Plant Reproduction

Annual Plant Reviews

A series for researchers and postgraduates in the plant sciences. Each volume in this annual series will focus on a theme of topical importance and emphasis will be placed on rapid publication.

Plant Reproduction

Edited by

SHARMAN D. O'NEILL
Section of Plant Biology
Division of Biological Science
University of California, Davis, USA

and

JEREMY A. ROBERTS
Plant Science Division
School of Biosciences
University of Nottingham, UK

CRC Press

AGR
QK
825
.P578x
2002

First published 2002
Copyright © 2002 Sheffield Academic Press

Published by
Sheffield Academic Press Ltd
Mansion House, 19 Kingfield Road
Sheffield S11 9AS, UK

ISBN 1-84127-226-4
ISSN 1460-1494

Published in the U.S.A. and Canada (only) by
CRC Press LLC
2000 Corporate Blvd., N.W.
Boca Raton, FL 33431, U.S.A.
Orders from the U.S.A. and Canada (only) to CRC Press LLC

U.S.A. and Canada only:
ISBN 0-8493-9791-X
ISSN 1097-7570

Printed on acid-free paper in Great Britain by
Antony Rowe Ltd, Chippenham, Wiltshire

British Library Cataloguing-in-Publication Data:
A catalogue record for this book is available from the British Library

Library of Congress Cataloging-in-Publication Data:
A catalog record for this book is available from the Library of Congress

Preface

Reproduction is the final stage in the life cycle of a plant. The developmental changes occurring during this event result in the production of the inflorescence and floral meristem, floral organ primordia, ovules and pollen, embryos and seeds. Each of these reproductive tissues is produced sequentially along the floral axis, reflecting the activity of the reproductive meristem as cell division affects the penultimate outcome of the genetic programme to reproduce sexually.

This volume provides an overview of our understanding of plant reproduction in a sequence that recapitulates the process of reproduction itself. The initial chapters focus on the transition to flowering and synthesize state-of-the-art knowledge of this topic from physiological, molecular and genetic perspectives. The sequence of events leading to pollen and ovule development is then described and the mechanisms by which gamete fusion leads to formation of the embryo and endosperm are considered. Self-incompatibility mechanisms are evaluated and formation of embryos in the absence of fertilization is discussed. The final chapter details mechanisms that regulate floral organ senescence.

The volume combines the interdisciplinary perspectives of cell biology, biochemistry, physiology, molecular biology, genetics and developmental biology. Each chapter is authored by a chosen expert in the field. The next frontier will be to understand the cellular basis of developmental changes. For example, we will need to understand the molecular events that occur during cell division and the signaling events that drive developmental change, resulting in the differentiation of ovules, pollen, gametes, embryos and seeds. These frontier areas will require the application of interdisciplinary research, using advanced genomic technologies and approaches, and will require participation and collaboration of the international community of plant scientists in investigating the functions of gene networks responsible for reproductive development. Finally, with advances in the field of evolution and phylogeny, plant scientists will begin to view reproductive development from an evolutionary perspective. This volume is a strong foundation for these future endeavors.

Sharman D. O'Neill
Jeremy A. Roberts

Contributors

Professor Richard Amasino   Department of Biochemistry, University of Wisconsin, 433 Babcock Drive, Madison, WI 53706-1544, USA

Professor Georges Bernier   Laboratory of Plant Physiology, Institute of Botany, University of Liège, B22 Sart Tilman, B-4000 Liège, Belgium

Dr Roy C. Brown   Department of Biology, University of Louisiana at Lafayette, Lafayette, LA 70504-2451, USA

Mr David Chevalier   Institute of Plant Biology, University of Zürich, Zollikerstrasse 107, CH-8008 Zürich, Switzerland

Mr Mark Doyle   Department of Biochemistry, University of Wisconsin, 433 Babcock Drive, Madison, WI 53706-1544, USA

Dr Anna M. Koltunow   CSIRO Plant Industry, Horticulture Unit, PO Box 350, Glen Osmond, South Australia, 5064, Australia

Dr Betty E. Lemmon   Department of Biology, University of Louisiana at Lafayette, Lafayette, LA 70504-2451, USA

Dr Jinhong Li   Centre for Plant Sciences, University of Leeds, Leeds LS2 9JT, UK

Dr Ed Newbigin   Plant Cell Biology Research Centre, School of Botany, University of Melbourne, Victoria 3010, Australia

Ms Hong Nguyen   Department of Biology, University of Louisiana at Lafayette, Lafayette, LA 70504-2451, USA

Mr Nicholas Paech   CSIRO Plant Industry, Horticulture Unit, PO Box 350, Glen Osmond, South Australia, 5064, Australia

Dr Luis Perez-Grau   J.R. Simplot Company, Plant Sciences Division, 5369 W. Irving Street, Boise, Idaho 83706, USA

Dr Claire Périlleux   Laboratory of Plant Physiology, Institute of Botany, University of Liège, B22 Sart Tilman, B-4000 Liège, Belgium

Dr Kay Schneitz   Institute of Plant Biology, University of Zürich, Zollikerstrasse 107, CH-8008 Zürich, Switzerland

Mr Si-Bum Sung   Department of Biochemistry, University of Wisconsin, 433 Babcock Drive, Madison, WI 53706-1544, USA

Mr Patrick Sieber   Institute of Plant Biology, University of Zürich, Zollikerstrasse 107, CH-8008 Zürich, Switzerland

Mr Matthew R. Tucker   Department of Plant Science, Waite Campus, University of Adelaide, P.M.B., 1 Glen Osmond, South Australia 5064, Australia.

Professor David Twell   Department of Biology, University of Leicester, University Road, Leicester LE1 7RH, UK

Dr Adam Vivian-Smith   CSIRO Plant Industry, Horticulture Unit, PO Box 350, Glen Osmond, South Australia, 5064, Australia

Professor William R. Woodson Department of Horticulture and Landscape Architecture, Purdue University, West Lafayette, IN 47907-1165, USA

Contents

6 Endosperm development   193
ROY C. BROWN, BETTY E. LEMMON and HONG NGUYEN

7 The central role of the ovule in apomixis and parthenocarpy   221
ANNA M. KOLTUNOW, ADAM VIVIAN-SMITH,
MATTHEW R. TUCKER and NICHOLAS PAECH

1 The control of flowering: do genetical and physiological approaches converge?

Claire Périlleux and Georges Bernier

1.1 Introduction

Since the early 1990s, *Arabidopsis thaliana* has been studied as the species of choice of 'flowering geneticists'. This species is a facultative long-day, vernalization-requiring plant in natural conditions. Both natural variation (ecotypes) and induced mutagenesis were exploited by several research groups (initially Maarten Koornneef and associates) to unravel the genetic control of flowering. Several pathways that either repress or promote flowering have been identified on the basis of (1) the flowering response of different genotypes to environmental factors (vernalization and photoperiod), (2) epistastis analyses, and (3) expression patterns of cloned genes in various backgrounds. Models attempting to include all information have been proposed repeatedly during the last decade (Koornneef *et al.*, 1991, 1998b; Martínez-Zapater *et al.*, 1994; Levy and Dean, 1998; Piñeiro and Coupland, 1998; Yang *et al.*, 1999; Samach *et al.*, 2000; Sheldon *et al.*, 2000), and their complexity is increasing with the bulk of data.

In this review, we shall attempt to integrate into this genetical framework the physiological knowledge accumulated on a variety of plants by generations of researchers (Halevy, 1985–1989) before the almost overnight breakthrough of *Arabidopsis*. During the first half of the twentieth century, physiologists uncovered the environmental signals used by plants to repress or promote flowering in natural conditions (Lang, 1965). With the advent of phytotrons in the 1940s, it became straightforward to grow plants in very varied environments, and these studies then revealed the incredible flexibility of plant development. By designing experimental systems where flowering could be strictly controlled, it became possible to subdivide the whole process into successive steps: the 'induction' mechanisms determining flowering time must precede floral 'evocation' or 'commitment' taking place in the shoot apical meristem (SAM) and occurring itself before the initiation of floral meristems, or floral 'morphogenesis' (Evans, 1969). In photoperiodic species, there is strong evidence that signals moving from the leaves to the SAM are involved as a link in between the two steps of 'induction' and 'evocation' and movement of these substances could be timed precisely in appropriate experiments (Bernier *et al.*, 1981a). In the 1960s, physiologists also started to analyse the changes occurring

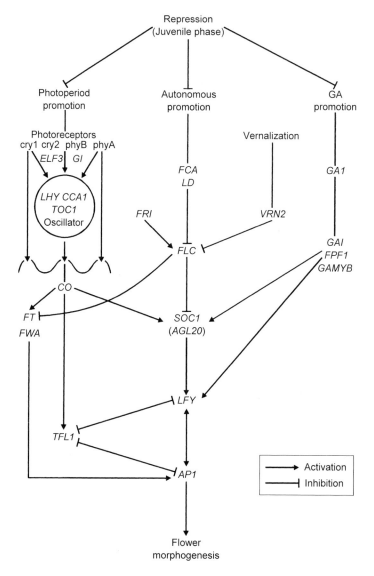

Figure 1.1 A simplified model of the possible interactions between genes and pathways controlling flowering time in *Arabidopsis*.

in the SAM during and after floral evocation, a field that had previously been investigated only by plant anatomists.

Genetic analyses allow us to place the action of genes in specific pathways; however, such studies provide little information about the temporal and spatial expression of the genes, leading to difficulties in the allocation of the precise roles of the gene products in the 'induction' or 'evocation' steps. The interpretation of their function is further complicated by the possibility that key signalling molecules may be transported from one site to another. In this review we have chosen to use a 'genetical skeleton' to organize our material, but because of the increasing complexity of existing schemes, we have generated our own (figure 1.1), omitting genes for which little information is available. Before explaining our model, we shall first summarize the evocation and early morphogenesis processes since the SAM is the final target where all floral pathways described below must converge.

1.2 Floral meristem identity genes

At floral transition, the shift from leaf to flower formation in *Arabidopsis* requires that the cells of the peripheral zone of the SAM undergo a change in identity. Several genes have been identified that are responsible for the identity of the floral meristems, among which *LEAFY* (*LFY*) and *APETALA1* (*AP1*)—encoding transcription factors—are prominent (Weigel, 1995).

Mutation at either the *LFY* or *AP1* locus causes the transformation of flowers into shoot-like structures but does not affect flowering time, or very weakly (the *ap1* mutant is slightly early-flowering; Page *et al.*, 1999). In contrast, overexpression of either gene not only converts all shoot meristems into single flowers but also accelerates flowering. On this basis, both genes are considered as necessary and sufficient for the initiation of individual flowers (Ng and Yanofsky, 2000), and they appear as potential targets for flowering time genes (figure 1.1).

LFY and *AP1* are repressed in the SAM during the vegetative phase, although *LFY* is expressed at a low level (Blásquez *et al.*, 1997), and both are markedly upregulated during floral transition, especially in floral anlagen. *LFY* is upregulated first (Hempel *et al.*, 1997). Other genes that are upregulated early in the transitional SAM, and also encode transcription factors, are *SUPPRESSOR OF OVEREXPRESSION OF CO1* (*SOC1*) and *AGAMOUS-LIKE 8* (*AGL8*) renamed *FRUITFULL* (*FUL*) (Hempel *et al.*, 1997; Samach *et al.*, 2000). We will return to the function of *SOC1* below.

Some genes, in contrast, are highly expressed in the SAM during the vegetative phase and are downregulated at floral transition, e.g. *KNOTTED-LIKE FROM ARABIDOPSIS THALIANA 1* (*KNAT1*) and *SHORT VEGETATIVE PHASE* (*SVP*) (Lincoln *et al.*, 1994; Hartmann *et al.*, 2000). Inactivation of these genes may be necessary if the transition is to occur, as indicated by the early-flowering phenotype of the *svp* mutant. Interestingly, both *KNAT1* and *SVP* also encode transcription factors.

In wild-type (WT) *Arabidopsis*, the inflorescence is indeterminate, implying that some mechanism inhibits the production of a terminal flower. Exclusion of *LFY* and *AP1* expression from the tip or central zone of the SAM requires the function of an antagonistic gene: *TERMINAL FLOWER 1* (*TFL1*) (Bradley *et al.*, 1997). When *TFL1* is mutated, heterochronic and ectopic expression of *LFY* and *AP1* in the central zone causes flowering to occur early and a terminal flower to be formed. In WT plants, *TFL1* is expressed just below the SAM and its transcription in the apex periphery is prevented by *LFY/AP1* (Ratcliffe *et al.*, 1999). *TFL1*, like *LFY*, is expressed weakly during the vegetative phase and upregulated at floral transition.

These meristem identity genes act in a transcriptional cascade leading to activation of the floral homeotic genes whose discovery has led to the ABC model of floral morphogenesis highlighted by Elliot Meyerowitz. These further regulatory interactions are outside the scope of this chapter, but have been reviewed (Ng and Yanofsky, 2000).

Physiologically, the SAM is found to be committed to flower, i.e. florally evoked, quite soon after the arrival of mobile floral signals generated in the leaves as a result of photoperiodic induction. This has been shown by determining when excised apices of induced individuals of the long-day plant (LDP) *Lolium temulentum* or the short-day plant (SDP) *Pharbitis nil* have become capable of initiating inflorescence morphogenesis *in vitro* (McDaniel *et al.*, 1991) or after grafting on a vegetative stock plant (Larkin *et al.*, 1990). For Hempel *et al.* (1997), evocation time occurs in *Arabidopsis* between the upregulation of *LFY* and *AP1*, and it would be critical to know whether this proposal is confirmed in well-controlled physiological systems. In *Lolium*, upregulation of the *AP1* homologue occurs early, at about the time of or just after evocation, supporting Hempel's claim, but curiously precedes by several days the upregulation of the *LFY* homologue, contrary to what is found in *Arabidopsis* (Gocal *et al.*, 2001).

It is important to recognize here that evocation seems to occur in many species well before the increase in size and the change in shape of the SAM. Usually, these changes start later, from one to several days after evocation is completed (Bernier *et al.*, 1981b). Early changes observed in switching SAMs are large increases in the rates of energy metabolism and cell division (Bernier, 1989). In some plants at least (e.g. *Sinapis alba*), these increases start so early that they are actually detectable before the transport of the floral signals is completed, thus well before the apex is fully evoked. Circumstantial evidence suggested that these changes are essential events of the floral switch at the SAM, and these inferences received some support from recent studies using transgenic plants. Transgenic potatoes with a reduced citrate synthase activity, and thus a decreased energy metabolism, have a late-flowering phenotype, although their vegetative growth is not impaired (Landschütze *et al.*, 1995). On the other hand, tobacco plants overexpressing the *AtCYCLIN D2* gene exhibit a higher rate of cell division in the SAM and flower earlier than WT (Cockcroft *et al.*, 2000).

1.3 Floral repression genes and pathway

Mutations causing an early flowering phenotype in *Arabidopsis* were found and hypothesized to concern genes whose normal function is to repress flowering. Central elements of this 'repression pathway' are the *EMBRYONIC FLOWER* (*EMF*) genes: mutations of these genes cause plants to bypass the rosette vegetative phase and produce a small inflorescence immediately upon germination (Sung *et al.*, 1992). *emf* is epistatic to most flowering-time mutants (Haung and Yang, 1998), suggesting that a vegetative phase takes place only if flowering is repressed by *EMF* genes. Thus *EMF* genes should act earlier in development than the other flowering-time genes by repressing flowering, which then appears as the 'default' program (Schultz and Haughn, 1993). The repression activity of *EMF* in WT plants is believed to decrease with age, since plants will eventually flower. Since the floral meristem identity gene *AP1* is precociously and ectopically expressed in *emf* seedlings (Chen *et al.*, 1997), repression of flowering by *EMF* genes could be due to their inhibitory action on the expression of floral meristem identity genes. This model may account for *EMF1* only, since *EMF2* has been shown recently to be expressed at all major developmental stages of *Arabidopsis* (Yoshida *et al.*, 2000).

For physiologists, the 'juvenile' phase of growth during which plants are unable to flower regardless of the growing conditions is the result of several systems repressing flowering and acting in varied combinations in different plant species (Bernier *et al.*, 1981a). Prominent among these systems are: (a) the influence of roots that maintain the SAM in a vegetative state during early growth after germination, (b) the inability of first-formed leaves to generate enough mobile floral-promoting signals whatever the photoperiod, and (c) the relative insensitivity of young SAMs to these floral promoters (SAM incompetence). In day-neutral tobaccos, for example, the genotype-specific number of nodes produced before flowering is primarily a function of foliar ability to produce floral signals, and of SAM competence (McDaniel, 1996).

Another critical factor repressing flowering in early stages of growth is insufficient photosynthetic activity due to insufficient leaf area and/or poor light conditions. High light, permitting high photosynthetic rates, has been found to shorten the juvenile phase in most kinds of plants—annuals, biennials and perennials—irrespective of their further photoperiodic and/or vernalization requirements (Bernier *et al.*, 1981a). Thus, insufficient carbohydrate availability appears as one major factor maintaining juvenility. Finally, maximal reduction in the number of leaves preceding flowering is obtained in several plants, including *Arabidopsis*, by combining light conditions optimal for flowering with extreme mineral (nitrogen) starvation (Lang, 1965; Bernier *et al.*, 1981a). Nitrogen nutrition is thus another factor controlling juvenility, at least in some plants.

At this stage, it is hard to relate these physiological observations with genetic data. Activity of *EMF* genes associates intuitively with juvenility and it is tempting to speculate that, provided other conditions are satisfied, flowering can occur when the SAM is 'relieved' from repression. One could hypothesize that *EMF* genes, together with other genes, such as the downregulated *KNAT1* and *SVP* and the upregulated *TFL1*, are involved in regulating SAM competence.

1.4   Floral promotion genes and pathways

Arabidopsis late-flowering mutants are used to identify WT genes acting in the promotion of flowering. It is striking that single mutants that are totally unable to flower have never been found, indicating that there is no single gene or pathway that controls the whole process. In fact, mutants have been classified into three independent pathways (figure 1.1). The first pathway includes mutants whose sensitivity to photoperiod is reduced or lost, and genes defective in these mutants that flower late in long days (LDs) are thought to act in a 'photoperiod-promotion' pathway. Mutants of the second pathway are delayed in LDs and in short days (SDs), and even show an increased sensitivity to photoperiod. Thus, the corresponding genes are believed to act in an 'autonomous promotion' pathway. Interestingly, vernalization may completely substitute for this pathway and rescue mutational effects. Finally, the third pathway includes mutants of gibberellin (GA) biosynthesis and response.

The concept that plants can use multiple pathways to achieve flowering is, of course, not new. Indeed if, in natural conditions, plants of temperate zones use essentially vernalization and/or photoperiod to synchronize their flowering to the changing season, experiments in phytotrons have disclosed that they can also be brought into the flowering stage by a variety of 'alternative' environmental pathways (Bernier *et al.*, 1981a; Bernier, 1988). For example, many obligate photoperiodic species, either LDPs or SDPs, become progressively independent of photoperiod when temperature drops from $\sim 20°C$ towards the vernalization range. Conversely, the natural vernalization requirement of several species can be bypassed by exposure to SDs at normal temperature (Lang, 1965; Heide, 1994). In other words, many plants can use alternatively photoperiod or vernalization. Additional environmental cues for flowering are high irradiance, high temperature ($\geqslant 30°C$), poor mineral nutrition and elevated CO_2 level, and it is not uncommon to find three to five different pathways individually leading to flowering in a single species, for example in the LDP *Silene armeria* or the SDP *Pharbitis* (Bernier, 1988).

Another interesting observation is that, in many species, the photoperiodic or vernalization requirements decrease with age. Thus these plants will ultimately flower, albeit slowly and weakly, in continuous 'unfavourable' conditions (Lang, 1965; Bernier *et al.*, 1981a). Clearly, with age, a majority of plants shift—for the promotion of their flowering—from environmentally controlled pathways to an autonomous pathway.

1.4.1 Photoperiod promotion pathway

Although major progress towards understanding photoperiodism in LDPs has emerged recently from genetic analyses in *Arabidopsis*, photoperiodism has been the subject of physiological investigation since Garner and Allard introduced the term in 1920, and there is a huge literature pertaining to all aspects of the subject (reviewed in Thomas and Vince-Prue, 1997).

Photoperiodism involves the coupling of the capacity to sense light with a timekeeping mechanism, in fact an endogenous clock or 'oscillator'. The biological clock is a ubiquitous mechanism driving circadian rhythms that are entrainable to environmental cues of light and temperature but continue in constant conditions. The underlying mechanisms are compensated such that the period length remains approximately the same within a 'physiological range' of ambient temperatures.

Erwin Bünning, in the 1930s, was one of the pioneers who suggested that photoperiodic timekeeping involves a regular oscillation of phases with different sensitivities to light. The most direct approach has been to combine a short photoperiod with a very long dark period and to scan the latter with a night-break of varying duration, light intensity and quality. In these experiments, maximum inhibition of flowering in SDPs or promotion of flowering in LDPs did occur at circadian intervals.

A model for circadian timekeeping in SDPs has been proposed in *Pharbitis*, from conclusive experiments in which dark-grown seedlings received a photoperiod of various durations followed by an inductive dark period interrupted at various times by a night break (Lumsden *et al.*, 1982). The time of the maximum sensitivity to a night-break was constant from light-on (15 h) with photoperiods of < 6 h duration and was constant from light-off (9 h) with photoperiods of > 6 h. Thus a rhythm of sensitivity to light is thought (1) to be initiated by transfer to light at dawn; (2) to run in continuous light for 6 h; (3) to be suspended thereafter as long as the plants stay in the light; and (4) to be released at dusk so that the maximum sensitivity to a night-break occurs 9 h later. The rhythm runs in constant darkness with a periodicity of ~24 h.

How LDPs measure time appeared to be more difficult to analyse: the effect of the night-break is less clear than in SDPs, unless its duration is increased up to a few hours, or it is superposed to marginally inductive conditions. However, a circadian rhythm in response to a short light period interrupting protracted darkness has been reported in LDPs such as *Hyoscyamus niger*, *Lolium*, and *Sinapis*. Hence flowering can be induced in LDPs by giving a short period of light 'at the right time'. From these observations the idea derived that, although a standard SD regime keeps LDPs vegetative, shifting one SD within a 24 h cycle may bring it into phase with the endogenous rhythm of sensitivity to light, and so induce flowering (Périlleux *et al.*, 1994). This 'displaced SD' (DSD) system is indeed used as an alternative treatment to induce flowering that helps discriminate between photosynthetic and photoperiodic effect of LDs.

Much of the physiological work aiming at the identification of photoreceptors involved in photoperiodism is based on action spectroscopy of night-break responses, where very similar action maxima are found in the red (R) for the inhibition of flowering in SDPs and the promotion of flowering in LDPs. Reversal of the effect of R by far red (FR) further identified phytochromes as the major photoreceptors involved. In the LDPs Cruciferae ('mustards'), blue (B) light perceived by cryptochromes was also found to promote flowering. Other experiments attempted to obtain action spectra for the effects of day-extensions in LDPs and a promotive effect of FR was consistently found, with maximal inductive responses to extensions with R plus FR. Thus incandescence is the selected light source for daylength extension experiments and, since high fluence is not required, this way of inducing flowering offers the conceptual advantage that photosynthetic contribution to the extension period is far reduced.

Various species used as physiological models in the investigation of photoperiodism, because they can be induced by a single LD or SD (*Lolium, Sinapis, Pharbitis*), were studied at the molecular level with the objective of identifying the genes whose expression in the leaves is modified during the induction of flowering. However careful these studies were, and despite the use of improved techniques, they did not uncover qualitative changes in gene expression in induced versus noninduced tissues, but repeatedly identified quantitative modifications in the rhythmic patterns of gene expression (Cremer *et al.*, 1991; O'Neill, 1993; Périlleux *et al.*, 1996). The work on *Pharbitis* is being continued with transcripts that seem to be associated with photoperiodic events since their abundance is photoperiod-dependent and also responds to a night-break treatment (Sage-Ono *et al.*, 1998; Zheng *et al.*, 1998).

These days, the understanding of photoperiodism in LDPs is taking great steps forward with the genetic analyses in *Arabidopsis*. The photoperiod promotion pathway includes genes (and their products) involved in the perception of daylength (i.e. photoreceptors, clock components, and transduction elements) but also genes acting downstream (figure 1.1).

1.4.1.1 Photoreceptors

Light-quality effects on *Arabidopsis* transition to flowering have been investigated both physiologically and genetically (Lin, 2000). When plants were grown under SDs (white light) and received a 1 h night-break in the middle of the night, flowering was hastened, though to a smaller extent than in continuous LDs (Goto *et al.*, 1991). Light effectiveness for the night-break was highest for FR and B light, and lowest for R and white light, as in other mustards. With daylength extension experiments, the promotion by FR was also observed and found to be rhythmic when superposed to continuous illumination (Deitzer, 1984). The characterization of mutants that carry lesions in individual photoreceptors has enabled the function of these pigments in the flowering process to be investigated.

Phytochrome A (phyA) is responsible for the FR promotion of flowering. Plants deficient in this photoreceptor are less sensitive to a night-break (Reed *et al.*, 1994) and are poorly responsive to inductive daylength extensions of low-fluence-rate incandescent light (Johnson *et al.*, 1994).

PhyB is the main R-sensing photoreceptor in the green plant. When it is absent, flowering is accelerated and the response to photoperiod is reduced (Goto *et al.*, 1991). This has led to the hypothesis that phyB has an inhibitory function in floral induction.

Cryptochrome 2 (cry2) is the primary B-light receptor regulating flowering time: the *cry2* mutant flowers significantly later than WT in LDs but not in SDs (Guo *et al.*, 1998). The peculiar light dependence of the late-flowering phenotype of *cry2* has led to the hypothesis that cry2 has two roles: a R-light-dependent function that would be to counteract phyB action, and a direct B-light function that would be to promote floral initiation redundantly with cry1 (Mockler *et al.*, 1999).

1.4.1.2 The biological clock

In *Arabidopsis*, as in other plants, the biological clock controls not only the induction of flowering by LDs but also leaf movements, opening of stomata, hypocotyl growth, and the transcription of a number of genes, such as those encoding chlorophyll *a/b* binding proteins (*CAB* genes). The search for individuals in which these rhythms are abnormal is the most common strategy of genetic screens aiming at the identification of clock components. This approach has been successful using as screening phenotypes the flowering response to photoperiod and the rhythmic expression of *CAB* genes, monitored via the activity of a *CAB:LUCIFERASE* (*CAB:LUC*) construct (Millar *et al.*, 1995a).

Conceptually, biological clocks consist of three parts: (1) a central oscillator that generates the periodicity; (2) input pathways that provide the temporal information from the environment to the oscillator; and (3) output pathways that transmit the periodicity to target processes or 'overt rhythms'. There is no single criterion that enables an arrhythmic mutant to be unequivocally classified as deficient in an input, output or oscillator component (Somers, 1999). However simplistic they may be, assumptions were made from the simplest models that provided the conceptual framework from which most clock research has proceeded (see the reviews of McClung (2000) and Samach and Coupland (2000) for more details and nuances about the functions of the genes described below). Mutations in the input pathways are expected to cause an environment-dependent phenotype and mutations in the output pathways should affect a limited number of downstream processes. Regarding the central core, evidence has been building for many years, in model organisms such as *Neurospora*,

Drosophila and mouse, in support of a model for an oscillator comprising, at least in part, a transcription/translation-based negative feedback loop wherein clock genes are rhythmically expressed, giving rise to cycling levels of clock transcripts and proteins (Dunlap, 1999). The proteins then feed back—after a lag—to depress the level of their own transcripts. Based on this 'autoregulatory feedback loop' model, genetic components of the circadian oscillator are expected to show rhythmic expression, negative feedback regulation, sensitivity to signals that reset the clock, and disruption of multiple circadian rhythms by misexpression. More recent evidence shows that, in fact, 'negative elements' do not feed back directly but do it by blocking the action of 'positive elements' whose role is to activate the clock genes (Dunlap, 1999).

The oscillator. The strategies and criteria described above have led to the isolation of three putative clock-components in *Arabidopsis*. These are: *LATE ELONGATED HYPOCOTYL (LHY), CIRCADIAN CLOCK ASSOCIATED 1 (CCA1)*, and *TIMING OF CAB 1 (TOC1)*. The *LHY* gene has been cloned from an overexpressing mutant whose flowering is delayed under LDs, and insensitive to photoperiod (Schaffer *et al.*, 1998). *LHY* encodes a MYB transcription factor very similar to CCA1, a protein that binds well-characterized regions of circadian- and phytochrome-regulated *CAB* promoters. Both *LHY* and *CCA1* transcripts exhibit a circadian rhythm and overexpression of the genes disrupts all rhythmic outputs examined (Schaffer *et al.*, 1998; Wang and Tobin, 1998). Moreover, overexpression of *LHY* or *CCA1* represses not only their own but also each other's expression, suggesting that these two genes are part of a feedback regulatory loop. Recent characterization of a *cca1* loss-of-function mutant demonstrated that *CCA1* is required for circadian clock regulation with a normal period and that, although *CCA1* and *LHY* are closely related, they are not completely redundant (Green and Tobin, 1999).

   TOC1 has been cloned recently from a short *CAB:LUC*-period mutant whose phenotype is light-independent and flowers as early in SDs as in LDs in certain ecotypes (Somers *et al.*, 1998b; Strayer *et al.*, 2000). *TOC1* was shown to encode a transcription factor containing a pseudo-receiver domain; it is circadianly expressed and its periodicity is shortened in the *toc1* mutant, suggesting an autoregulation.

The light input pathway. Light is a universally active stimulus in setting both the phase and the free-running period of the clock: light pulses control the phase of the clock, mediating entrainment to the day–night cycle, whereas continuous photoreceptor activation modulates the period. This latter kind of experiment showed that both R and B lights shorten the period of *CAB:LUC* to almost the same extent as continuous white light (Millar *et al.*, 1995b). Using photoreceptor-deficient mutants, Somers *et al.* (1998a) concluded that phyA, phyB and cry1 are involved in clock entrainment in specific conditions of light quality and intensity, while cry2 plays a minor role. However, it was recently reported that a quadruple *phyA;phyB;cry1;cry2* mutant still exhibits light entrainment of the clock, thus one more photoreceptor—at least—is at work (Yanofsky *et al.*, 2000). Although phytochromes other than A and B may be involved (Lin, 2000), new candidates are encoded by *ZEITLUPE* (*ZTL*) and *FLAVIN-BINDING, KELCH REPEAT, F-BOX 1* (*FKF1*) (Nelson *et al.*, 2000; Somers *et al.*, 2000). The products of these two genes, which show high similarity, contain conserved domains suggesting that ZTL and FKF1 may function in a novel B-light signalling pathway—maybe as photoreceptors—targeting proteins for ubiquitin-mediated degradation. Since mutations in *ZTL* or *FKF1* perturb several clock-controlled processes, an exciting hypothesis would be that the targets of ZTL and FKF1 are the clock-proteins themselves.

Information about the light signalling to the clock also comes from flowering-time mutants *gigantea* (*gi*) and *early flowering 3* (*elf3*) that show a reduced sensitivity to photoperiod, but 'inverse' phenotypes: *gi* mutations cause late flowering under LDs (Koornneef *et al.*, 1991), while *elf3* flowers early in SDs (Zagotta *et al.*, 1996).

The *gi* mutations modify the period of the rhythm of a *CAB:LUC* reporter gene (Park *et al.*, 1999) and affect the R-light fluence effect on the free-running period of this rhythm, suggesting that *GI* functions in a light input pathway. Indeed, it has been shown recently that GI is a nuclear protein involved in phyB signalling (Huq *et al.*, 2000). *GI* may function in an outer feedback loop in between photoreceptor(s) and the clock since (1) its expression is circadian-regulated and (2) defects in *GI* expression reduce the amplitude—and possibly alter the period—of the putative clock-components *LHY* and *CCA1* (Fowler *et al.*, 1999; Park *et al.*, 1999). Interestingly, *GI* expression—timing, height and duration of the daily peak—is influenced by daylength (Fowler *et al.*, 1999).

The *elf3* mutation causes defective circadian rhythms only during and imme-diately after light exposure, suggesting that *ELF3* is involved in an interaction between phototransduction and the circadian system (Hicks *et al.*, 1996). Both B- and R-light signals can cause *elf3* mutant to exhibit arrhythmic *CAB:LUC* transcription and the identities of the photoreceptor species involved in the *elf3* phenotypes is still controversial (Zagotta *et al.*, 1996; Reed *et al.*, 2000). *ELF3* seems to regulate clock-components, such as *GI* and *LHY* expression in the light (Schaffer *et al.*, 1998; Fowler *et al.* 1999).

The output pathway. Theoretically, circadian control may be mediated either by a direct output from the circadian clock to the immediate effectors, or indirectly by regulating other signalling pathways. Direct output is suggested by the fact that putative components of the central oscillator are themselves transcription factors. For example, both LHY and CCA1 proteins bind to the *CAB* gene promoter and as a consequence may induce its circadian expression (Green and Tobin, 1999). However, some evidence suggests that the cascades leading to floral induction are more complex. A good example comes from *COLD CIRCADIAN REGULATED 2* (*CCR2*), a gene that is often used as an 'indicator' of clock function (Kreps and Simon, 1997). It has turned out that *CCR2* is identical to *AtGRP7*, a gene encoding an RNA-binding glycine-rich peptide. *AtGRP7* is the *Arabidopsis* homologue of *SaGRP*, a gene identified in *Sinapis* as being putatively involved in the photoperiodic induction of flowering (Staiger and Heintzen, 1999). *CCR2/AtGRP7* transcript oscillation is not only linked to the biological clock, it is also the consequence of an autoregulatory feedback exerted by its own protein (Heintzen *et al.*, 1997). However, *CCR2/AtGRP7* could hardly function as a central clock component since its overexpression does not affect widespread circadian-regulated processes. A model was thus proposed where this genetic loop forms a 'slave' oscillator.

It is obvious from this short survey that the conceptual model of biological clocks is getting more complex. Not only does the central oscillator generate rhythmicity by autoregulated loops, but input and output pathways no longer appear as 'linear' and exert feedback controls. We will return later to working models of how photoreceptors and the biological clock may control the transition to flowering.

1.4.1.3 Downstream genes

Acting downstream of photoreceptors and clock components, the *CONSTANS* (*CO*) gene is of primary importance in the photoperiod promotion pathway. Mutations in *CO* result in late flowering in LDs, while overexpression of the same gene causes very precocious flowering (Putterill *et al.*, 1995; Simon *et al.*, 1996; Onouchi *et al.*, 2000). In WT plants, *CO* transcript is found in all plant parts, including the SAM, and is more abundant in LDs than in SDs, consistent with the idea that *CO* promotes flowering under LDs only. The critical role of *CO* in mediating the floral response to photoperiod in SDPs was recently disclosed by analyses of *CO* homologues in *Pharbitis* (Liu *et al.*, 2001) and rice (*Oryza sativa*) (Yano *et al.*, 2000). In both *Arabidopsis* and *Pharbitis*, *CO* transcript levels are under circadian clock regulation and increase in inductive photoperiods: in *Arabidopsis* LDs broaden the daily peak of accumulation (Suárez-López *et al.*, 2001), while in *Pharbitis CO* transcripts accumulate in the dark (Liu *et al.*, 2001). However, the exact nature of the common mechanisms involved in photoperiod promotion of flowering by *CO* in both LDPs and SDPs remains elusive. Important information will likely come from the analysis of posttranscriptional events and their control by light.

CO encodes a zinc finger transcription factor and several genes have been found to be upregulated quite early, in a matter of hours, following induction of its activity (figure 1.1). One of them is *TFL1*, a repressor of the floral meristem identity genes *LFY/AP1* (see above). Two others are *SOC1* (formerly named *AGL20*) and *FLOWERING LOCUS T* (*FT*) (Samach *et al.*, 2000). Mutations at either of these two loci cause late flowering and, in addition, reduce the earliness of plants overexpressing *CO*. On the other hand, overexpression of either *SOC1* or *FT* results in an early-flowering phenotype similar to *CO* overexpressors (Kardailsky *et al.*, 1999; Kobayashi *et al.*, 1999; Bonhomme *et al.*, 2000; Borner *et al.*, 2000; Onouchi *et al.*, 2000; Samach *et al.*, 2000).

Interestingly, both *SOC1* and *FT* are proposed to activate floral meristem identity genes. *SOC1* encodes a transcription factor that is expressed in the whole plant and accumulates in LDs. It is activated in the SAM at an early stage during floral transition, at a time that clearly precedes upregulation of *LFY*. *SOC1* is therefore seen as part of a transcriptional cascade within the SAM that ultimately results in the activation of *LFY* (Samach *et al.*, 2000). *FT*, like *CO* and *SOC1*, is expressed in the whole plant but has been proposed to activate another floral meristem identity gene: *AP1* (Ruiz-García *et al.*, 1997). This activation might not be direct since the semidominant late-flowering *fwa* mutation, which does not affect *FT* expression, partially suppresses the early-flowering phenotype caused by overexpressing *FT* (Kardailsky *et al.*, 1999; Kobayashi *et al.*, 1999). Thus, *FWA* apparently acts downstream of *FT* and upstream of *AP1*, and the *FT/FWA* function could be viewed as acting redundantly with *LFY* to regulate *AP1* activity (Nilsson *et al.*, 1998). The role of *FWA*, however, is not clear, since Soppe *et al.* (2000) found that the late-flowering phenotype of the *fwa* mutant correlates with overexpression of the gene due to its demethylation, while *FWA* has no inhibitory effect, and may have no function, in WT plants.

Clearly, *CO* has a pivotal role in the photoperiod promotion pathway by activating at least one subpathway (*TFL1*) that represses *LFY/AP1* functions and two redundant subpathways (*SOC1*, *FT/FWA*) that promote the same functions. Interestingly, *TFL1* and *FT* encode related proteins and their antagonistic effects on flowering could be balanced to fine tune the response to signals that promote floral induction (Kardailsky *et al.*, 1999; Kobayashi *et al.*, 1999).

1.4.1.4 Physiology of photoperiodic promotion of flowering—floral signals
A major unresolved problem of the genetic analysis approach is that it does not elucidate where in the plant the expression of the genes must take place to enable the photoperiodic pathway to proceed. Physiological evidence has demonstrated that, in intact plants, the most effective perception site for photoperiodic floral induction is the expanded leaves (Lang, 1965). From this, it was inferred that photoinduction of flowering involves transmission of floral signals from expanded leaves to the SAM. However, excised shoot apices of several species can be caused to flower *in vitro* by exposure to an appropriate daylength, provided that a carbohydrate is supplied in the medium (Bernier *et al.*, 1981a). Thus sufficient carbon in the SAM is essential and apices, including immature leaves, are apparently capable of providing the necessary floral signals. The predominant role attributed classically to expanded leaves in photoperiod perception can thus be explained by the fact that they are the sole plant part capable

of producing enough carbohydrates and other floral signals at the same time. Supporting evidence for the participation of immature leaves in the control of flowering time can be found in recent work on maize (Irish and Jegla, 1997; Colasanti *et al.*, 1998).

On this basis, we can hypothesize that the genes involved upstream of *CO* in the photoperiod pathway (figure 1.1) are active in all plant parts exposed to light/dark cycles. On the other hand, activity of *CO* and downstream genes could be essential only in the SAM, despite their widespread expression within the plant. It should be stressed here that in *Sinapis* the *SOC1* homologue (*SaMADSA*) is only expressed in the SAM at floral transition (Bonhomme *et al.*, 2000; Borner *et al.*, 2000). A similar situation has been found with the expression pattern of *FUL* (*SaMADSB*) and *AP1* (*SaMADSC*), which is highly discrete in *Sinapis* (Menzel *et al.*, 1996; F. Bonhomme, personal communication) but more diffuse in *Arabidopsis* (Hempel *et al.*, 1997). Diffuse gene expression in *Arabidopsis* could be due to small plant size, and absence of a clear spatial separation of roots, leaf bases, nodes and SAM in the rosette core.

Another question, central in the physiological work but not addressed in genetic studies in *Arabidopsis*, concerns the nature of the leaf-generated mobile signals that control the floral shift at the SAM. Grafting experiments with various photoperiodic species disclosed that these signals may be of a promotive or inhibitory nature, depending on whether they are produced by induced or noninduced leaves (Cleland, 1978; Bernier *et al.*, 1981a; O'Neill, 1992). Moreover, as far as the experimental evidence goes, it seems that species differ in the relative importance of these positive and negative signals. In the LDPs *Nicotiana sylvestris*, *Hyoscyamus*, *Lolium* and pea, both promoters and inhibitors are produced and transmitted in favourable and unfavourable conditions, respectively. In contrast, in the SDP Maryland Mammoth tobacco, promoters seem predominant since inhibitors are not detected in grafting experiments. The opposite situation seems to prevail in the SDP strawberry, in which flowering time was found to be essentially controlled by transmissible inhibitors produced by noninduced leaves. Until now, it is unknown which situation prevails in *Arabidopsis*. Movement of promoting substances out of mature leaves has been demonstrated (Corbesier *et al.*, 1996), but the existence of inhibitors has not been tested.

The problem of the nature of floral signals has been vigorously pursued since the 1940s. However, all attempts at identifying specific floral promoter(s) and inhibitor(s) have failed (Cleland, 1978). It was often observed that substances transmitted through grafts affect not only the flowering of the receptor but also its growth habit, casting doubts about the paradigm of specific florigenic and antiflorigenic substances (Bernier, 1988).

Long-distance signalling has been comprehensively investigated in the adult mustard *Sinapis* induced to flower by one LD or one DSD (Bernier *et al.*, 1993, 1998; Havelange *et al.*, 2000; Corbesier *et al.*, 2001). This work identified several putative floral signals that move sequentially within the plant: the earliest signal is sucrose, moving both upwards and downwards; later signals are cytokinins (CKs) and nitrogenous compounds (glutamine or asparagine and putrescine), all moving essentially upwards (figure 1.2). Sucrose and nitrogenous

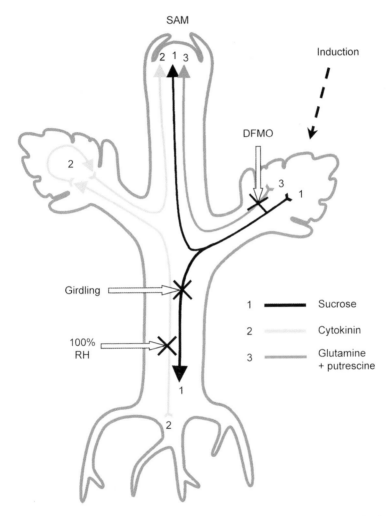

Figure 1.2 Diagram showing the sequence of long-distance movements of floral signals in plants of *Sinapis alba* induced to flower by one LD. Three treatments were applied that interrupt these movements and, at the same time, inhibit the floral response to the LD. (a) Girdling consists in the removal of a ring of superficial living tissues, including phloem; when done below the lower leaves, it stops movement of the phloem sap from shoot to roots. (b) Growing plants temporarily in air with 100% relative humidity (RH) stops the transpiration stream and, consequently, the movement of the xylem sap from roots to shoot. (c) DFMO (difluoromethylornithine), an inhibitor of putrescine biosynthesis, was sprayed on the leaves.

compounds arise in expanded leaves, while CKs initially originate from root tissues. Interestingly, the increased export of CKs from the roots is actually caused by extra sucrose coming from the leaves. This two-way traffic of chemicals is essential for flowering as, if movement is interrupted, the floral response of the plants to one LD is considerably reduced (figure 1.2). Moreover, it was found that the sucrose/CKs signalling loop (shoot-to-root-to-shoot) must be operative during a narrow time window, 8–12 h after start of the LD (Havelange *et al.*, 2000). Within the SAM, the extra sucrose and CKs cause events that were described above as essential for evocation: activation of the energy metabolism by sucrose, activation of *SaMADSA* (*SOC1* homologue) by CKs, and stimulation of cell division (presumably via activation of D-type *CYCLIN* genes) by both sucrose and CKs (Bernier *et al.*, 1993; Bonhomme *et al.*, 2000). Although the effects of sucrose and CKs at the SAM are identical in time and space to those produced by one LD, nevertheless, these two chemicals are unable to trigger flowering of vegetative plants even when applied in combination. The effects of nitrogenous compounds at the SAM remain to be investigated. Strangely, however, although excess nitrogen (nitrate) in the substrate delays flowering in *Sinapis*, as in many other plants (see above), photoinduction of flowering in this species causes an increase in the SAM supply of nitrogenous compounds. The inhibition of flowering by high nitrate could be due to some signalling role of this anion—independent of its nutritional role—as recently shown in the case of root system architecture (Zhang *et al.*, 1999).

Similar observations concerning movement of sucrose and CKs from leaves to SAM were made in the SDP *Xanthium strumarium* induced by one SD (Bernier *et al.*, 1998). Increased sucrose is observed shortly after the extended night, suggesting that it arises from starch mobilization.

In *Arabidopsis*, there is evidence that the transmissible floral signals are the same as those disclosed in *Sinapis* (Corbesier *et al.*, 1998, 2001; L. Corbesier, personal communication), but few studies have explored the floral behaviour of mutants and transgenics affected in the metabolism, transport or transduction of these signals. New issues may, however, arise from molecular and genetic analyses. A good example of such a feedback from genetics to physiology can be seen in maize. The *indeterminate* (*id1*) mutant of maize is unable to undergo a normal transition to flowering and produces many more leaves than WT plants (Colasanti *et al.*, 1998). Strikingly, the *ID1* transcript is restricted to young leaves, more precisely to the portion of these leaves where the transition from sink to source tissues is occurring (Colasanti and Sundaresan, 2000). This correlation suggests that the product of *ID1* could be involved in controlling the transmission or loading of some floral signal; such a function has never been conceived for flowering-time genes isolated in *Arabidopsis*.

At this point we may conclude that the long-distance floral signalling is complex, is multifactorial, and involves 'housekeeping' compounds such as assimilates and hormones (Bernier, 1988). This may explain why many chemicals, although insufficient to trigger flowering when applied on plants grown in noninductive conditions, may have promotive effects in threshold inductive conditions. Only in this latter case would the other necessary signals be nonlimiting.

1.4.2 Autonomous promotion pathway

Representative genes of this pathway are *FCA* and *LUMINIDEPENDENS* (*LD*) (figure 1.1). These genes have no role in the flowering response to photoperiod, since mutants are still photoperiodic, flowering later than WT under both LDs and SDs. The *FCA* gene is expressed at low levels in all parts of WT plants, and at all developmental stages (Macknight *et al.*, 1997), whereas *LD* is mainly expressed in shoot and root tips (Aukerman *et al.*, 1999). Thus, these two genes may have different functions in the autonomous promotion pathway (Koornneef *et al.*, 1998b), although both have been shown to regulate—probably indirectly—the floral meristem identity genes *LFY* and *AP1*.

A common characteristic of mutants belonging to the autonomous promotion pathway is their increased sensitivity to vernalization that can almost completely suppress their late-flowering phenotype (Koornneef *et al.*, 1991). Clearly, vernalization makes the autonomous pathway unnecessary, suggesting that this pathway may converge downstream with the product(s) of the vernalization process.

Convergence apparently occurs at the level of the *FLOWERING LOCUS C* (*FLC*) gene (also named *FLF*) (figure 1.1). A dominant allele at this locus, together with a dominant allele at the *FRIGIDA* (*FRI*) locus, is responsible for the extreme late flowering of some natural ecotypes (Lee and Amasino, 1995). *FRI* is in fact a positive regulator of *FLC*, and the product of *FLC* (a transcription factor) is a strong repressor of flowering (Michaels and Amasino, 1999; Sheldon *et al.*, 1999). *FLC* expression is detectable in all plant parts, but there is circumstantial evidence that this gene acts primarily in the SAM. *FLC* expression is upregulated in the mutants belonging to the autonomous promotion pathway. This observation could explain the late-flowering phenotype of these mutants since overexpression of *FLC* has been found to delay markedly flowering in transgenic lines. The action of *FLC* may be to suppress the activity of the *SOC1* and *FT* genes, as the expression of these two genes is reduced in a *fca* mutant (Samach *et al.*, 2000). This cascade could account for the regulation of the downstream genes *LFY* and *AP1* by the autonomous pathway (figure 1.1).

The expression of *FLC* is downregulated to undetectable levels by a vernalization treatment, explaining the high cold sensitivity of ecotypes and mutants containing high levels of *FLC* product. In contrast, mutants of the photoperiod promotion pathway do not have increased *FLC* expression (Sheldon *et al.*, 1999), which accounts for their insensitivity to vernalization. This interpretation has recently been challenged by the observation that loss-of-function *flc* mutants retain sensitivity to vernalization, thus vernalization promotes flowering via

FLC-dependent and *FLC*-independent mechanisms (Michaels and Amasino, 2001).

Mutations at several loci named *VERNALIZATION* (*VRN*) suppress the response of the *fca* mutant to vernalization (Dean *et al.*, 1999). Thus the *VRN* genes appear to be involved in perception or transduction of the vernalization signal. One of these genes, *VRN2*, appears to be a putative suppressor of *FLC* (Sheldon *et al.*, 1999).

The convergence of the vernalization effect with the autonomous pathway, as revealed by genetic studies, is rather puzzling for physiologists. As already stressed above, autonomous flowering is of common occurrence, since many plants progressively lose their environmental requirements for promotion of flowering with age. To our knowledge, there is no evidence that vernalization would have a promotive effect in all these autonomously flowering plants, for example in old individuals of *Xanthium* or *Lolium* (strain Ceres) flowering slowly in noninductive photoperiods or in plants of day-neutral tobaccos or tomato (Halevy, 1985–1989). Thus it remains to be demonstrated whether the effect of vernalization as shown in figure 1.1 is peculiar to some plants or is of general applicability.

One clear outcome of the physiological work is that vernalization, to be effective in intact plants, must include the chilling of the mitotically active SAM (Lang, 1965; Bernier *et al.*, 1981a). This observation agrees with the above inference that *FLC* activity is primarily important at the SAM. Isolated shoot apices can also be vernalized, provided a carbohydrate is supplied in the culture medium *in vitro*. The vernalized state is cell autonomous, being transmitted by cell division only, and it was proposed that vernalization is partly mediated by genomic demethylation (Sheldon *et al.*, 2000). Evidence to support this hypothesis comes from a study of an *Arabidopsis METHYLTRANSFERASE 1* antisense line that exhibits a reduced level of DNA methylation and *FLC* activity, and flowers earlier than untransformed controls. However, hypomethylation may in other cases cause late flowering, indicating that the relationship between DNA methylation and vernalization is not simple (Soppe and Koornneef, 2000). The vernalized state, however stable it may be, can be suppressed in some plants, including *Arabidopsis*, by further growth in inadequate environmental conditions, such as high temperature, low irradiance or SDs (Lang, 1965; Bernier *et al.*, 1981a). Whether these so-called 'devernalizing' treatments do affect the levels of DNA methylation and gene activities remains to be investigated.

1.4.3   GA promotion pathway

Flowering of many ecotypes and late-flowering mutants of *Arabidopsis* is promoted by exogenous gibberellins (GAs) and delayed by inhibitors of GA biosynthesis (Napp-Zinn, 1969; Chandler and Dean, 1994). Moreover, mutants that are either deficient in GAs like *ga1*, or insensitive to GAs like *gibberellin insensitive* (*gai*), are late-flowering, especially in SDs (Wilson *et al.*, 1992). Promotion of flowering by GAs is also demonstrated by overexpressing the gene encoding the biosynthetic enzyme GA20-oxidase, or overexpressing components of the GA-signalling pathway, such as *FLOWERING PROMOTING FACTOR 1* (*FPF1*): transgenics look like plants repeatedly treated with GAs and flower early (Kania *et al.*, 1997; Coles *et al.*, 1999).

Application of a GA activates expression of both *SOC1* and *LFY*, but does not affect *FLC* expression (Sheldon *et al.*, 1999; Blásquez and Weigel, 2000; Borner *et al.*, 2000). Activation of *LFY* is through a cis-acting element in the promoter of the gene that resembles the consensus binding site of some MYB transcription factors. On the other hand, some GA-controlled germination-related processes in cereals are mediated by a *GAMYB* gene (Gubler *et al.*, 1995). Thus a *GAMYB*-like gene may provide a transactivation link between GAs and flowering, as shown in *Lolium* by Gocal *et al.* (1999). These authors observed that, after the plants have been induced to flower by one LD, there are increases in the GA contents of leaves and SAM, followed by upregulation in the SAM first of a *GAMYB* gene then of the *LFY* homologue.

The role of GAs in the control of the floral shift has been a controversial issue among physiologists (Zeevaart, 1983; Bernier, 1988; Metzger, 1995). This was mainly because all kinds of responses to exogenous GAs were obtained, from clear promotion in many rosetted LDPs or cold-requiring plants as well as in some caulescent SDPs (e.g. *Pharbitis*) to clear inhibition in other species, including the SDP Maryland Mammoth tobacco and many perennials (e.g. the SDP strawberry, the LDP *Fuchsia* and the day-neutral tomato). In between these opposite effects, GAs are simply ineffective in many other species of various environment-requirement types. This diversity might be related to the nature of the GAs tested. In the LDP *Lolium*, where a single application of GA_3 can substitute for the LD requirement, biochemical dissection has been pursued to design the most 'florigenic' GA, which at the highest specialization may have growth-retardant effects (Mander *et al.*, 1998). On the other hand, exogenous treatments with hormones are known to perturb the relationships between competing sinks, and this was proposed to explain, for example, the inhibitory effect of GAs on flowering of *Bougainvillea* (Halevy, 1985–1989). The same conclusion was reached by King and Ben-Tal (2001), who demonstrated that GA inhibits flowering of *Fuchsia hybrida* by reducing import into the SAM of leaf-sourced assimilates. At this point, it is evident that more targeted studies on GA action are needed.

In *Sinapis*, in which exogenous GAs have no effect on flowering time, it has been found that a single application of GA to the SAM of noninduced plants upregulates the expression of the *SOC1* homologue, *SaMADSA* (Bonhomme *et al.*, 2000). Thus, GAs may be part of the floral signals in *Sinapis*. However, whether GAs are long-distance signalling molecules remains to be shown. Indeed, these hormones are commonly detected in phloem and xylem saps, and thus are translocated, but genes involved in their biosynthesis have been found to be active in most plant parts in *Arabidopsis* (Xu *et al.*, 1997), casting doubts on the necessity for their transport (Colasanti and Sundaresan, 2000). The shoot tip itself has been shown to be an important site of GA production (Xu *et al.*, 1997; Coles *et al.*, 1999) and it was recently observed in rice that GA catabolism is drastically altered around the shoot apex by the transition from vegetative to reproductive development (Sakamoto *et al.*, 2001).

1.5   Interactions between different genes and pathways: welcome to complexity

Independence of the three promotion pathways just described was deduced from the fact that double mutants blocked in two of them flower significantly later than their latest parent, as is the case for *co;fca*, *co;ga1* and *fca;ga1* double mutants (Putterill *et al.*, 1995; Koornneef *et al.*, 1998a; Chandler *et al.*, 2000). These three pathways are also thought to act in parallel, since each may compensate for the inefficiency of the others: when one promotion pathway is blocked, increasing the activity of another may partially or completely rescue the late-flowering phenotype. Such a situation has been found when *fca* mutation was weakened by overexpressing *CO*, or when a GA was applied to *co* or *fca* mutants (Chandler and Dean, 1994; Piñeiro and Coupland, 1998). According to this scheme, only a triple mutant blocked in all promotion pathways should be unable to flower. A triple *co;fca;ga1* mutant was indeed found to be unable to flower in LDs, provided plants are not vernalized (Reeves and Coupland, 2001). Vernalization rescues flowering of the triple mutant by overcoming the block in the autonomous promotion pathway (figure 1.1). By adding a *vrn* mutation in this background, one might expect to prevent flowering in any environmental conditions.

At this point, the pathways promoting the floral transition to flowering in *Arabidopsis* appear in figure 1.1 as multiple, linear, parallel, and independent; these characteristics somehow overlap and deserve discussion.

1.5.1   Are the pathways linear?

The pathways are not as ideally linear as pictured in figure 1.1 and may, on the contrary, loop back at several levels. We have already mentioned 'secondary' loops within the input and output pathways of the biological clock in the photoperiod pathway. Another example comes from the autonomous pathway gene *FCA* that was shown above to positively regulate the downstream gene *AP1*. It was recently found that in turn *AP1* negatively regulates *FCA* function (Page *et al.*, 1999), establishing a feedback loop between these two genes.

1.5.2 Do the pathways act in parallel?

Although physiological studies have shown the existence of alternative pathways leading to flowering (see above), the situation depicted in *Arabidopsis* may be simplified. Indeed, many species flower only when environmental requirements are fulfilled in sequence. The most common examples are those many plants requiring first vernalization and only afterwards photoperiodic induction (most often LDs) (Lang, 1965). This was interpreted as indicating that vernalization renders the SAM competent to respond to the photoperiod-dependent floral signals. If we refer to figure 1.1, this might actually mean that suppression of *FLC* activity by vernalization is a prerequisite for *SOC1* and *FT* to be activated by CO or a GA. As far as GAs are concerned, it was found that they are able to accelerate flowering in *FLC* overexpressors, which are very late, and that this effect occurs without downregulating this gene (Sheldon *et al.*, 2000). However, the GA dose that is required is much higher than is normally required in WT plants or many late-flowering mutants. Thus the *FLC* gene seems indeed to antagonize the promotive effect of GAs in the SAM.

1.5.3 Are the pathways independent?

For physiologists, the postulated independence of the GA pathway appears quite puzzling, since these hormones have long been hypothesized to be involved in photoperiod or vernalization signalling. It is well-documented that exposure of many plants, including *Arabidopsis*, to LDs (photoperiod pathway) increases the endogenous GA contents and upregulates some genes of the GA biosynthesis pathway (Metzger, 1995; Xu *et al.*, 1997). Similarly, vernalization causes the specific induction of a GA biosynthetic enzyme and subsequent increases in GAs in the SAM of *Thlaspi arvense* and *Brassica napus* (Metzger, 1995; Zanewich and Rood, 1995). If the same situation could be shown in *Arabidopsis*, then all three promotion pathways could be linked by GAs. However, recent results suggest that GAs are not required for vernalization in *Arabidopsis*, since the double *fca;ga1* mutant—very deficient in GAs—responds as strongly to vernalization as the *fca* single mutant (Chandler *et al.*, 2000). The main control may actually be exerted downstream, in the signalling cascade. Vernalization was indeed found to increase the GA sensitivity of both the *fca* mutant of *Arabidopsis* and *Eustoma grandiflorum* (Oka *et al.*, 2001), and late flowering mutants previously allocated to the autonomous pathway (*fpa*) were identified in a gibberellin-response screen (Meier *et al.*, 2001).

Another simplification of genetic schemes as depicted in figure 1.1 that is open to criticism is to confine light and the biological clock to the photoperiod promotion pathway: their key functions there are not questioned, but their ubiquitous involvement in plant metabolism may suggest a more widespread action (McClung, 2000). Before going to complexity, a question that remains to be debated here is how photoreceptors and the biological clock regulate flowering time. Two models have long been opposed and are reassessed in recent reviews (Lin, 2000; Samach and Coupland, 2000). The 'internal coincidence' model proposes that photoperiod changes the 'relative phasing' of underlying circadian rhythms, with floral induction occurring only when specific phase points in critical multiple rhythms coincide. Support for this hypothesis comes from the fact that several genes that are required for the photoperiodic response of *Arabidopsis* are clock-regulated and the phase of their rhythmic expression changes in LDs compared to SDs (e.g. *GI*, *CO* and *FT*; Suárez-López *et al.*, 2001). Other 'internal' rhythms that could be important in the coincidental induction of flowering could concern 'housekeeping' metabolites acting as modulators or 'floral signals'. The 'external coincidence' model, originally proposed by Bünning, postulates that induction depends on the coincidence of an external signal (light) with an internal light-sensitive phase of a circadian rhythm. In more fashionable terms, the biological clock 'gates' light-regulated processes and, conceptually, signal transduction can proceed only at the phase when the gate is open. *ELF3* may have a critical role in this 'gating' of the sensitivity to light (McWatters *et al.*, 2000). Thus the functions of photore-ceptors in this model are twofold: they regulate the circadian clock, as seen above, and also mediate signal transductions that directly affect floral initiation. This hypothesis could explain why phyA and cry2, although they affect clock entrainment only under specialized low-fluence conditions, have marked effects on the photoperiodic control of flowering in normal conditions. At the molecular level, a direct effect of photoreceptors on floral initiation is suggested, for example, by the fact that the early-flowering phenotype of a *phyB* mutant is independent of *CO* and *FT* expression, that is of the photoperiod promotion pathway (Blásquez and Weigel, 1999). Conceptually, the 'internal' and 'external' coincidence models are not antagonistic since light, which is regarded as an 'external' factor, regulates many aspects of plant metabolism whose products could in turn be regarded as 'internal' effectors. Thus there is increasing evidence that light does not act through the biological clock only, and it is also known that circadian rhythms run in etiolated seedlings (i.e. in complete darkness) and may be synchronized by seed imbibition (Zhong *et al.*, 1998). If some independent actions of photoreceptors and the biological clock are accepted, there is no obvious reason to include them in the photoperiod promotion pathway only, and experimental evidence is accumulating along this line. For example, *ELF3* apparently functions upstream of the autonomous pathway too (Chou and Yang, 1999), and mutations in the autonomous pathway repressor *FLC* shorten circadian period (Swarup *et al.*, 1999).

1.5.4 Where do the pathways converge?

We described above that, although all pathways promoting flowering have been shown to regulate the activity of the floral meristem identity genes *LFY/AP1*, convergence may occur upstream of these genes, towards *SOC1* and *FT* (figure 1.1). Interestingly, the *SOC1* homologue of *Sinapis* (*SaMADSA*) can be induced by both a CK, which had been previously identified as a floral signal in photoperiodic induction (figure 1.2), and a GA (Bonhomme *et al.*, 2000). In addition, the constitutive expression of *SOC1* in *Arabidopsis* is sufficient for photoperiod-independent flowering. This bypassing of the photoperiodic control is also observed, and this is still more remarkable, when *SaMADSA* is overexpressed in the obligate SDP Maryland Mammoth tobacco (Borner *et al.*, 2000). Involvement of *SOC1* in the autonomous/vernalization pathway was clearly demonstrated by Lee *et al.* (2000), who screened for *FRI/FLC* suppressors and found plants overexpressing *SOC1*. Thus *SOC1* is an important integrator of floral promotion pathways.

Another 'meeting point' of the pathways leading to flowering that may be a common effector is carbohydrate supply. Sufficient carbohydrate availability at the SAM has been found on many occasions to be critical for its switch to the floral transition. Importantly, a sucrose supply to the SAM of *Arabidopsis* plants grown *in vitro*, either in LDs or in continuous darkness, suppresses the late-flowering phenotype of many mutants belonging to either the photoperiod or the autonomous promotion pathways (Roldán *et al.*, 1999). Only mutants *ft* and *fwa* are not rescued by sucrose. It seems, therefore, that the promotion by sucrose operates downstream of the photoperiod and autonomous pathways, but upstream of *FT /FWA*. One target of sucrose is *LFY* (Blásquez *et al.*, 1998): activation of the *LFY* promoter by a GA (see above) does indeed require the presence of sucrose. Another proposed effect of sucrose is to decrease *FLC* expression, but this remains to be documented (Sheldon *et al.*, 2000). On the other hand, in *Sinapis*, sucrose alone does not affect the expression of the *SOC1* homologue (Bonhomme *et al.*, 2000).

According to the growth conditions and the photosynthetic capacity of the plants, sucrose involved in the developmental switch of the SAM may come either directly from photosynthetic sites or from storage sources. For example, *phosphoglucomutase* (*pgm*) and *phospho-glucose isomerase* (*pgi*) mutants of *Arabidopsis*, which are unable to synthesize starch, have a late-flowering phenotype in SDs only, which can be rescued by sucrose application (Corbesier *et al.*, 1998; Yu *et al.*, 2000). In LDs, the extended daily periods suffice to supply enough sucrose to the SAM, and the starch mutants flower as early as WT. This suggests that carbohydrates are involved in the photoperiod promotion pathway. However, since these mutants remains photoperiod-sensitive and can be rescued by vernalization (Bernier *et al.*, 1993; Corbesier *et al.*, 1998), sucrose has been implicated in the autonomous pathway too.

Surprisingly, the *sucrose-uncoupled* (*sun6*) mutant, which is insensitive to sucrose, has an early-flowering phenotype (Dijkwel *et al.*, 1997). This could be due to a high carbohydrate level, since the regulation of carbohydrate home-ostasis may be altered in this mutant. However, this mutation was also found to be allelic with *abscisic acid insensitive 4* (*abi4*) (Smeekens, 2000). Clas-sically, abscisic acid (ABA) is considered as a general 'background' inhibitor of flowering (Bernier, 1988). In *Arabidopsis*, ABA-deficient mutants flower earlier than WT (Martínez-Zapater *et al.*, 1994) and combining *abi* mutations to *fca* partially compensates the late-flowering phenotype of the latter parent (Chandler *et al.*, 2000). It is thus surprising that, up to now, ABA has received little consideration in the control of flowering in *Arabidopsis*. Recent evidence showing that a protein identical to TOC1 has the ability to interact with the gene product of *ABI3* should motivate new interest in that direction (Kurup *et al.*, 2000).

Sucrose may act directly at the SAM but has also been shown in *Sinapis* to trigger CK production by the roots (see figure 1.2). Information about the putative role of this class of hormones in *Arabidopsis* is scarce but conclusive. It has been shown that indeed exogenous application of a CK accelerates flowering, but only when the light irradiance is low, supporting the idea that CK effect is dependent on the carbohydrate level (Dennis *et al.*, 1996). Interestingly, genetic enrichment in CKs by the *altered meristem program* (*amp1*) mutation has been found to hasten flowering (Chaudhury *et al.*, 1993) and to rescue the late-flowering phenotype of the *gi* mutant belonging to the photoperiod promotion pathway (Dennis *et al.*, 1996). These observations suggest that CKs act downstream of *GI*. On the other hand, *amp1* is unable to rescue the late-flowering phenotype of *fca*, suggesting that CKs are not sufficient to overcome *FLC* inhibition (figure 1.1). These results are consistent with the observation that, in *Sinapis*, a CK upregulates the *SOC1* homologue but does not, alone, affect flowering time (Bonhomme *et al.*, 2000).

As suggested by Dielen *et al.* (2001), similar chemicals may be implicated in the SAM floral transition in autonomous plants: in the late-flowering *uniflora* mutant of tomato, transition *in vitro* requires sucrose and CKs, together with nitrogenous nutrients, at optimal concentrations.

It is evident from the previous discussion that the mechanisms leading to floral commitment are complex, and that the SAM integrates signals coming from the different—interconnected—pathways. This was already suspected from physiological experiments showing that a plant is able to sum whatever changes are caused by alternative inductive treatments (Bernier, 1988). Indeed, it has been observed that, in plants capable of flowering in response to a variety of environmental pathways, flowering can be reached by sequential exposure to different inductive treatments each at a subthreshold levels (Bernier, 1988). Strikingly, inductive factors of a totally opposite nature may have additive effects, such as low and high temperatures, or continuous darkness and LDs.

1.6 Conclusions

It is no longer possible to produce a straightforward scheme to explain how flowering is regulated. Evidence is rapidly accumulating that the so-called 'independent' pathways interact in complex ways and the reason for their conceptual separation may be purely experimental. Indeed researchers, either geneticists or physiologists, design experiments in such a way that some pathways are brought to the forefront while others are placed in the background. In such a selective and somewhat artificial approach, the search for limiting factors such as floral signals may ignore nonlimiting components. As a consequence, the latter may be inappropriately classified as 'nonessential' or even as 'inhibitory' if additional supply results in supraoptimal levels. On the other hand, their implication could be highlighted in other growth conditions, leading to controversy. However, since the final issue of all pathways is the activation of floral initiation genes, it is conceivable that signals are common and probably act in balance between 'repressive' and 'promotive' pathways; we presented data supporting this view. Carbohydrates and hormones may play such a role, in part and in concert, and there is an increasing need to combine their analysis with genetic networks. Another gap to fill between the genetic and physiological approaches is a careful examination of the spatial and temporal expression patterns of the isolated genes. Finally, one cannot ignore the exciting discoveries taking place in our understanding of circadian mechanisms in plants, and it can be expected that these new developments will throw new light on the flowering process. Indeed, if many genes are known to exhibit circadian oscillations, and the importance of some of these has been stressed, the circadian regulation of hormone production and responsiveness can no longer be ignored.

The labyrinthine system of control of flowering time is probably critical for the successful reproduction of plants in the contrasting and unpredictable environments that have been experienced throughout evolution. Extending the multidisciplinary approach of flowering to as many plant species as possible will undoubtedly confirm the incredible plasticity of this process (see Weller *et al.*, 1997). In that respect, it may be expected that any experimental approach will start with simple schemes and end up, as in *Arabidopsis*, with the need for the third dimension!

Acknowledgements

We thank Professor J.-M. Kinet for critical reading of the manuscript. This review is based on work supported by Pôles d'Attraction Interuniversitaires Belges (Service du Premier Ministre, Services Fédéraux des Affaires Scientifiques, Techniques et Culturelles, P4/15).

References

Aukerman, M.J., Lee, I., Weigel, D. and Amasino, R.M. (1999) The *Arabidopsis* flowering-time gene *LUMINIDEPENDENS* is expressed primarily in regions of cell proliferation and encodes a nuclear protein that regulates *LEAFY* expression. *Plant J.*, **18**, 195-203.

Bernier, G. (1988) The control of floral evocation and morphogenesis. *Annu. Rev. Plant Physiol. Plant Mol. Biol.*, **39**, 175-219.

Bernier, G. (1989) Events of the floral transition of meristems, in *Plant Reproduction: From Floral Induction to Pollination* (eds. E. Lord and G. Bernier), Symposium Series vol. I, The American Society of Plant Physiologists, Rockville, MD, pp. 42-50.

Bernier, G., Kinet, J.-M. and Sachs, R.M. (1981a) *The Physiology of Flowering*, vol. I, CRC Press, Boca Raton, FL.

Bernier, G., Kinet, J.-M. and Sachs, R.M. (1981b) *The Physiology of Flowering*, vol. II, CRC Press, Boca Raton, FL.

Bernier, G., Havelange, A., Houssa, C., Petitjean, A. and Lejeune, P. (1993) Physiological signals that induce flowering. *Plant Cell*, **5**, 1147-1155.

Bernier, G., Corbesier, L., Périlleux, C., Havelange, A. and Lejeune, P. (1998) Physiological analysis of the floral transition, in *Genetic and Environmental Manipulation of Horticultural Crops* (eds. K.E. Cockshull, D. Gray, G.B. Seymour and B. Thomas), CAB International, Wallingford, pp. 103-109.

Blázquez, M.A. and Weigel, D. (1999) Independent regulation of flowering by phytochrome B and gibberellins in *Arabidopsis. Plant Physiol.*, **120**, 1025-1032.

Blázquez, M.A. and Weigel, D. (2000) Integration of floral inductive signals in *Arabidopsis. Nature*, **404**, 889-892.

Blázquez, M.A., Soowal, L.N., Lee, I. and Weigel, D. (1997) *LEAFY* expression and flower initiation in *Arabidopsis. Development*, **124**, 3835-3844.

Blázquez, M.A., Green, R., Nilsson, O., Sussman, M.R. and Weigel, D. (1998) Gibberellins promote flowering of *Arabidopsis* by activating the *LEAFY* promoter. *Plant Cell*, **10**, 791-800.

Bonhomme, F., Kurz, B., Melzer, S., Bernier, G. and Jacqmard, A. (2000) Cytokinin and gibberellin activate *SaMADS A*, a gene apparently involved in regulation of the floral transition in *Sinapis alba. Plant J.*, **24**, 103-111.

Borner, R., Kampmann, G., Chandler, J., Gleißner, R., Wisman, E., Apel, K. and Melzer, S. (2000) A MADS domain gene involved in the transition to flowering in *Arabidopsis. Plant J.*, **24**, 591-599.

Bradley, D., Ratcliffe, O., Vincent, C., Carpenter, R. and Coen, E. (1997) Inflorescence commitment and architecture in *Arabidopsis. Science*, **275**, 80-83.

Chandler, J. and Dean, C. (1994) Factors influencing the vernalization response and flowering time of late flowering mutants of *Arabidopsis thaliana* (L.) Heynh. *J. Exp. Bot.*, **45**, 1279-1288.

Chandler, J., Martínez-Zapater, J.M. and Dean, C. (2000) Mutations causing defects in the biosynthesis and response to gibberellins, abscisic acid and phytochrome B do not inhibit vernalization in *Arabidopsis fca-1. Planta*, **210**, 677-682.

Chaudhury, A.M., Letham, S., Craig, S. and Dennis, E.S. (1993) *amp1*—a mutant with high cytokinin levels and altered embryonic pattern, faster vegetative growth, constitutive photomorphogenesis and precocious flowering. *Plant J.*, **4**, 907-916.

Chen, L., Cheng, J.-C., Castle, L. and Sung, Z.R. (1997) *EMF* genes regulate *Arabidopsis* inflorescence development. *Plant Cell*, **9**, 2011-2024.

Chou, M.-L. and Yang, C.-H. (1999) Late-flowering genes interact with early-flowering genes to regulate flowering time in *Arabidopsis thaliana. Plant Cell Physiol.*, **40**, 702-708.

Cleland, C.F. (1978) The flowering enigma. *BioScience*, **28**, 265-269.

Cockcroft, C.E., den Boer, B.G.W., Healy, J.M.S. and Murray, J.A.H. (2000) Cyclin D control of growth rate in plants. *Nature*, **405**, 575-579.

Colasanti, J. and Sundaresan, V. (2000) 'Florigen' enters the molecular age: long-distance signals that cause plants to flower. *Trends Biochem. Sci.*, **25**, 236-240.

Colasanti, J., Yuan, Z. and Sundaresan, V. (1998) The *indeterminate* gene encodes a zinc finger protein and regulates a leaf-generated signal required for the transition to flowering in maize. *Cell*, **93**, 593-603.

Coles, J.P., Phillips, A.L., Croker, S.J., García-Lepe, R., Lewis, M.J. and Hedden, P. (1999) Modification of gibberellin production and plant development in *Arabidopsis* by sense and antisense expression of gibberellin 20-oxidase genes. *Plant J.*, **17**, 547-556.

Corbesier, L., Gadisseur, I., Silvestre, G., Jacqmard, A. and Bernier, G. (1996) Design in *Arabidopsis thaliana* of a synchronous system of floral induction by one long day. *Plant J.*, **9**, 947-952.

Corbesier, L., Lejeune, P. and Bernier, G. (1998) The role of carbohydrates in the induction of flowering in *Arabidopsis thaliana*: comparison between the wild type and a starchless mutant. *Planta*, **206**, 131-137.

Corbesier, L., Havelange, A., Lejeune, P., Bernier, G. and Périlleux, C. (2001) N-content of phloem and xylem exudates during the transition to flowering in *Sinapis alba* and *Arabidopsis thaliana*. *Plant, Cell Environ.*, **24**, 367-375.

Cremer, F., Van de Walle, C. and Bernier, G. (1991) Changes in mRNA level rhythmicity in the leaves of *Sinapis alba* during a lengthening of the photoperiod which induces flowering. *Plant Mol. Biol.*, **17**, 465-473.

Dean, C., Dijkwel, P., Duroux, M., *et al.* (1999) Molecular analysis of flowering time and vernalization response in *Arabidopsis*. *Flowering Newslett.*, **28**, 6-11.

Deitzer, G.F. (1984) Photoperiodic induction in long-day plants, in *Light and the Flowering Process* (eds. D. Vince-Prue, B. Thomas and K.E. Cockshull), Academic Press, London, pp. 51-63.

Dennis, E.S., Finnegan, E.J., Bilodeau, P., *et al.* (1996) Vernalization and the initiation of flowering. *Semin. Cell Dev. Biol.*, **7**, 441-448.

Dielen, V., Lecouvet, V., Dupont, S. and Kinet, J.-M. (2001) *In vitro* control of floral transition in tomato (*Lycopersicon esculentum* Mill.), the model of autonomously flowering plants, using the late flowering *uniflora* mutant. *J. Exp. Bot.*, **52**, 715-723.

Dijkwel, P.P., Huijser, C., Weisbeek, P.J., Chua, N.-H. and Smeekens, S.C.M. (1997) Sucrose control of phytochrome A signaling in *Arabidopsis*. *Plant Cell*, **9**, 583-595.

Dunlap, J. (1999) Molecular bases for circadian clocks. *Cell*, **96**, 271-290.

Evans, L.T. (1969) The nature of flower induction, in *The Induction of Flowering—Some Case Histories* (ed. L.T. Evans), The Macmillan Company of Australia, Melbourne, pp. 457-480.

Fowler, S., Lee, K., Onouchi, H., *et al.* (1999) *GIGANTEA*: a circadian clock-controlled gene that regulates photoperiodic flowering in *Arabidopsis* and encodes a protein with several possible membrane-spanning domains. *EMBO J.*, **18**, 4679-4688.

Gocal, G.F.W., Poole, A.T., Gubler, F., Watts, R.J., Blundell, C. and King, R.W. (1999) Long-day up-regulation of a *GAMYB* gene during *Lolium temulentum* inflorescence formation. *Plant Physiol.*, **119**, 1271-1278.

Gocal, G.F.W., King, R.W., Blundell, C.A., Schwartz, O.M., Andersen, C.H. and Weigel, D. (2001) Evolution of floral meristem identity genes. Analysis of *Lolium temulentum* genes related to *APETALA1* and *LEAFY* of *Arabidopsis*. *Plant Physiol.*, **125**, 1788-1801.

Goto, N., Kumagai, T. and Koornneef, M. (1991) Flowering responses to light breaks in photomorphogenic mutants of *Arabidopsis thaliana*, a long-day plant. *Physiol. Plant.*, **83**, 209-215.

Green, R.M. and Tobin, E.M. (1999) Loss of the circadian clock-associated protein 1 in *Arabidopsis* results in altered clock-regulated gene expression. *Proc. Natl. Acad. Sci. USA*, **96**, 4176-4179.

Gubler, F., Kalla, R., Roberts, J.K. and Jacobsen, J.V. (1995) Gibberellin-regulated expression of a *MYB* gene in barley aleurone cells: evidence for MYB transactivation of a high-pI α-amylase gene promoter. *Plant Cell*, **7**, 1879-1891.

Guo, H., Yang, H., Mockler, T.C. and Lin, C. (1998) Regulation of flowering time by *Arabidopsis* photoreceptors. *Science*, **279**, 1360-1363.

Halevy, A.H. (1985–1989). *Handbook of Flowering*, vols. I to VI, CRC Press, Boca Raton, FL.

Hartmann, U., Höhmann, S., Nettesheim, K., Wisman, E., Saedler, H. and Huijser, P. (2000) Molecular cloning of *SVP*: a negative regulator of the floral transition in *Arabidopsis*. *Plant J.*, **21**, 351-360.

Haung, M.-D. and Yang, C.-H. (1998) *EMF* genes interact with late-flowering genes to regulate *Arabidopsis* shoot development. *Plant Cell Physiol.*, **39**, 382-393.

Havelange, A., Lejeune, P. and Bernier, G. (2000) Sucrose/cytokinin interaction in *Sinapis alba* at floral induction: a shoot-to-root-to-shoot physiological loop. *Physiol. Plant.*, **109**, 343-350.

Heide, O.M. (1994) Control of flowering and reproduction in temperate grasses. *New Phytol.*, **128**, 347-362.

Heintzen, C., Nater, M., Apel, K. and Staiger, D. (1997) *At*GRP7, a nuclear RNA-binding protein as a component of a circadian-regulated negative feedback loop in *Arabidopsis thaliana*. *Proc. Natl. Acad. Sci. USA*, **94**, 8515-8520.

Hempel, F.D., Weigel, D., Mandel, M.A., *et al.* (1997) Floral determination and expression of floral regulatory genes in *Arabidopsis*. *Development*, **124**, 3845-3853.

Hicks, K.A., Millar, A.J., Carré, I.A., *et al.* (1996) Conditional circadian dysfunction of the *Arabidopsis early-flowering 3* mutant. *Science*, **274**, 790-792.

Huq, E., Tepperman, J.M. and Quail, P.H. (2000) GIGANTEA is a nuclear protein involved in phytochrome signalling in *Arabidopsis*. *Proc. Natl. Acad. Sci. USA*, **97**, 9789-9794.

Irish, E. and Jegla, D. (1997) Regulation of extent of vegetative development of the maize shoot meristem. *Plant J.*, **11**, 63-71.

Johnson, E., Bradley, M., Harberd, N.P. and Whitelam, G.C. (1994) Photoresponses of light-grown *phyA* mutants of *Arabidopsis*. Phytochrome A is required for the perception of daylength extensions. *Plant Physiol.*, **105**, 141-149.

Kania, T., Russenberger, D., Peng, S., Apel, K. and Melzer, S. (1997) *FPF1* promotes flowering in *Arabidopsis*. *Plant Cell*, **9**, 1327-1338.

Kardailsky, I., Shukla, V.K., Ahn, J.H., *et al.* (1999) Activation tagging of the floral inducer *FT*. *Science*, **286**, 1962-1965.

King, R.W. and Ben-Tal, Y. (2001) A florigenic effect of sucrose in *Fuchsia hybrida* is blocked by gibberellin-induced assimilate competition. *Plant Physiol.*, **125**, 488-496.

Kobayashi, Y., Kaya, H., Goto, K., Iwabuchi, M. and Araki, T. (1999) A pair of related genes with antagonistic roles in mediating flowering signals. *Science*, **286**, 1960-1962.

Koornneef, M., Hanhart, C.J. and van der Veen, J.H. (1991) A genetic and physiological analysis of late flowering mutants in *Arabidopsis thaliana*. *Mol. Gen. Genet.*, **229**, 57-66.

Koornneef, M., Alonso-Blanco, C., Blankestijn-de Vries, H., Hanhart, C.J. and Peeters, A.J.M. (1998a) Genetic interactions among late-flowering mutants of *Arabidopsis*. *Genetics*, **148**, 885-892.

Koornneef, M., Alonso-Blanco, C., Peeters, A.J.M. and Soppe, W. (1998b) Genetic control of flowering time in *Arabidopsis*. *Annu. Rev. Plant Physiol. Plant Mol. Biol.*, **49**, 345-370.

Kreps, J.A. and Simon, A.E. (1997) Environmental and genetic effects on circadian-clock-regulated gene expression in *Arabidopsis*. *Plant Cell*, **9**, 297-304.

Kurup, S., Jones, H.D. and Holdsworth, M.J. (2000) Interactions of the developmental regulator ABI3 with proteins identified from developing *Arabidopsis* seeds. *Plant J.*, **21**, 143-155.

Landschütze, V., Willmitzer, L. and Müller-Röber, B. (1995) Inhibition of flower formation by antisense repression of mitochondrial citrate synthase in transgenic potato plants leads to a specific disintegration of the ovary tissues of flowers. *EMBO J.*, **14**, 660-666.

Lang, A. (1965) Physiology of flower initiation, in *Encyclopedia of Plant Physiology*, vol. XV, part 1 (ed. W. Ruhland), Springer Verlag, Berlin, pp. 1380-1536.

Larkin, J.C., Felsheim, R. and Das, A. (1990) Floral determination in the terminal bud of the short-day plant *Pharbitis nil*. *Dev. Biol.*, **137**, 434-443.

Lee, H., Suh, S.-S., Park, E., *et al.* (2000) The AGAMOUS-LIKE 20 MADS domain protein integrates floral inductive pathways in *Arabidopsis*. *Genes Dev.*, **14**, 2366-2376.

Lee, I. and Amasino, R.M. (1995) Effect of vernalization, photoperiod, and light quality on the flowering phenotype of *Arabidopsis* plants containing the *FRIGIDA* gene. *Plant Physiol.*, **108**, 157-162.

Levy, Y.Y. and Dean, C. (1998) The transition to flowering. *Plant Cell*, **10**, 1973-1989.

Lin, C. (2000) Photoreceptors and regulation of flowering time. *Plant Physiol.*, **123**, 39-50.

Lincoln, C., Long, J., Yamaguchi, J., Serikawa, K. and Hake, S. (1994) A *knotted1*-like homeobox gene in *Arabidopsis* is expressed in the vegetative meristem and dramatically alters leaf morphology when overexpressed in transgenic plants. *Plant Cell*, **6**, 1859-1876.

Liu, J., Yu, J., McIntosh, L., Kende, H. and Zeevaart, J.A.D. (2001) Isolation of a *CONSTANS* ortholog from *Pharbitis nil* and its role in flowering. *Plant Physiol.*, **125**, 1821-1830.

Lumsden, P.J., Thomas, B. and Vince-Prue, D. (1982) Photoperiodic control of flowering in dark-grown seedlings of *Pharbitis nil* Choisy. The effect of skeleton and continuous light photoperiods. *Plant Physiol.*, **70**, 277-282.

Macknight, R., Bancroft, I., Page, T., *et al.* (1997) *FCA*, a gene controlling flowering time in *Arabidopsis*, encodes a protein containing RNA-binding domains. *Cell*, **89**, 737-745.

Mander, L.N., Sheburn, M., Camp, D., King, R.W., Evans, L.T. and Pharis, R.P. (1998) Effects of D-ring modified gibberellins on flowering and growth in *Lolium temulentum*. *Phytochemistry*, **49**, 2195-2206.

Martínez-Zapater, J.M., Coupland, G., Dean, C. and Koornneef, M. (1994) The transition to flowering in *Arabidopsis*, in *Arabidopsis* (eds. E.M. Meyerowitz and C.R. Somerville), Cold Spring Harbor Press, Cold Spring Harbor, NY, pp. 403-433.

McClung, C.R. (2000) Circadian rhythms in plants: a millennial view. *Physiol. Plant.*, **109**, 359-371.

McDaniel, C.N. (1996) Developmental physiology of floral initiation in *Nicotiana tabacum* L. *J. Exp. Bot.*, **47**, 465-475.

McDaniel, C.N., King, R.W. and Evans, L.T. (1991) Floral determination and *in-vitro* floral differentiation in isolated shoot apices of *Lolium temulentum* L. *Planta*, **185**, 9-16.

McWatters, H.G., Bastow, R.M., Hall, A. and Millar, A.J. (2000) The *ELF3 zeitnehmer* regulates light signalling to the circadian clock. *Nature*, **408**, 716-720.

Meier, C., Bouquin, T., Nielsen, M.E., *et al.* (2001) Gibberellin response mutants identified by luciferase imaging. *Plant J.*, **25**, 509-519.

Menzel, G., Apel, K. and Melzer, S. (1996) Identification of two MADS box genes that are expressed in the apical meristem of the long-day plant *Sinapis alba* in transition to flowering. *Plant J.*, **9**, 399-408.

Metzger, J.D. (1995) Hormones and reproductive development, in *Plant Hormones* (ed. P.J. Davies), 2nd edn, Kluwer Academic, Dordrecht, pp. 617-648.

Michaels, S.D. and Amasino, R.M. (1999) *FLOWERING LOCUS C* encodes a novel MADS domain protein that acts as a repressor of flowering. *Plant Cell*, **11**, 949-956.

Michaels, S.D. and Amasino, R.M. (2001) Loss of *FLOWERING LOCUS C* activity eliminates the late-flowering phenotype of *FRIGIDA* and autonomous pathway mutations but not responsiveness to vernalization. *Plant Cell*, **13**, 935-941.

Millar, A.J., Carré, I.A., Strayer, C.A., Chua, N.-H. and Kay, S.A. (1995a) Circadian clock mutants in *Arabidopsis* identified by luciferase imaging. *Science*, **267**, 1161-1163.

Millar, A.J., Straume M., Chory, J., Chua N.-H. and Kay, S.A. (1995b) The regulation of circadian period by phototransduction pathways in *Arabidopsis*. *Science*, **267**, 1163-1166.

Mockler, T.C., Guo, H., Yang, H., Duong, H. and Lin, C. (1999) Antagonistic actions of *Arabidopsis* cryptochromes and phytochrome B in the regulation of floral induction. *Development*, **126**, 2073-2082.

Napp-Zinn, K. (1969) *Arabidopsis thaliana* (L.) Heynh, in *The Induction of Flowering. Some Case Histories* (ed. L.T. Evans), The Macmillan Company of Australia, Melbourne, pp. 291-304.

Nelson, D.C., Lasswell, J., Rogg, L.E., Cohen, M.A. and Bartel, B. (2000) *FKF1*, a clock-controlled gene that regulates the transition to flowering in *Arabidopsis*. *Cell*, **101**, 331-340.

Ng, M. and Yanofski, M.F. (2000) Three ways to learn the ABCs. *Curr. Opin. Plant Biol.*, **3**, 47-52.

Nilsson, O., Lee, I., Blázquez, M.A. and Weigel, D. (1998) Flowering-time genes modulate the response to *LEAFY* activity. *Genetics*, **150**, 403-410.

Oka, M., Tasaka, Y., Iwabuchi, M. and Mino, M. (2001) Elevated sensitivity to gibberellin by vernalization in the vegetative rosette plants of *Eustoma grandiflorum* and *Arabidopsis thaliana. Plant Sci.*, **160**, 1237-1245.

O'Neill, S.D. (1992) The photoperiodic control of flowering: progress toward understanding the mechanism of induction. *Photochem. Photobiol.*, **56**, 789-801.

O'Neill, S.D. (1993) Changes in gene expression associated with floral induction and evocation, in *The Molecular Biology of Flowering* (ed. B.R. Jordan), CAB International, Wallingford, pp. 69-92.

Onouchi, H., Igeño, M.I., Périlleux, C., Graves, K. and Coupland, G. (2000) Mutagenesis of plants overexpressing *CONSTANS* demonstrates novel interactions among *Arabidopsis* flowering-time genes. *Plant Cell*, **12**, 885-900.

Page, T., Macknight, R., Yang, C.-H. and Dean, C. (1999) Genetic interactions of the *Arabidopsis* flowering time gene *FCA*, with genes regulating floral initiation. *Plant J.*, **17**, 231-239.

Park, D.H., Somers, D.E., Kim, Y.S., *et al.* (1999) Control of circadian rhythms and photoperiodic flowering by the *Arabidopsis GIGANTEA* gene. *Science*, **285**, 1579-1582.

Périlleux, C., Bernier, G. and Kinet, J.-M. (1994) Circadian rhythms and the induction of flowering in the long-day grass *Lolium temulentum* L. *Plant, Cell Environ.*, **17**, 755-761.

Périlleux, C., Ongena, P. and Bernier, G. (1996) Changes in gene expression in the leaf of *Lolium temulentum* L. Ceres during the photoperiodic induction of flowering. *Planta*, **200**, 32-40.

Piñeiro, M. and Coupland, G. (1998) The control of flowering time and floral identity in *Arabidopsis. Plant Physiol.*, **117**, 1-8.

Putterill, J., Robson, F., Lee, K., Simon, R. and Coupland, G. (1995) The *CONSTANS* gene of *Arabidopsis* promotes flowering and encodes a protein showing similarities to zinc finger transcription factors. *Cell*, **80**, 847-857.

Ratcliffe, O.J., Bradley, D.J. and Coen, E.S. (1999) Separation of shoot and floral identity in *Arabidopsis. Development*, **126**, 1109-1120.

Reed, J.W., Nagatanii, A., Elich, T.D., Fagan, M. and Chory, J. (1994) Phytochrome A and phytochrome B have overlapping but distinct function in *Arabidopsis* development. *Plant Physiol.*, **104**, 1139-1149.

Reed, J.W., Nagpal, P., Bastow, R.M., *et al.* (2000) Independent action of ELF3 and phyB to control hypocotyl elongation and flowering time. *Plant Physiol.*, **122**, 1149-1160.

Reeves, P.H. and Coupland, G. (2001) Analysis of flowering time control in *Arabidopsis* by comparison of double and triple mutants. *Plant Physiol.*, **126**, 1085-1091.

Roldán, M., Gómez-Mena, C., Ruiz-Garcia, L., Salinas, J. and Martínez-Zapater, J.M. (1999) Sucrose availability on the aerial part of the plant promotes morphogenesis and flowering of *Arabidopsis* in the dark. *Plant J.*, **20**, 581-590.

Ruiz-García, L., Madueño, F., Wilkinson, M., Haughn, G., Salinas, J. and Martínez-Zapater, J.M. (1997) Different roles of flowering-time genes in the activation of floral initiation genes in *Arabidopsis. Plant Cell*, **9**, 1921-1934.

Sage-Ono, K., Ono, M., Harada, H. and Kamada, H. (1998) Accumulation of a clock-regulated transcript during flower-inductive darkness in *Pharbitis nil. Plant Physiol.*, **116**, 1479-1485.

Sakamoto, T., Kobayashi, M., Itoh, H., *et al.* (2001) Expression of a gibberellin 2-oxidase gene around the shoot apex is related to phase transition in Rice. *Plant Physiol.*, **125**, 1508-1516.

Samach, A. and Coupland, G. (2000) Time measurement and the control of flowering in plants. *BioEssays*, **22**, 38-47.

Samach, A., Onouchi, H., Gold, S.E., *et al.* (2000) Distinct roles of CONSTANS target genes in reproductive development of *Arabidopsis. Science*, **288**, 1613-1616.

Schaffer, R., Ramsay, N., Samach, A., *et al.* (1998) The *late elongated hypocotyl* mutation of *Arabidopsis* disrupts circadian rhythms and the photoperiodic control of flowering. *Cell*, **93**, 1219-1229.

Schultz, E.A. and Haughn, G.W. (1993) Genetic analysis of the floral initiation process (FLIP) in *Arabidopsis. Development*, **119**, 745-765.

Sheldon, C.C., Burn, J.E., Perez, P.P., *et al.* (1999) The *FLF* MADS box gene: a repressor of flowering in *Arabidopsis* regulated by vernalization and methylation. *Plant Cell*, **11**, 445-458.

Sheldon, C.C., Finnegan, E.J., Rouse, D.T., *et al.* (2000) The control of flowering by vernalization. *Curr. Opin. Plant Biol.*, **3**, 418-422.

Simon, R., Igeño, M.I. and Coupland, G. (1996) Activation of floral meristem identity genes in *Arabidopsis*. *Nature*, **384**, 59-62.

Smeekens, S. (2000) Sugar-induced signal transduction in plants. *Annu. Rev. Plant Physiol. Plant Mol. Biol.*, **51**, 49-81.

Somers, D.E. (1999) The physiology and molecular bases of the plant circadian clock. *Plant Physiol.*, **121**, 9-19.

Somers, D.E., Devlin, P.F. and Kay, S.A. (1998a) Phytochromes and cryptochromes in the entrainment of the *Arabidopsis* circadian clock. *Science*, **282**, 1488-1490.

Somers, D.E., Webb, A.A.R., Pearson, M. and Kay, S.A. (1998b) The short-period mutant, *toc1-1*, alters circadian clock regulation of multiple outputs throughout development in *Arabidopsis thaliana*. *Development*, **125**, 485-494.

Somers, D.E., Schultz, T.F., Milnamow, M. and Kay, S.A. (2000) *ZEITLUPE* encodes a novel clock-associated PAS protein from *Arabidopsis*. *Cell*, **101**, 319-329.

Soppe, W.J.J. and Koornneef, M. (2000) Flowering initiation and methylation of DNA in *Arabidopsis*. *Flowering Newslett.*, **30**, 3-7.

Soppe, W.J.J., Jacobsen, S.E., Alonso-Blanco, C., *et al.* (2000) The late flowering phenotype of *fwa* mutants is caused by gain-of-function epigenetic alleles of a homeodomain gene. *Mol. Cell*, **6**, 791-802.

Staiger, D. and Heintzen, C. (1999) The circadian system of *Arabidopsis thaliana*: forward and reverse genetic approaches. *Chronobiol. Int.*, **16**, 1-16.

Strayer, C., Oyama, T., Schultz, T.F., *et al.* (2000) Cloning of the *Arabidopsis* clock gene *TOC1*, an autoregulatory response regulator homolog. *Science*, **289**, 768-771.

Suárez-López, P., Wheatley, K., Robson, F., Onouchi, H., Valverde, F. and Coupland, G. (2001) *CONSTANS* mediates between the circadian clock and the control of flowering in *Arabidopsis*. *Nature*, **410**, 1116-1120.

Sung, Z.R., Belachew, A., Shunong, B. and Bertrand-Garcia, R. (1992) *EMF*, an *Arabidopsis* gene required for vegetative shoot development. *Science*, **258**, 1645-1647.

Swarup, K., Alonso-Blanco, C., Lynn, J.R., *et al.* (1999) Natural allelic variation identifies new genes in the *Arabidopsis* circadian system. *Plant J.*, **20**, 67-77.

Thomas, B. and Vince-Prue, D. (1997) *Photoperiodism in Plants*, 2nd edn, Academic Press, London.

Wang, Z.-Y. and Tobin, E.M. (1998) Constitutive expression of the *CIRCADIAN CLOCK ASSOCIATED 1 (CCA1)* gene disrupts circadian rhythms and suppresses its own expression. *Cell*, **93**, 1207-1217.

Weigel, D. (1995) The genetics of flower development: from floral induction to ovule morphogenesis. *Annu. Rev. Genet.*, **29**, 19-39.

Weller, J.L., Reid, J.B., Taylor, S.A. and Murfet, I.C. (1997) The genetic control of flowering in pea. *Trends Plant Sci.*, **2**, 412-418.

Wilson, R.N., Heckman, J.W. and Somerville, C.R. (1992) Gibberellin is required for flowering in *Arabidopsis thaliana* under short days. *Plant Physiol.*, **100**, 403-408.

Xu, Y.-L., Gage, D.A. and Zeevaart, J.A.D. (1997) Gibberellins and stem growth in *Arabidopsis thaliana*—effects of photoperiod on expression of the *GA4* and *GA5* loci. *Plant Physiol.*, **114**, 1471-1476.

Yang, C.-H., Chou, M.-L. and Haung, M.-D. (1999) Genetic interactions between genes in regulating flowering time, flower initiation, and flower formation in *Arabidopsis thaliana*. *Flowering Newslett.*, **28**, 12-19.

Yano, M., Katayose, Y., Ashikari, M., *et al.* (2000) *Hd1*, a major photoperiod sensitivity quantitative trait locus in rice, is closely related to the *Arabidopsis* flowering time gene *CONSTANS*. *Plant Cell*, **12**, 2473-2483.

Yanofsky, M.J., Mazzela, M.A. and Casal, J.J. (2000) A quadruple photoreceptor mutant still keeps track on time. *Curr. Biol.*, **10**, 1013-1015.

Yoshida, N., Yanai, Y., Chen, L., Sung, Z.R. and Takahashi, S., 2000. Functional analysis of *EMBRYONIC FLOWER 2* gene in *Arabidopsis*. *11th International Conference on Arabidopsis Research*, Madison, abstract 332.

Yu, T.-S., Lue, W.-L., Wang, S.-M. and Chen, J. (2000) Mutation of *Arabidopsis* plastid phosphoglucose isomerase affects leaf starch synthesis and floral initiation. *Plant Physiol.*, **123**, 319-325.

Zagotta, M.T., Hicks, K.A., Jacobs, C.I., Young, J.C., Hangarter, R.P. and Meeks-Wagner, D.R. (1996) The *Arabidopsis ELF3* gene regulates vegetative photomorphogenesis and the photoperiodic induction of flowering. *Plant J.*, **10**, 691-702.

Zanewich, K.P. and Rood, S.B. (1995) Vernalization and gibberellin physiology of winter canola— endogenous gibberellin (GA) content and metabolism of $[^3H]GA_1$ and $[^3H]GA_{20}$. *Plant Physiol.*, **108**, 615-621.

Zeevaart, J.A.D. (1983) Gibberellins and flowering, in *The Biochemistry and Physiology of Gibberellins*, vol. II (ed. A. Crozier), Praeger, New York, pp. 333-374.

Zhang, H., Jennings, A., Barlow, P.W. and Forde, B.G. (1999) Dual pathways for regulation of root branching by nitrate. *Proc. Natl. Acad. Sci. USA*, **96**, 6529-6534.

Zheng, C.C., Porat, R., Lu, P. and O'Neill, S.D. (1998) *PNZIP* is a novel mesophyll-specific cDNA that is regulated by phytochrome and a circadian rhythm and encodes a protein with a leucine zipper motif. *Plant Physiol.*, **116**, 27-35.

Zhong, H.H., Painter, J.E., Salome, P.A., Straume, M. and McClung, C.R. (1998) Imbibition, but not release from stratification, sets the circadian clock in *Arabidopsis* seedlings. *Plant Cell*, **10**, 2005-2018.

2 The genetic control of flowering time

Mark Doyle, Si-Bum Sung and Richard Amasino

2.1 Introduction

The floral transition is a major developmental event in the lifecycle of flowering plants whereby plants switch from a phase of vegetative growth to one of reproductive growth. The timing of this event is governed by many factors. Proper timing of seed production is critical for optimizing reproductive fitness in a given species.

Developing plants first pass through a juvenile phase in which flowering is repressed. During this phase, photosynthetic tissue is produced to help meet the high energetic demands of reproduction. The length of this juvenile phase varies greatly among species. Once the juvenile phase is completed, many plants require additional environmental cues to trigger the switch from vegetative to reproductive growth. Thus, in many species there is an integration of developmental and environmental information that leads to the floral transition.

Daylength and cold temperatures are common environmental cues that can induce flowering. Some plants absolutely require specific environmental signals in order to flower; this is known as an obligate response. This is in contrast to a facultative response in which a signal is not absolutely required but acts to hasten the floral transition. Changes in both daylength and temperature are excellent indicators of season. The geographical distribution of many species, however, spans a large range of latitudes and habitats that can include large variations in seasonal daylength and temperature changes. Owing to this environmental variability, the underlying sensing mechanism must be highly adaptable. For example, the cocklebur, *Xanthium strumarium*, displays an obligate response to daylength. Although all known strains of *X. strumarium* have a daylength requirement for flowering, the actual required daylength varies considerably between strains found in different climates and at different latitudes (Evans, 1975). This attests to the adaptability of the environment-sensing components of flowering-time control.

The induction of flowering in response to the relative length of day and night is known as photoperiodism. Plants that flower in response to short days and long nights are known as short-day plants (SDP). These plants are 'dark dominant' and flower when the dark period *exceeds* a critical value. Thus, SDP typically flower in autumn as days shorten and nights lengthen. Long-day plants (LDP) flower in the spring when days are lengthening. Some plants, referred to as day-neutral plants (DNP), do not respond to photoperiod. It is important to mention that what differentiates a 'long' day from a 'short' one varies greatly between species and sometimes between different ecotypes of the same species, as in *X. strumarium*. Also there are known species in which intermediate daylengths or successive daylength changes, such as long days followed by short days, are required for flowering (Thomas and Vince-Prue, 1997).

Much of what is currently known about photoperiodism comes from the study of plants that respond to a single inductive photoperiod. Such plants allow for tight control of floral induction under laboratory conditions. Leaves are the site of photoperiod perception. Upon induction, a signal of unknown composition is produced in leaves and transported to the shoot apical meristem (SAM) where the floral transition occurs (Zeevaart, 1984; Aukerman and Amasino, 1998). The elusive flowering signal, known as florigen, can be transmitted through graft junctions. Grafting of induced leaves can lead to flowering in noninduced plants. Both intra- and interspecific transmission of a flowering signal have been demonstrated. This does not, however, demonstrate that florigen is the same in all plants because grafts are only successful between closely related species.

Another environmental cue that affects flowering in many plants is prolonged exposure to near-freezing temperatures (typically 0–7°C for 4–12 weeks). The acquisition or acceleration of the ability to flower by a chilling treatment is known as vernalization (Chouard, 1960). The requirement for prolonged exposure to cold has adaptive value in that it enables plants to become established prior to the onset of winter and to flower rapidly in the spring, thus avoiding competition. An important requirement for this strategy to be successful is that short periods of cold in the autumn season followed by warmer conditions do not lead to premature flowering.

Unlike daylength, vernalization is sensed at the shoot apex. Also, vernalization does not lead to the production of a flowering signal. Rather vernalization is signified by the attainment of a reproductively competent state. Thus, vernalization alone is not always sufficient to induce flowering, but is instead a process that renders the meristem reproductively competent to flower once other requirements for flowering such as exposure to inductive photoperiods have been met.

Genetic analysis of flowering time is greatly enhancing our understanding of this complex process. Mendel first discovered the basic tenets of genetic inheritance using pea. Variation in flowering time was a trait studied by Mendel; however, his data were never fully presented (Weller *et al.*, 1997a). Today, the genetics of flowering time regulation in pea and other systems is better understood. The floral transition is governed by a complex network of pathways. These pathways act to integrate internal developmental cues with environmental cues to determine the optimal time for flowering. Using genetics, it is possible to dissect the various pathways involved in flowering and study the interactions between these different pathways.

W.W. Garner and H.A. Allard were the first to use a flowering-time mutant to study the physiology of flowering. They found a late-flowering tobacco mutant named Maryland Mammoth and observed that this mutant only flowered in a greenhouse during the winter. With the aid of a light-tight 'doghouse' that made it possible to expose field-grown plants to well-defined photoperiods, Garner and Allard concluded that these plants were responding to daylength and flowered only when exposed to conditions of short days and long nights (Evans, 1975).

This obligate SDP, Maryland Mammoth, arose from a strain of tobacco that does not respond to photoperiod—a DNP. In the case of Maryland Mammoth, the parental line, which does not respond to photoperiod, is converted to an obligate SDP by a single recessive mutation. Thus, the parental day-neutral tobacco maintained all of the machinery for photoperiod reception throughout domestication and cultivation. The Maryland Mammoth phenotype is a fairly common one as it has arisen several times independently; however, a molecular explanation for this phenotype and the nature of the affected gene is not yet known.

The Maryland Mammoth story provides an illustration of the relationship between agriculture and the control of flowering time. When one is trying to cultivate crops in a nonindigenous environment, flowering-time regulation is an important trait for selection. Environmental requirements for flowering such as daylength and prolonged exposure to cold temperatures place constraints on the geographical range in which crop plants can be successfully cultivated. Thus, extensive breeding in crop plants often leads to an attenuated ability to respond to environmental signals. It is hoped that an understanding of the molecular regulators of flowering will make it possible to engineer plants with specific flowering-time responses, which could further increase the range in which crops can be cultivated.

Much of what is known about the genes involved in flowering-time regulation comes from the study of natural variation and mutations in *Arabidopsis thaliana*. Accessions of *Arabidopsis* can be found growing in a variety of environments throughout the world. Thus, *Arabidopsis* has optimized its regulation of flowering time under many environmental conditions, providing natural variation that can be studied genetically.

Certain accessions of *Arabidopsis* respond to both daylength and cold temperatures and display a facultative response to both stimuli. Gibberellins (GAs), a class of plant hormones, have also been implicated in the control of flowering time in *Arabidopsis*; mutants that are deficient in GAs exhibit an obligate requirement for inductive photoperiods (Wilson *et al.*, 1992). *Arabidopsis* is a LDP and thus flowers more rapidly when grown in long days. In the experiments described below, long days are typically 16 hours of light and 8 hours of darkness. Short days are typically 8 hours of light and 16 hours of darkness. Thus, in addition to the advantage of *Arabidopsis* as a system for genetics, this species can be used to study a number of flowering-time control mechanisms.

In the following pages, we present a detailed account of genes that are currently known to affect flowering time in *Arabidopsis*. Mutations in these genes can cause early or late flowering relative to wild type. Flowering time can be measured in days; however, most studies also use a developmental measure and define 'flowering time' as the number of leaves produced by the primary meristem prior to flowering. Work done in other species is discussed with an emphasis on possible parallels with what is known from studies of *Arabidopsis*.

2.2 *FLOWERING LOCUS C* and *FRIGIDA* confer a vernalization requirement in *Arabidopsis*

The natural variation in flowering time among many accessions of *Arabidopsis* has been used for the identification of genes that regulate the timing of the floral transition. Certain accessions of *Arabidopsis* are very late-flowering unless exposed to prolonged periods of cold temperatures; i.e. these accessions require vernalization for rapid flowering (figure 2.1a). These accessions of *Arabidopsis* have a facultative response to vernalization and behave as winter annuals (Karlsson *et al.*, 1993; Napp-Zinn, 1979). There are also summer-annual accessions of *Arabidopsis* that only require inductive photoperiods to flower. These summer-annual varieties are most commonly used in *Arabidopsis* research.

Genetic analysis of crosses between a number of winter annual and summer annual varieties of *Arabidopsis* revealed that two loci were primarily responsible for the late-flowering vernalization response, *FRIGIDA* (*FRI*) and *FLOWERING LOCUS C* (*FLC*). Initial studies found that vernalization-responsive late flowering mapped to the *FRI* locus (Burn *et al.*, 1993b; Clarke and Dean, 1994; Lee *et al.*, 1993; Sanda and Amasino, 1996). Additional studies demonstrated that *FRI* acted synergistically with a second locus, *FLC*, to render plants extremely late flowering unless vernalized (figure 2.1a) (Lee *et al.*, 1994b; Koornneef *et al.*, 1994).

FLC has been cloned and encodes a MADS-box-type transcription factor (Michaels and Amasino, 1999; Sheldon *et al.*, 1999). *FLC* mRNA is found at high levels in the presence of *FRI*. Overexpression of *FLC* from the 35S CaMV promoter leads to late flowering in the absence of *FRI*. Thus, *FLC* appears to be the gene that has a direct role in preventing flowering, and the role of *FRI* appears to be elevation of *FLC* expression to levels sufficient to block flowering. The fact that *FRI* does not impact flowering in the absence of *FLC* supports this model (Michaels and Amasino, 1999). In *Arabidopsis*, introduction of extra copies of *FLC* is sufficient to create an obligate response to vernalization (figure 2.1d) (Michaels and Amasino, 2000). Thus, plants may use *FLC* expression level as a rheostat for the control of flowering time (Michaels and Amasino, 2000).

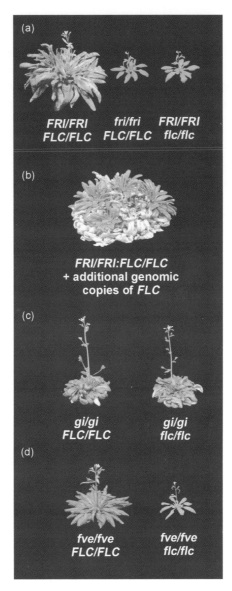

Figure 2.1 Representative flowering phenotypes in *Arabidopsis*. (a) The vernalization-responsive winter-annual habit is conferred by the presence of dominant alleles at two loci, *FRIGIDA* (*FRI*) and *FLOWERING LOCUS C* (*FLC*). Mutations at either locus confer a rapid-flowering, summer-annual habit. (b) Increasing the dosage of *FLC* in the presence of *FRI* creates plants with an extreme block to flowering as shown. This late flowering is completely overcome by vernalization. Thus, increasing *FLC* dosage creates an obligate requirement for cold in order to flower. (c) Late-flowering mutants in the photoperiod pathway delay flowering in an *FLC*-independent manner. *gigantea* (*gi*) mutants are late-flowering in the presence or absence of *FLC*. (d) Mutations in autonomous-pathway genes are another route to vernalization-responsive late flowering. On the left is an unvernalized *fve* mutant. The late-flowering phenotype is completely *FLC*-dependent as *fve/flc* double mutants flower early (right).

FRI has recently been cloned (Johanson *et al.*, 2000). The predicted protein shows no significant similarity to other known genes, making it difficult to formulate a hypothesis as to how *FRI* upregulates *FLC*. The *FRI* gene was sequenced from many early-flowering *Arabidopsis* accessions (Johanson *et al.*, 2000). Mutations leading to a loss of *FRI* function were found in several accessions including Ler, Ws and Col, indicating that many accessions of *Arabidopsis* have lost the vernalization requirement for rapid flowering through mutations in *FRI*. The discovery of different mutations in *FRI* indicates that a conversion from winter to summer annual occurred at least twice (Johanson *et al.*, 2000). Thus, *Arabidopsis* was originally a winter annual. The summer-annual forms commonly used by many laboratories have been derived from this winter annual by loss-of-function mutations in *FRI*.

The appearance of early-flowering, summer-annual varieties of *Arabidopsis* indicates that there has been selection against the vernalization requirement for flowering (Johanson *et al.*, 2000). Mutations in dominant genes such as *FRI* and *FLC* would make it possible for the winter-annual *Arabidopsis* to inhabit more moderate climates where temperatures would not become cold enough for vernalization to occur. Given that *FLC* is a more direct repressor of flowering, it is intriguing that in commonly studied summer-annual varieties *FRI* is the target for the selection. A *FRI*-independent phenotype of *FLC* in SD as well as the slight effect of *flc* mutations on circadian rhythms (Swarup *et al.*, 1999) suggests that *FLC* may have roles in pathways unaffected by cold treatment. These additional functions of *FLC* could have made *FRI* a more likely target for selection, as mutations in *FRI* may be less disruptive than *FLC* mutations to plant development.

2.3   Genes in the autonomous pathway of floral promotion

In addition to naturally occurring variation in flowering time, induced late-flowering mutants have also been isolated from several early-flowering, summer-annual accessions of *Arabidopsis* (Redei, 1962; Vetrilova, 1973; Koornneef *et al.*, 1991). Genetic and physiological studies place these late-flowering mutations into two distinct groups. One group, the photoperiod-pathway mutations, includes *constans* (*co*), *fha*, *ft*, *fwa* and *gigantea* (*gi*) (figure 2.1c). Flowering times of these mutants are delayed under LD but not under SD; i.e., these mutations render a plant 'blind' to photoperiod. The late-flowering phenotype in this group cannot be fully suppressed by vernalization (Koornneef *et al.*, 1991).

The other group of late-flowering mutants, the autonomous-pathway mutants, includes *luminidependens* (*ld*), *fca*, *fld*, *fpa*, *fve*, and *fy* (figure 2.1d). Late flowering in these mutants can be fully suppressed by vernalization (Koornneef *et al.*, 1991). Mutations in these genes cause later flowering than wild type under both LD and SD; furthermore, even though flowering is delayed, this group of mutants still responds to inductive photoperiods. Thus, this group of genes is believed to promote flowering in an 'autonomous' or photoperiod-independent manner that is involved in monitoring the developmental stage of the plant. Phenotypic differences between the juvenile and adult phases include differences in leaf morphology and trichome distribution (Poethig, 1997). The autonomous-pathway genes may be involved in controlling the length of a juvenile stage of development because certain autonomous-pathway mutants display an extended juvenile phase (Telfer *et al.*, 1997).

FLC mRNA has been shown to accumulate in the autonomous mutants *fca*, *fpa*, *ld* and *fve*, suggesting that these genes normally act as *FLC* repressors (Michaels and Amasino, 1999; Sheldon *et al.*, 1999). Further studies revealed that late-flowering in autonomous-pathway mutants is *FLC* dependent; i.e., late-flowering in *fca*, *fpa*, *fve* and *ld* is eliminated in the presence of a null allele of *FLC* (Michaels and Amasino, 2001). Thus autonomous-pathway genes likely act upstream of *FLC* and promote flowering by suppressing *FLC* expression. The L*er* allele of *FLC* is weak and partially suppresses late flowering in autonomous-pathway mutants (Lee *et al.*, 1994b; Koornneef *et al.*, 1994; Sanda and Amasino, 1996).

The relationship between some autonomous-pathway genes has been studied genetically (Koornneef *et al.*, 1998). Epistasis analysis placed the autonomous-pathway genes into two subgroups. The *fy* mutant is epistatic with *fca*, and the *fpa* mutant is epistatic with *fve*. An *fy/fpa* double mutant could not be produced, suggesting that these autonomous-pathway genes may have an essential but redundant role in plant development (Koornneef *et al.*, 1998).

The behavior of *fpa* mutants differs from that of other autonomous-pathway mutants. When overexpressed, *FPA* induces plants to flower earlier than *flc* mutants (Schomburg and Amasino, 2001). Such a response has not been found with other cloned autonomous-pathway genes. This indicates that *FPA* is capable of expediting the vegetative to reproductive switch in an *FLC*-independent manner. *FPA* may play a role in GA action. Two *fpa* mutants were isolated in a screen for plants with increased *GA-5::LUC* activity (Meier *et al.*, 2001). Higher *GA5* promoter activity seems to be caused by reduced GA signaling in *fpa* mutant backgrounds. Moreover, *fpa* mutants exhibit compact dark green rosettes in SD, a phenotype commonly associated with defects in GA action.

Although a GA-like phenotype, similar to that seen in *fpa*, has not been reported in *fy* mutants, it is possible that failure to find an *fy/fpa* double mutant might be the result of severe GA insensitivity. The existence of a possible redundant but essential function shared by *FPA* and *FY* is supported by the fact that single mutations in either *fpa* or *fy* lead to a delay in the juvenile–adult phase transition (Telfer *et al.*, 1997; Simpson *et al.*, 1999).

To date, three autonomous-pathway genes have been cloned: *LUMINIDEPENDENS* (*LD*), *FCA* and *FPA* (Lee *et al.*, 1994a; Macknight *et al.*, 1997; Schomburg *et al.*, 2001). *FPA* and *FCA* contain regions predicted to act as RNA-binding domains, and *FCA* has been shown to bind RNA *in vitro*. The presence of RNA-binding motifs in these two genes raises the possibility that posttranscriptional regulation has an important function in the autonomous pathway. *LD* encodes a protein with a homeodomain at its N-terminus and two putative bipartite nuclear localization signals (Lee *et al.*, 1994a). *LD* is targeted to the nucleus and may function as a DNA-binding protein (Aukerman *et al.*, 1999). However, given the fact that *FCA* and *FPA* are putative RNA-binding proteins, it is also possible that the *LD* homeodomain-like region serves as a RNA-binding motif similar to the homeodomain in the *Drosophila* protein Bicoid (Dubnau and Struhl, 1996).

Studies of *FRI*-containing plants and autonomous-pathway mutants indicate that *FLC* is a major player in flowering-time regulation in *Arabidopsis*. As described above, two possible routes lead to vernalization-responsive late flowering. Plants containing a functional copy of *FRI* are late-flowering as are plants with mutations in the autonomous pathway. In both cases late flowering is caused almost exclusively by the upregulation of one gene, *FLC* (Michaels and Amasino, 2001).

2.4 *FLC* as a key regulator for the competence to flower

The high levels of *FLC* expression seen in *FRI*-containing lines and autonomous-pathway mutants are irreversibly turned off by vernalization (Michaels and Amasino, 1999; Sheldon *et al.*, 1999). Transgenic plants expressing 35S*::FLC* are late-flowering and do not respond to vernalization, indicating that vernalization acts to suppress the activity of the native *FLC* promoter. As mentioned above, additional genomic copies of *FLC* result in plants that flower only after vernalization, demonstrating that it is possible to create an obligate requirement for vernalization by increasing *FLC* copy number and hence *FLC* expression level (Michaels and Amasino, 1999, 2000). Thus, *FLC* downregulation appears to be a necessary step in attaining reproductive competence in the meristem.

Many classical studies of flowering revealed that both vernalization and developmental state act to regulate meristem competence. In *Arabidopsis*, the developmental and vernalization pathways seem to converge on regulating *FLC* expression. Consistent with this model, *FLC* is expressed in shoot meristems where the floral transition takes place and is downregulated after vernalization. Autonomous-pathway genes are expressed in meristems. Both *FLC* and autonomous-pathway genes are also expressed in root tips and in developing leaf primordia (Aukerman *et al.*, 1999; Michaels and Amasino, 2000; Schomburg *et al.*, 2001); i.e., these genes are primarily expressed in regions that contain dividing cells. Other studies have shown that only actively dividing cells can be vernalized (Wellensiek, 1962; Michaels and Amasino, 2000).

Although vernalization in *Arabidopsis* acts largely through the suppression of *FLC*, it is important to note that a residual vernalization response is observed in *flc* mutant plants (Michaels and Amasino, 2001). Under SD, vernalized *flc* mutants flower earlier than wild type. This indicates that there is a *FLC*-independent component of the vernalization response in *Arabidopsis* although the effect of this component is minor compared to the effect of suppressing *FLC*.

2.5 The molecular basis of vernalization

The vernalized state in *Arabidopsis* is stable throughout the life of an individual plant but is reset in the next generation. As discussed above, *FLC* downregulation accounts for much of the vernalization response in *Arabidopsis*. After vernalization *FLC* is repressed for the remainder of the plant lifecycle. Thus, *FLC* expression, like the vernalization response as a whole, undergoes a stable epigenetic switch after cold treatment.

Epigenetic repression of genes is often associated with DNA methylation. Certain cytosine residues of eukaryotic DNA are methylated in inactive regions of the genome (Tilghman, 1999). It has been proposed that such DNA methylation may account for the flowering promotion by vernalization (Dennis *et al.*, 1997; Finnegan *et al.*, 1998). For example, cold treatment of cultured tobacco and *Arabidopsis* cells leads to reduced DNA methylation (Burn *et al.*, 1993a; Finnegan *et al.*, 1998). Furthermore, flowering is promoted either by the introduction of a DNA methylase antisense construct (Finnegan *et al.*, 1998) or after exposure to the demethylating agent 5-azacytidine (Burn *et al.*, 1993a). This suggests that DNA methylation might be involved in the regulation of *FLC* expression by vernalization (Sheldon *et al.*, 1999). However, demethylating agents or antisense constructs do not promote flowering as strongly as vernalization. Also, reduced methyltransferase activity in plants causes pleiotropic effects including ectopic expression of floral homeotic genes such as *AGAMOUS* (*AG*) (Finnegan *et al.*, 1996) that can directly promote flowering (Mizukami and Ma, 1997). Thus, only the identification of components of the vernalization system that are differentially regulated by DNA methylation will provide direct evidence for this model.

Genetic identification of the molecular components of the vernalization pathway is in the early stages. Most mutagenesis of *Arabidopsis* populations has been done on early-flowering strains. Because of the very subtle effects of vernalization on these strains, mutants are not easily identified from these populations. As discussed above, analyses of natural variation have revealed genes that alter the vernalization requirement for flowering. However, these studies have given little insight into the mechanism of vernalization itself.

Two *Arabidopsis* mutants showing an altered response to vernalization, *vrn1* and *vrn2*, have been described (Chandler *et al.*, 1996). These mutants came from a population of mutagenized *fca* plants and were identified as having an attenuated vernalization response. *vrn2* mutations partially block the decrease in *FLC* expression that normally occurs during vernalization (Sheldon *et al.*, 2000). However, plants containing these mutations still show approximately 50% of the wild-type vernalization response. Furthermore, *vrn2* causes later flowering than the parental line without vernalization and thus may not be entirely specific to the vernalization process.

2.6 Possible parallels between *Arabidopsis* and other species

As discussed above, genetic studies between winter-annual and summer-annual accessions of *Arabidopsis* revealed that the dominant *FRI* gene accounts for the winter-annual flowering response in many late-flowering strains. Recessive mutations in the autonomous pathway also lead to a vernalization-responsive, late-flowering phenotype similar to plants containing *FRI*. In both *FRI*-containing plants and autonomous-pathway mutants, late flowering is due largely to an upregulation of *FLC*. Similar mechanisms may account for the dominant and recessive nature of this response in other species.

In brassicas, the most closely related crop plants to *Arabidopsis*, two loci have been identified that are responsible for biennial behavior. One of these loci corresponds to a *Brassica* homologue of *FLC* (Osborn *et al.*, 1997; Kole *et al.*, 2000). Furthermore, *FLC* seems to be downregulated by vernalization in brassicas (Kole *et al.*, 2000). At least three *FLC* homologues have been cloned from *B. rapa* and two of them significantly delay flowering in *Arabidopsis* when constitutively expressed (S. Sung and R. Amasino, unpublished). Thus, quite similar mechanisms most likely govern the vernalization response in *Arabidopsis* and brassicas.

In diploid wheat and barley, crosses between winter and spring varieties have identified two loci that contribute greatly to the control of flowering time. Dominant alleles of *Vrn-Am2* from wheat and *Vrn-H2* from barley increase the requirement for vernalization and thus promote a winter growth habit (Tranquilli and Dubcovsky, 2000). This is similar to many natural late-flowering variations of *Arabidopsis* in which a dominant *FRI* allele governs vernalization requirement. There also exist dominant alleles of *Vrn-Am1* and *Vrn-H1* from wheat and barley, respectively, that *reduce* the requirement for vernalization, resulting in plants that behave like spring varieties (Tranquilli and Dubcovsky, 2000). These loci could correspond to weak alleles of an *FLC* homologue that are not responsive to *Vrn-Am2* or *Vrn-H2*.

A single dominant locus confers the biennial habit of *Hyoscyamus. niger* varieties (reviewed in Lang, 1986). This locus may represent a *FRI* homologue.

However, in beets a single recessive allele accounts for the vernalization response (reviewed in Michaels and Amasino, 2000). Similarly to *Arabidopsis* it is possible that an autonomous pathway mutation is responsible for this trait in beet. A recessive, autonomous-pathway mutation that enhanced *FLC* expression may have been selected for during the domestication of beets.

Pisum sativum, garden pea, has also been used as a genetic system for the study of flowering time. The analysis of natural variation and induced mutations has led to the identification of many flowering-time loci (reviewed in Weller *et al.*, 1997b). An advantage of studying flowering in pea is that mutants can be readily used in grafting experiments. Grafting experiments revealed that the *LATE FLOWERING* (*LF*) gene acts in the shoot either at or after perception of a floral stimulus. *VEGETATIVE-1* (*VEG-1*) also seems to act in the shoot and to be involved in establishing reproductive competence. *veg-1* is epistatic to *LF*. Variations in the two loci do not have an effect on the photoperiod response. Since these two loci are responsible for providing competence to flower upon receiving the floral stimulus, it is possible that these genes may operate in a similar way to the *FRI-FLC* system in *Arabidopsis*. The induced late-flowering *gi* mutation leads to behavior that is similar to autonomous-pathway mutants in *Arabidopsis* in that the *gi* mutant has an enhanced response to vernalization and retains a photoperiod response (Weller *et al.*, 1997b).

2.7   Photoreceptors and flowering time

Photoperiodism begins with light perception, and plants have evolved a range of photoreceptors that respond to different wavelengths and intensities of light. Mutants showing altered photoperiodic flowering responses in several species have been shown to have mutations in genes encoding photoreceptors. A brief account of photoreceptor mutants and their potential interactions with other flowering-time genes and pathways is presented below.

Several types of light receptors have been identified in plants. The first type of photoreceptors to be characterized were the phytochromes, which are proteins that interconvert between red (R) and far-red (FR) absorbing forms, Pr and Pfr respectively. Pfr is believed to be the biologically active form for most responses. *Arabidopsis* contains five phytochrome genes, *PHYA–E*. *PhyA* is light labile and mediates responses to FR light. PhyB–E are light stable and function as R light receptors (reviewed in Quail *et al.*, 1995).

Phytochromes have been well documented to have an impact on flowering. Several flowering-time mutants are defective in phytochrome biosynthesis. Such mutants reveal the net effect of phytochromes on flowering. In rice, a SDP, mutations in the *PHOTOPERIOD SENSITIVITY 5* (*SE5*) gene cause early flowering and impair the ability of plants to respond to photoperiod (Izawa *et al.*, 2000). This gene has recently been cloned and encodes the rice homologue of

Arabidopsis HY1, a gene involved in phytochrome chromophore biosynthesis (Davis *et al.*, 1999). Thus, *se5* mutants are phytochrome deficient, and early flowering in these mutants indicates that the net effect of phytochromes on flowering is inhibitory in rice. This is also seen in the LDP *Arabidopsis*, where mutations in the phytochrome biosynthesis genes *HY1* and *HY2* both lead to early flowering (Goto *et al.*, 1991).

In *Arabidopsis PHYA*, a mediator of FR signals, acts as a promotor of flowering. This is consistent with physiological studies linking FR with floral promotion in LDP. *phyA* mutants flower late in response to daylength extensions. Plants overexpressing *PHYA* are photoperiod insensitive and flower early (Johnson *et al.*, 1994; Bagnall *et al.*, 1995). In garden pea, a LDP, the *phyA* mutant *fun1* is photoperiod insensitive and flowers late, clearly demonstrating that *PHYA* is the major phytochrome required for the promotion of flowering by inductive photoperiods (Weller *et al.*, 1997b).

PHYTOCHROME B (*PHYB*) acts as a floral repressor in many species. Mutants in the *Arabidopsis PHYB* (*HY3*) gene display a reduced photoperiod response and flower early in SD (Goto *et al.*, 1991; Bagnall *et al.*, 1995). A similar phenotype is seen in the BMDR-1 genotype of barley, which contains a light labile form of *PHYB* (Hanumappa *et al.*, 1999). Early flowering due to a lack of light-stable *PHYB* has also been shown in the *ein* mutants of *Brassica rapa* (Devlin *et al.*, 1992), the *lh* mutant in cucumber (Lopez-Juez *et al.*, 1992) and the *ma3* mutant in sorghum, a SDP (Childs *et al.*, 1997). Similar early-flowering phenotypes seen in such a variety of species suggests that the role of *PHYB* as an inhibitor of flowering is largely conserved throughout the plant kingdom.

In *Arabidopsis* the early-flowering phenotype of *phyB* mutants is enhanced by mutations in *phyD* and *phyE*, suggesting some redundancy in the ability of these phytochromes to inhibit flowering. *phyD* and *phyE* single mutants have little effect on flowering, indicating that *PHYB* has a more substantial role in the control of flowering time (Aukerman *et al.*, 1997; Devlin *et al.*, 1998, 1999). A mutation in *PHYC* has not yet been reported.

The cryptochrome blue light receptors also affect flowering. The photoperiod-insensitive *fha* mutant is late flowering in long days (Koornneef *et al.*, 1991) and has been shown to result from a mutation in *CRYPTOCHROME 2* (*CRY2*) (Guo *et al.*, 1998). An interesting link between *CRY2* and photoperiodism comes from work with recombinant inbred lines (RILs) between the L*er* and Cape Verde Islands (Cvi) ecotypes. Several of these RILs were day-neutral (Alonso-Blanco *et al.*, 1998). A major quantitative trait locus (QTL) leading to day neutrality was identified as *CRY2*. (M. Koornneef, personal communication). Mutants in *CRYPTOCHROME1* (*CRY1*) have a less severe effect on flowering than *cry2*; however, under certain light conditions *CRY1* plays a slight role in the promotion of flowering (Bagnall *et al.*, 1996; Mockler *et al.*, 1999).

Analyses of single and double photoreceptor mutants have begun to reveal a model for photoreceptor action. *cry2* mutants flower at the same time as wild type under blue light, but flower late under red + blue light. This led to the hypothesis that *CRY2* may function to block an inhibitory signal created by *PHYB* under red light (Guo *et al.*, 1998). Genetic characterization provides further support to this hypothesis as *phyB/cry2* double mutants show that *PHYB* is largely epistatic to *CRY2* (Mockler *et al.*, 1999). The *PHYB* and *CRY2* proteins have recently been shown to interact in the nucleus in a light-dependent manner (Mas *et al.*, 2000).

2.8 Photoperiodism and circadian clocks

Physiological studies of flowering have revealed a clear link between the photoperiodic induction of flowering and circadian rhythms. Extensive genetic analyses in *Arabidopsis* support this link as all known genes involved in photoperiod perception either affect or are affected by the circadian clock. Below is a brief overview of circadian rhythms. Genes involved in photoperiodism are then discussed in the context of their predicted role in the circadian network.

Circadian rhythms can be subdivided into three main components: input pathways, central oscillator components, and output rhythms. Input pathways mediate the interaction between the environment and the clock. These pathways sense environmental signals such as daily fluctuations in light and temperature and act to reset the clock each day, a process known as entrainment. In the absence of entrainment signals, clocks continue to run with a periodicity that is close to but not exactly 24 hours. Thus, daily resetting via input pathways is necessary to keep the clock synchronous with the external environment.

The central oscillator generates rhythms. Genetic analysis in other systems has uncovered components of the central oscillator (reviewed in Dunlap, 1999). These genes include the *WHITE COLLAR* (*WC*) and *FREQUENCY* (*FRQ*) genes in *Neurospora crassa*, the *CLK*, *PER* and *TIM* genes of *Drosophila melanogaster*, and the *KAI* genes in *Synechococcus*. These genes encode positive and negative components of a feedback loop that cycles with a periodicity of approximately 24 hours.

In addition to a function in feedback regulation, oscillator components, many of which encode transcription factors, also control the expression of clock-controlled genes (CCGs). As levels of these transcription factors oscillate, so too do levels of CCGs, creating output rhythms. A hallmark of circadian rhythms is the maintenance of the oscillator under constant conditions; i.e. in the absence of input signals. Thus, when the clock is intact, CCG expression, as well as higher-order outputs such as daily leaf movements, continues to cycle in the absence of input signals.

A survey of 8000 *Arabidopsis* genes ($\sim 1/3$ of the genome) revealed that approximately 6% have a circadian pattern of expression in mRNA abundance (Harmer *et al.*, 2000). These genes are involved in a diversity of plant processes including flowering-time regulation and photosynthesis. One class of photosynthesis genes, the *Lhcb* or *CAB* genes, shows robust cycling of mRNA. Proper cycling of these genes is commonly used as an indicator of clock activity (Millar *et al.*, 1995) and will be referred to below.

2.8.1   Input genes

The primary entrainment signal for the circadian clock in *Arabidopsis* is light, although temperature can also act to entrain the clock. Light entrains the clock through its interaction with photoreceptors. In *Arabidopsis phyA, phyB, cry1* and *cry2* mutants display changes in period length under different light treatments (Somers *et al.*, 1998a). *phyA* mutants entrain to new photoperiods more slowly than wild-type plants (Somers *et al.*, 1998a). A similar role of *PHYA* in entrainment has been shown in potato (Yanovsky *et al.*, 2000a). Although entrainment is typically affected under specific light conditions in phytochrome and cryptochrome mutants, functional copies of these genes are not required for entrainment. Entrainment and robust cycling have been demonstrated in *Arabidopsis phyA/phyB/cry1/cry2* quadruple mutants (Yanovsky *et al.*, 2000b).

In many cases the efficacy of light signals is dependent on the phase of the rhythm at the time of the signal. In many SDP a brief R light pulse, called a night-break, given at specific times during an inductive night period, is sufficient to keep plants in a vegetative state (Thomas and Vince-Prue, 1997). The time of maximum effectiveness of such a treatment is under circadian control. In LDP the ability of FR light to promote flowering is also greatest during a particular phase of the rhythm. The ability of the clock to control when a signal is effective is known as gating. The promotion of flowering by FR light in long-day plants is a gated response occurring only when the light signal is present during a window of receptivity dictated by the clock (Deitzer, 1984; Thomas and Vince-Prue, 1997). Interestingly, the transcripts of photoreceptors *PHYB, CRY1* and *CRY2* cycle in *Arabidopsis* (Bognar *et al.*, 1999; Harmer *et al.*, 2000). Cycling of *PHYB* mRNA and protein has also been demonstrated in tobacco (Bognar *et al.*, 1999). Despite the evidence that effective light signals must be coincident with a receptive phase of the rhythm to be effective, no correlation has been made between the timing of gated responses to light and the expression level of known photoreceptors (Thomas and Vince-Prue, 1997).

ZEITLUPE (ZTL) (Somers *et al.*, 2000) and *FKF1* (flavin-binding, kelch repeat, F-box 1) (Nelson *et al.*, 2000) contain flavin-binding domains making them potential blue light receptors. These genes also affect circadian rhythms and flowering time. As the name *FKF1* implies, these proteins also contain kelch repeats and an F-box suggesting a potential role in protein degradation via the ubiquitin pathway (Nelson *et al.*, 2000). Thus, *FKF1* and other members of this small gene family may provide light input to the clock by targeting central oscillator components for degradation (Nelson *et al.*, 2000).

Both *fkf1* and *ztl* mutants are late-flowering in long days but not short days; however, the *fkf1* flowering phenotype is more severe (Nelson *et al.*, 2000; Somers *et al.*, 2000). *fkf1* mutants have a slightly altered pattern of *CAB* expression (Nelson *et al.*, 2000). The *ztl-1* mutation, which lengthens the periods of *CAB* expression by about 3 hours, has a greater affect on rhythms than *fkf1* mutations (Somers *et al.*, 2000). Thus, *FKF1* appears to play a larger role in flowering-time regulation than *ZTL* but a lesser role in clock function.

EARLY FLOWERING 3 (*ELF3*) was first identified in a screen for *Arabidopsis* mutants that flowered early in SD (figure 2.2) (Zagotta *et al.*, 1996). *elf3* mutants resemble *Arabidopsis phyB* mutants; however, *ELF3* and *PHYB* appear to be in separate pathways because (i) *elf3/phyB* double mutants have longer hypocotyls and flower earlier than either single mutant, and (ii) mutations in *ELF3* do not impede the effect of *PHYB* overexpression on hypocotyl elongation (Reed *et al.*, 2000).

Other studies of *elf3* revealed a defect in circadian rhythms. Rhythmic expression of *CAB* appears normal in mutants grown in SD. *CAB* expression progressively deviates from wild type as daylength increases. *elf3* grown in continuous light shows arrhythmicity in *CAB* expression and daily leaf movements,

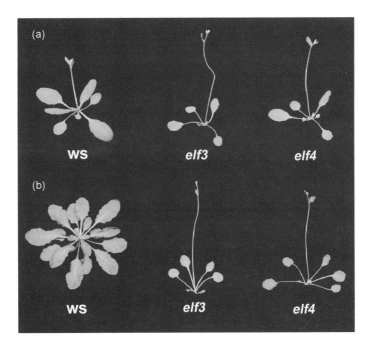

Figure 2.2 Early-flowering mutants of *Arabidopsis*. *Arabidopsis* is a facultative LDP: it flowers more rapidly in LD than in SD. (a) Wild type (WS) and two early-flowering mutants, *elf3* and *elf4* grown in LD. (b) WS, *elf3* and *elf4* grown in SD. *elf3* and *elf4* have an attenuated response to photoperiod.

suggesting that *ELF3* has a global effect on circadian rhythms (Hicks *et al.*, 1996). This result, however, does not explain the *elf3* flowering phenotype. *ELF3* mutants have the most extreme flowering phenotype in SD, the photoperiod in which *elf3* rhythms of *CAB* expression and leaf movement look most like those in wild type (Hicks *et al.*, 1996).

Another interesting aspect of *elf3* mutants is that *CAB* expression continues to cycle in continuous darkness. This eliminates *ELF3* as a component of the central oscillator and has lead to the idea that *ELF3* plays a role in light input to the clock (Hicks *et al.*, 1996). It has been shown that mutations in *elf3* do not disable the clock in continous light, but mask detectable outputs of the oscillator. The oscillator in *elf3* mutants, though undetectable in continuous light (LL), can still be entrained with temperature cycles (McWatters *et al.*, 2000).

Typically in SDP, rhythms that control flowering are suspended when the daylength exceeds a given length. Rhythms then proceed at the onset of darkness, and plants begin to measure nightlength, the determinant of flowering in SDP (Thomas and Vince-Prue, 1997). Interestingly, arrhythmicity seen in *ELF3* mutants under LL resembles the daily suspension of rhythms seen in SDP. When 'suspended' *ELF3* mutants are shifted to darkness, a peak in *CAB* expression is seen 8–11 hours later, regardless of when the light-to-dark transfer occurs (McWatters *et al.*, 2000). This becomes even more intriguing when considering that *ELF3* mutants have been shown to act like SDP, flowering earlier in SD than in LD (Zagotta *et al.*, 1996), Y.S. Noh and R. Amasino, unpublished data).

2.8.2   *Central oscillator components*

Several genes have been isolated that may encode components of the central oscillator. Not surprisingly these genes also play a role in the control of flowering time. Two potential central clock components in *Arabidopsis*, *CIRCADIAN CLOCK ASSOCIATED 1 (CCA1)* and *LATE ELONGATED HYPOCOTYL (LHY)* encode Myb transcription factors. The *CCA1* protein was initially isolated owing to its ability to bind a promoter element of the *Lhcb1*3* gene, a member of the *CAB* family (Wang *et al.*, 1997). The *cca1-1* null mutant shows altered circadian gene expression in LL; the periods of *CAB* and other circadian regulated genes in *cca1-1* are approximately 3 hours shorter than in wild type (Green and Tobin, 1999). The effect of a *cca1-1* mutation on flowering time has not been reported.

The transcripts of *CCA1* and *LHY* cycle in circadian fashion. Peak levels of expression occur around dawn for both genes (Schaffer *et al.*, 1998; Wang and Tobin, 1998). *CCA1* has also been shown to cycle at the protein level (Wang and Tobin, 1998). Overexpression of either *CCA1* or *LHY* leads to late flowering and the loss of detectable rhythms, including the rhythmic patterns of the endogenous *CCA1* and *LHY* transcripts (Wang and Tobin, 1998; Fowler *et al.*, 1999). This suggests a role for these genes in an autoregulatory feedback loop that may comprise the central oscillator. The role of *CCA1* in *CAB* induction suggests that these genes may act as positive components of the oscillator. The high similarity between these genes in addition to their similarity in temporal expression patterns could explain why *CCA1* mutants still maintain a functional, though altered clock. A loss-of-function allele of *LHY* has not yet been reported. It will be interesting to see whether *lhy* mutants behave like *cca1* mutants and whether a *cca1/lhy* double mutant abolishes clock function.

Other genes that control rhythms and flowering time have been revealed through further study of *CCA1*. A yeast two-hybrid screen (Sugano *et al.*, 1999) identified *CKB3*, a β subunit of *CASEIN KINASE 2* (*CK2*), as a protein that interacts with *CCA1*. The *CK2* holoenzyme is composed of two α subunits containing kinase domains and two β subunits that regulate kinase activity. It was subsequently shown that both α and β subunits of *CK2* can physically interact with *CCA1 in vitro*. Interaction of *CCA1* with *CKB3* facilitates the binding of *CCA1* to the promoter of *Lhcb1*1*, a CAB gene. Work done with whole plant extracts suggests that *CK2* is responsible for *CCA1* phosphorylation and that phosphorylation of *CCA1* is required for the formation of DNA–protein complexes (Sugano *et al.*, 1998). *CK2* has also been shown to interact and phosphorylate *LHY* (Sugano *et al.*, 1999).

Overexpression of *CKB3* alters rhythms and flowering time. Like the *cca1-1* mutant, *CKB3* overexpression (*CKB3*ox) shortens the period of the circadian rhythms and impairs the light induction of *CAB* gene expression (Sugano *et al.*, 1999). The periods of mRNA expression for *CCA1*, *LHY*, and other circadian-regulated genes are shortened by ∼4 hours in *CKB3*ox plants. Unlike plants overexpressing *CCA1* and *LHY*, *CKB3*ox plants flower earlier than wild type. This effect on flowering time is enhanced in short days (Sugano *et al.*, 1999). This work indicates a role for *CK2* in the ability of the clock to control flowering time. Phosphorylation has been shown to be important in the circadian feedback loops of other systems. Casein kinases in particular may play a role in insect clocks. The recently cloned *Drosophila* gene *DOUBLE-TIME* (*DT*) encodes a protein that resembles other known casein kinases (Kloss *et al.*, 1998).

timing of cab 1 (*toc1*) mutants were originally isolated in a screen for period-length mutants. The circadian period is shortened in *toc1* under constant conditions. This mutant is also early-flowering in SD and shows a reduced photoperiod response (Somers *et al.*, 1998b). Interestingly, the response to photoperiod is fully restored when *toc1* mutants are grown in 21-hour photoperiods (LD 14L:7D, SD 7L:14D) (Strayer *et al.*, 2000). This indicates that the early-flowering phenotype in *TOC1* mutants can be fully attributed to defects in circadian rhythms.

TOC1 has recently been cloned (Strayer *et al.*, 2000). The amino-terminus contains a motif similar to that of receiver domains in bacterial two-component response regulator systems. *TOC1* also contains a distinctive COOH-terminal motif found in several *Arabidopsis* proteins including the flowering-time gene *CONSTANS* (*CO*) (see below). This motif, referred to as a CCT motif

(*CONSTANS, CONSTANS-LIKE*, and *TOC1*), is a basic stretch of ~ 45 amino acids containing a putative nuclear localization sequence. The C-terminal end of *TOC1* contains an acidic motif common to transcriptional activators. This, along with the fact that *toc1* mutants show shortened periods of *TOC1* gene cycling, suggests a possible role for *TOC1* in a regulatory feedback loop (Strayer *et al.*, 2000).

Redei first described the late-flowering *gigantea* (*gi*) mutant in 1962 (Redei, 1962). *gi* mutants are photoperiod insensitive, but, unlike the SD early-flowering phenotype seen in *elf3* and *toc1*, *gi* mutants flower late in LD and form about the same number of leaves as wild-type plants in SD (Koornneef *et al.*, 1991; Araki and Komeda, 1993). *GI* encodes a novel protein containing up to 11 potential transmembrane domains. No homologous proteins exist in *Arabidopsis*; however, a relative was found in a rice EST database (Fowler *et al.*, 1999; Park *et al.*, 1999). Despite predicted transmembrane domains, *GI* is consitutively localized to the nucleus (Huq *et al.*, 2000). This was discovered after a *GI* mutant was identified in a screen for long-hypocotyl mutants in red light, suggesting that *GI* plays a role in *PHYB* signaling.

Although all *gi* alleles show late-flowering in LD, the effects of *gi* mutants on other circadian outputs are allele-specific. *gi-1* and *gi-2* lengthened and shortened the period of *CAB* expression, respectively (Park *et al.*, 1999). However, periods of leaf movement were shortened in both *gi-1* and *gi-2*, indicating uncoupling of these two outputs. The *GI* transcript cycles, and mutations in *GI* reduce the amplitudes of *CCA1*, *LHY* and *GI* mRNA expression, suggesting that *GI* may be a positive component in a feedback loop (Fowler *et al.*, 1999; Park *et al.*, 1999).

2.9 Photoperiod pathway genes downstream of the clock

The genes described above regulate flowering at least in part through their role in global regulation of the circadian machinery. Floral induction is one of many circadian outputs and it is the effect of these genes on this output that defines their role in flowering time regulation. A gene in *Arabidopsis* has been identified that seems to be an important component of this flowering-specific output.

CONSTANS (*CO*) encodes a putative zinc finger transcription factor with similarity to the GATA1 class of transcription factors (Putterill *et al.*, 1995). *co* mutants are photoperiod insensitive and late-flowering in long days (Redei, 1962; Koornneef *et al.*, 1991). *CO* is expressed at higher levels in LD than in SD (Putterill *et al.*, 1995) and shows a circadian pattern of expression in both *Arabidopsis* (Suarez-Lopez *et al.* 2001) and morning glory Liu *et al.* (2001). Constitutive expression of *CO* causes extreme early flowering and photoperiod insensitivity (Onouchi *et al.*, 2000). *co* mutants do not appear have a global impact on rhythms (Suarez-Lopez *et al.*, 2001), suggesting that regulation of *CO* is a flowering-specific output of the clock. In fact, *CO*, a promoter of flowering, may be the primary floral regulator in this system. Other flowering and circadian mutants display levels of *CO* expression that are consistent with their flowering phenotype; i.e., late-flowering mutants such as *lhy* and *gi* mutants display reduced levels of *CO* and the early-flowering *elf3* mutant displays elevated levels of *CO* expression (Suarez-Lopez *et al.*, 2001).

2.10   Clocks and flowering—summary

The study of flowering-time mutants has further established a link between photoperiodism and circadian rhythms. As mentioned above, there is much evidence that light signals interact with a receptive phase of the rhythm to either promote or repress flowering. A model in which an environmental input must occur simultaneously with a circadian phase in order to get a flowering response is known as a coincidence model. Bünning first proposed such a model in 1936. An extension of this model involves the interaction of two outputs from the clock where at least one output is differentially phased under different photoperiods (Thomas and Vince-Prue, 1997; Samach and Coupland, 2000). For such a system to operate, there must exist genes with different patterns of expression in LD and SD. Several flowering-time genes including *GI*, *CO* (see below) and *ELF4* (figure 2.2) exhibit this behavior (Fowler *et al.*, 1999; Suarez-Lopez *et al.*, 2001; M. Doyle and R. Amasino, unpublished data). These genes could then interact with other circadian-regulated genes only under specific daylengths, leading to a promotion or inhibition of flowering.

Despite a large number of cloned genes that function in this pathway, a clear model for how these genes interact to control flowering time is not yet apparent. Mutant analysis thus far has revealed clock mutants that both repress and promote flowering. However, no correlations can currently be made between the global effect of a mutation on rhythms and its effect on flowering time. Almost all clock mutants in plants are from *Arabidopsis*, a LDP. A future challenge will be to determine the extent to which photoperiod sensing mechanisms are conserved between LDP and SDP and to determine any differences between these two response types.

2.11   Genes downstream of the major flowering pathways

Recent evidence indicates that the different flowering pathways converge prior to the transition to flowering (figure 2.3). This was revealed in part by studies done using plants that overexpressed *CO*. Using a dexamethasone-inducible system, *CO* has been shown to promote the expression of floral meristem-identity genes as well as *FT* (see below) (Simon *et al.*, 1996; Kobayashi *et al.*, 1999). Using

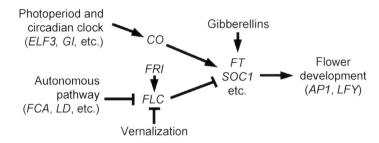

Figure 2.3 An outline of flowering pathways in *Arabidopsis*. This model lists only a few genes from each pathway for reference. Information from several pathways including the photoperiod, autonomous, vernalization, and gibberellin pathways control the expression of downstream genes such as *SOC1* and *FT*. These genes in turn influence the expression of meristem-identity genes that specify the formation of floral meristems and ultimately of flowers.

the same line, genes directly downstream of *CO* were found by simultaneously applying dexamethasone and the protein synthesis inhibitor cycloheximide and monitoring changes in mRNA levels. This work identified *FT* as an early target gene of *CO*. *SUPPRESSOR OF OVEREXPRESSION OF CO 1* (*SOC1*) was also identified as a direct downstream target of *CO* in this screen (Samach *et al.*, 2000).

Further support for some of these genetic interactions came from a screen for suppressor mutations of *35S::CO* plants. Mutations in *ft* and *soc1* partially suppress the early-flowering phenotype of *35S::CO* (Onouchi *et al.*, 2000). Onouchi *et al.* also identified another known flowering-time gene, *FWA* (Koornneef *et al.*, 1991), in this screen (Onouchi *et al.*, 2000). *SOC1* and *FWA* also act downstream of the autonomous pathway. Thus, these genes appear to act at a point of convergence between the photoperiod and autonomous pathways.

The *FT* family of genes consists of three members: *TERMINAL FLOWER* (*TFL*), *FT*, and *TWIN SISTER OF FT* (*TSF*). All of these genes have been shown to affect the control of flowering time. *ft* mutants flower late in both LD and SD (Koornneef *et al.*, 1991). The *FT* transcript is regulated in a circadian fashion with a peak time of expression in the evening (Harmer *et al.*, 2000; Suarez-Lopez *et al.*, 2001). Overexpression of *FT* leads to early flowering in long days and even earlier flowering in short days (Kardailsky *et al.*, 1999; Kobayashi *et al.*, 1999). Similar results have also been found in plants overexpressing *TSF*, the closest relative of *FT*. These early-flowering plants have a greatly reduced period of vegetative growth, forming about two leaves before flowering. *FT* expression is delayed in *co* mutants and overexpression of *FT* completely suppresses the late-flowering *co* phenotype. These data and the partial suppression of *35S::CO* effects discussed above are consistent with *FT* acting downstream of *CO* (Samach *et al.*, 2000).

terminal flower (*tfl*) mutants flower earlier than wild type in LD and SD (Shannon and Meeks-Wagner, 1991). As the name implies, these mutants have inflorescences that eventually form a terminal flower. *TFL* was cloned owing to its similarity with the *Antirrhinum* gene *CENTRORADIALIS* (*CEN*). Mutations in *CEN* lead to a terminal flower but do not affect flowering time (Bradley *et al.*, 1996). The role of *TFL* in both flowering-time regulation and inflorescence architecture in *Arabidopsis* is consistent with its pattern of expression. *TFL* is first expressed at low levels in the vegetative meristem, where it presumably acts as a repressor of flowering. Later in development *TFL* is expressed at higher levels in the inflorescence meristem, where it acts to maintain indeterminacy by inhibiting the expression of floral meristem-identity genes in the SAM (Bradley *et al.*, 1997; Liljegren *et al.*, 1999; Ratcliffe *et al.*, 1999).

35S::TFL plants flower late and display a variety of inflorescence phenotypes (Ratcliffe, 1998). Thus, overexpression of *TFL* causes the opposite phenotype from overexpression of its close relative *FT*. Although there is no clear-cut epistatic relationship between these two genes and their effect on flowering time, *ft;tfl* plants and *35S::FT;35S::TFL* plants more closely resemble *ft* and *35S::FT*, respectively (Ruiz-Garcia *et al.*, 1997; Kobayashi *et al.*, 1999).

SOC1 encodes a MADS-box transcription factor that acts as an activator of flowering. Three groups independently cloned this gene using different strategies. *SOC1* was first isolated as a suppressor of *35S::CO* (see above). A second group isolated *SOC1* owing to its homology with *SaMADSA* from *Sinapis alba* (Borner *et al.*, 2000). A third group isolated *SOC1* in a screen for gain-of-function mutants that suppress late flowering in plants containing *FRI* (Lee *et al.*, 2000). The ability to isolate mutants in *SOC1* by these different means attests to its importance in floral promotion downstream of both the photoperiod- and *FLC*-dependent pathways.

An increase in expression levels of *SOC1* and *SaMADSA* occurs rapidly upon exposure to an inductive LD in *Arabidopsis* (Borner *et al.*, 2000; Samach *et al.*, 2000) and *S. alba* (Menzel *et al.*, 1996), respectively. Late-flowering mutants such as *co*, *gi*, *fca*, and *fve* show a delay in the upregulation of *SOC1* (Lee *et al.*, 2000; Samach *et al.*, 2000). Autonomous-pathway mutants, however, show a greater delay in *SOC1* expression. This is consistent with a model in which *FLC* acts as a negative regulator of *SOC1* (Lee *et al.*, 2000). Plants overexpressing *SOC1* flower earlier than wild type but maintain a slight response to photoperiod (Borner *et al.*, 2000). *FRI* does not affect flowering in plants overexpressing *SOC1*, indicating that *SOC1* acts downstream of *FLC* (Lee *et al.*, 2000).

GAs are known to induce flowering in many species. *SOC1* (Borner *et al.*, 2000) and *SaMADSA* (Bonhomme *et al.*, 2000) expression is upregulated in response to GA. Thus, in addition to being downstream of the photoperiod- and *FLC*-dependent pathways, hormones can also regulate *SOC1*, suggesting a role for this gene downstream of yet another pathway controlling flowering time.

2.12 The transition to flowering—activation of floral meristem identity genes

The action of all genes involved in controlling the floral transition eventually leads to the activation of floral meristem-identity genes such as the *Arabidopsis* *LEAFY* (*LFY*) and *APETALA1* (*AP1*) genes. *LFY* and *AP1* are expressed in organ primordia flanking the SAM (Parcy *et al.*, 1998; Ratcliffe *et al.*, 1999) and act to establish the fate of these primordia as flowers. Thus, activation of these genes indicates that the floral transition has already taken place and that reproductive development is underway. Here we briefly discuss what is known about the regulation of these genes. An in-depth discussion of meristem-identity genes and their effect on flower development is beyond the scope of this review.

CO overexpression leads to early flowering and precocious upregulation of *LFY* (Simon *et al.*, 1996). The autonomous pathway also plays a role in promoting *LFY* expression (Ruiz-Garcia *et al.*, 1997; Aukerman *et al.*, 1999). *FT* and *FWA* promote *AP1* expression and seem to function in the same pathway (Koornneef *et al.*, 1998). *ft/lfy* and *fwa/lfy* double mutants reveal new inflorescence phenotypes similar to those seen in *ap1*, and these phenotypes are not seen in *fwa/ft* double mutants (Ruiz-Garcia *et al.*, 1997; Koornneef *et al.*, 1998). This suggests that two pathways control the expression of *AP1*. One pathway is through *LFY* and the other through *FWA* and *FT*.

Mutations in *lfy* and *ap1* have phenotypes associated with floral meristem development but do not affect the timing of the floral transition (Mandel *et al.*, 1992; Weigel *et al.*, 1992). Overexpression of *AP1*, however, leads to early flowering in LD and SD as well as formation of terminal flowers as seen in *tfl* (Mandel and Yanofsky, 1995). Overexpression of *LFY* also leads to the formation of terminal flowers and early flowering in SD (Weigel and Nilsson, 1995). Mutations in *ap1* suppress the early-flowering phenotype of *35S::LFY*, indicating that *LFY* acts upstream of *AP1* and positively regulates *AP1* expression (Weigel and Nilsson, 1995). *AP1* is also capable of activating *LFY* (Liljegren *et al.*, 1999). The fact each of *LFY* and *AP1* can positively regulate the expression of the other suggests that a *LFY/AP1* positive feedback loop may exist in *Arabidopsis*, which could account for the irreversible nature of the floral transition in this species.

2.13 Summary

Genetic analyses have identified a large number of genes that control flowering time in several species, particularly *Arabidopsis*. This genetic progress is leading to an understanding of flowering pathways at a molecular level. The progress has resulted from a combination of approaches that have been reviewed in this chapter. The approach of classical mutagenesis to identify induced mutations in flowering-time loci has revealed much about the photoperiod and autonomous pathways. The use of genetic screens designed to identify suppressors and enhancers of existing mutations has identified genes that operate at convergence points of flowering pathways (Lee *et al.*, 2000; Samach *et al.*, 2000). To overcome the limits on classical screens imposed by the redundancy of most plant genomes, activation-tagging mutagenesis has been used to create gain-of-function alleles of flowering-time loci (Lee *et al.*, 2000; Weigel *et al.*, 2000). Another approach is to exploit naturally occurring variation in flowering time. This has, for example, led to the identification of a gene involved in meristem competence (*FLC*) and has revealed that *Arabidopsis* was first a winter annual. Naturally occurring mutations that led to loss or reduction of gene activity gave rise to the summer-annual forms that are the popular varieties used in most laboratories. The analysis of natural variation is likely to be used more frequently in the future with *Arabidopsis*. With the development of efficient molecular marker technology, specific statistical methods and the complete genome sequence of *Arabidopsis*, it has become much easier to map and subsequently clone quantitative trait loci. An advantage of studying natural variation is that any allelic variation at flowering-time loci has survived selection. An interesting example of this is the recent demonstration of allelic variation at the CRY2 blue light photoreceptor discussed above (M. Koorneef, personal communication).

As we learn more about flowering-time regulation in *Arabidopsis* and other well-characterized systems such as pea, the challenge will be to explore the molecular basis of different flowering behaviors among a wide range of different plants. What were the first regulatory systems to evolve? How have these regulatory systems been modified throughout evolution to allow adaptation to specific climates and locations, and to permit successful competition with other plants in the same region? Impressive progress has been made in understanding the molecular control of the floral transition in some species, but much remains to be learned.

References

Alonso-Blanco, C., El-Assal, S.E., Coupland, G. and Koornneef, M. (1998) Analysis of natural allelic variation at flowering time loci in the Landsberg erecta and Cape Verde Islands ecotypes of Arabidopsis thaliana. *Genetics*, **149**, 749-764.

Araki, T. and Komeda, Y. (1993) Analysis of the role of the late-flowering locus, *GI*, in the flowering of *Arabidopsis thaliana*. *Plant J.*, **3**, 231-239.

Aukerman, M.J. and Amasino, R.M. (1998) Floral induction and florigen. *Cell*, **93**, 491-494.

Aukerman, M.J., Hirschfeld, M., Wester, L., *et al.* (1997) A deletion in the PHYD gene of the *Arabidopsis* Wassilewskija ecotype defines a role for phytochrome D in red/far-red light sensing. *Plant Cell*, **9**, 1317-1326.

Aukerman, M.J., Lee, I., Weigel, D. and Amasino, R.M. (1999) The *Arabidopsis* flowering-time gene *LUMINIDEPENDENS* is expressed primarily in regions of cell proliferation and encodes a nuclear protein that regulates *LEAFY* expression. *Plant J.*, **18**, 195-203.

Bagnall, D.J., King, R.W., Whitelam, G.C., Boylan, M.T., Wagner, D. and Quail, P.H. (1995) Flowering responses to altered expression of phytochrome in mutants and transgenic lines of *Arabidopsis thaliana* (L.) Heynh. *Plant Physiol.*, **108**, 1495-1503.

Bagnall, D.J., King, R.W. and Hangarter, R.P. (1996) Blue-light promotion of flowering is absent in *hy4* mutants of *Arabidopsis*. *Planta*, **200**, 278-280.

Bognar, L.K., Hall, A., Adam, E., Thain, S.C., Nagy, F. and Millar, A.J. (1999) The circadian clock controls the expression pattern of the circadian input photoreceptor, photychrome B. *Proc. Natl. Acad. Sci. USA*, **96**, 14652-14657.

Bonhomme, F., Kurz, B., Melzer, S., Bernier, G. and Jacqmard, A. (2000) Cytokinin and gibberellin activate *SaMADS A*, a gene apparently involved in regulation of the floral transition in *Sinapis alba*. *Plant J.*, **24**, 103-111.

Borner, R., Kampmann, G., Chandler, J., *et al.* (2000) A MADS domain gene involved in the transition to flowering in arabidopsis. *Plant J.*, **24**, 591-599.

Bradley, D., Carpenter, R., Copsey, L., Vincent, C., Rothstein, S. and Coen, E. (1996) Control of inflorescence architecture in *Antirrhinum*. *Nature*, **379**, 791-797.

Bradley, D., Ratcliffe, O., Vincent, C., Carpenter, R. and Coen, E. (1997) Inflorescence commitment and architecture in *Arabidopsis*. *Science*, **275**, 80-83.

Burn, J.E., Bagnall, D.J., Metzger, J.D., Dennis, E.S. and Peacock, W.J. (1993a) DNA methylation, vernalization, and the initiation of flowering. *Proc. Natl. Acad. Sci. USA*, **90**, 287-291.

Burn, J.E., Smyth, D.R., Peacock, W.J. and Dennis, E.S. (1993b) Genes conferring late flowering in *Arabidopsis thaliana*. *Genetica*, **90**, 147-155.

Chandler, J., Wilson, A. and Dean, C. (1996) Arabidopsis mutants showing an altered response to vernalization. *Plant J.*, **10**, 637-644.

Childs, K.L., Miller, F.R., Cordonnier-Pratt, M.M., Pratt, L.H., Morgan, P.W. and Mullet, J.E. (1997) The sorghum photoperiod sensitivity gene, *Ma3*, encodes a phytochrome B. *Plant Physiol.*, **113**, 611-619.

Chouard, P. (1960) Vernalization and its relations to dormancy. *Annu. Rev. Plant Physiol.*, **11**, 191-238.

Clarke, J.H. and Dean, C. (1994) Mapping *FRI*, a locus controlling flowering time and vernalization response in *Arabidopsis thaliana*. *Mol. Gen. Genet.*, **242**, 81-89.

Davis, S.J., Kurepa, J. and Vierstra, R.D. (1999) The *Arabidopsis thaliana* HY1 locus, required for phytochrome—chromophore biosynthesis, encodes a protein related to heme oxygenases. *Proc. Natl. Acad. Sci. USA*, **96**, 6541-6546.

Deitzer, G. (1984) Photoperiodic induction in long-day plants, in *Light and the Flowering Process* (eds. D. Vince-Prue, B. Thomas and K.E. Cockshul), Academic Press, New York, pp. 51-63.

Dennis, E.S., Bilodeau, P., Burn, J., *et al.* (1997) Methylation controls the low temperature induction of flowering in *Arabidopsis*, in *Control of Plant Development, Genes and Signals*, vol. 51 (eds. A.J. Greenland, E.M. Meyerowitz and M. Steer), Portland Press, London, pp. 97-103.

Devlin, P.F., Rood, S.B., Somers, D.E., Quail, P.H. and Whitelam, G.C. (1992) Photophysiology of the elongated internode (*ein*) mutant of *Brassica rapa*. The *ein* mutant lacks a detectable phytochrome B-like polypeptide. *Plant Physiol.*, **100**, 1442-1447.

Devlin, P.F., Patel, S.R. and Whitelam, G.C. (1998) Phytochrome E influences internode elongation and flowering time in *Arabidopsis*. *Plant Cell*, **10**, 1479-1487.

Devlin, P.F., Robson, P.R.H., Patel, S.R., Goosey, L., Sharrock, R.A. and Whitelam, G.C. (1999) Phytochrome D acts in the shade-avoidance syndrome in *Arabidopsis* by controlling elongation growth and flowering time. *Plant Physiol.*, **119**, 909-915.

Dubnau, J. and Struhl, G. (1996) RNA recognition and translational regulation by a homeodomain protein. *Nature*, **379**, 694-699.

Dunlap, J.C. (1999) Molecular bases for circadian clocks. *Cell*, **96**, 271-290.

Evans, L.T. (1975) *Daylength and the Flowering of Plants*, W.A. Benjamin, Menlo Park, CA.

Finnegan, E.J., Peacock, W.J. and Dennis, E.S. (1996) Reduced DNA methylation in *Arabidopsis thaliana* results in abnormal plant development. *Proc. Natl. Acad. Sci. USA*, **93**, 8449-8454.

Finnegan, E.J., Genger, R.K., Kovac, K., Peacock, W.J. and Dennis, E.S. (1998) DNA methylation and the promotion of flowering by vernalization. *Proc. Natl. Acad. Sci. USA*, **95**, 5824-5829.

Fowler, S., Lee, K., Onouchi, H., *et al.* (1999) *GIGANTEA*: a circadian clock-controlled gene that regulates photperiodic flowering in *Arabidopsis* and encodes a protein with several possible membrane-spanning domains. *EMBO J.*, **18**, 4679-4688.

Goto, N., Kumagai, T. and Koornneef, M. (1991) Flowering responses to light-breaks in photomorphogenic mutants of *Arabidopsis thaliana*, a long-day plant. *Physiol. Plant.*, **83**, 209-215.

Green, R.M. and Tobin, E.M. (1999) Loss of the circadian clock-associated protein 1 in *Arabidopsis* results in altered clock-regulated gene expression. *Proc. Natl. Acad. Sci. USA*, **96**, 4176-4179.

Guo, H., Yang, H., Mockler, T.C. and Lin, C. (1998) Regulation of flowering time by *Arabidopsis* photoreceptors. *Science*, **279**, 1360-1363.

Hanumappa, M., Pratt, L.H., Cordonnier-Pratt, M.M. and Deitzer, G.F. (1999) A photoperiod-insensitive barley line contains a light-labile phytochrome B. *Plant Physiol.*, **119**, 1033-1040.

Harmer, S.L., Hogenesch, J.B., Straume, M., *et al.* (2000) Orchestrated transcription of key pathways in *Arabidopsis* by the circadian clock. *Science*, **290**, 2110-2113.

Hicks, K.A., Millar, A.J., Carre, I.A., *et al.* (1996) Conditional circadian dysfunction of the *Arabidopsis early-flowering 3* mutant. *Science*, **274**, 790-792.

Huq, E., Tepperman, J.M. and Quail, P.H. (2000) GIGANTEA is a nuclear protein involved in phytochrome signaling in *Arabidopsis*. *Proc. Natl. Acad. Sci. USA*, **97**, 9789-9794.

Izawa, T., Oikawa, T., Tokutomi, S., Okuno, K. and Shimamoto, K. (2000) Phytochromes confer the photoperiodic control of flowering in rice (a short-day plant). *Plant J.*, **22**, 391-399.

Johanson, U., West, J., Lister, C., Michaels, S., Amasino, R. and Dean, C. (2000) Molecular analysis of *FRIGIDA*, a major determinant of natural variation in *Arabidopsis* flowering time. *Science*, **290**, 344-347.

Johnson, E., Bradley, M., Harberd, N.P. and Whitelam, G.C. (1994) Photoresponses of light-grown *phyA* mutants of *Arabidopsis*. *Plant Physiol.*, **105**, 141-149.

Kardailsky, I., Shukla, V.K., Ahn, J.H., *et al.* (1999) Activation tagging of the floral inducer FT. *Science*, **286**, 1962-1965.

Karlsson, B.H., Sills, G.R. and Nienhuis, J. (1993) Effects of photoperiod and vernalization on the number of leaves at flowering in 32 *Arabidopsis thaliana* (Brassicaceae) ecotypes. *Am. J. Bot.*, **80**, 646-648.

Kloss, B., Price, J.L., Saez, L., *et al.* (1998) The *Drosophila* clock gene double-time encodes a protein closely related to human casein kinase Iepsilon. *Cell*, **94**, 97-107.

Kobayashi, Y., Kaya, H., Goto, K., Iwabuchi, M. and Araki, T. (1999) A pair of related genes with antagonistic roles in mediating flowering signals. *Science*, **286**, 1960-1962.

Kole, C., Quijada, P., Michaels, S.D., Amasino, R.M. and Osborn, T.C. (2001) Evidence for homology of flowering-time genes *VFR2* from *Brassica rapa* and *FLC* from *Arabidopsis thaliana*. *Theor. Appl. Genet.*, **102**, 425-430.

Koornneef, M., Hanhart, C.J. and van der Veen, J.H. (1991) A genetic and physiological analysis of late flowering mutants in *Arabidopsis thaliana*. *Mol. Gen. Genet.*, **229**, 57-66.

Koornneef, M., Blankestijn-de Vries, H., Hanhart, C., Soppe, W. and Peeters, T. (1994) The phenotype of some late-flowering mutants is enhanced by a locus on chromosome 5 that is not effective in the landsberg *erecta* wild-type. *Plant J.*, **6**, 911-919.

Koornneef, M., Alonso-Blanco, C., Blankestijn-de Vries, H., Hanhart, C.J. and Peeters, A.J. (1998) Genetic interactions among late-flowering mutants of *Arabidopsis*. *Genetics*, **148**, 885-892.

Lang, A. (1986) *Hyoscyamus niger*, in *CRC Handbook of Flowering*, vol. V (ed. A.H. Halevy) CRC Press, Boca Raton, FL, pp. 144-186.

Lee, H., Suh, S.S., Park, E., *et al.* (2000) The AGAMOUS-LIKE 20 MADS domain protein integrates floral inductive pathways in *Arabidopsis*. *Genes Dev.*, **14**, 2366-2376.

Lee, I., Bleecker, A. and Amasino, R. (1993) Analysis of naturally occurring late flowering in *Arabidopsis thaliana*. *Mol. Gen. Genet.*, **237**, 171-176.

Lee, I., Aukerman, M.J., Gore, S.L., *et al.* (1994a) Isolation of *LUMINIDEPENDENS*—a gene involved in the control of flowering time in *Arabidopsis*. *Plant Cell*, **6**, 75-83.

Lee, I., Michaels, S.D., Masshardt, A.S. and Amasino, R.M. (1994b) The late-flowering phenotype of *FRIGIDA* and *LUMINIDEPENDENS* is suppressed in the Landsberg *erecta* strain of *Arabidopsis*. *Plant J.*, **6**, 903-909.

Liljegren, S.J., Gustafson-Brown, C., Pinyopich, A., Ditta, G.S. and Yanofsky, M.F. (1999) Interactions among APETALA1, LEAFY, and TERMINAL FLOWER1 specify meristem fate. *Plant Cell*, **11**, 1007-1018.

Liu, J., Yu, J., McIntosh, L., Kende, H. and Zeevaart, J.A. (2001) Isolation of a CONSTANS ortholog from *Pharbitis nil* and its role in flowering. *Plant Physiol.*, **125**, 1821-1830.

Lopez-Juez, E., Nagatani, A., Tomizawa, K.-I., *et al.* (1992) The cucumber long hypocotyl mutant lacks a light-stable PHYB-like phytochrome. *Plant Cell*, **4**, 241-251.

Macknight, R., Bancroft, I., Page, T., *et al.* (1997) *FCA*, a gene controlling flowering time in *Arabidopsis*, encodes a protein containing RNA-binding domains. *Cell*, **89**, 737-745.

Mandel, M.A. and Yanofsky, M.F. (1995) A gene triggering flower formation in Arabidopsis. *Nature*, **377**, 522-524.

Mandel, M.A., Gustafson-Brown, C., Savidge, B. and Yanofsky, M.F. (1992) Molecular characterization of the *Arabidopsis* floral homeotic gene APETALA1. *Nature*, **360**, 273-277.

Mas, P., Devlin, P.F., Panda, S. and Kay, S.A. (2000) Functional interaction of phytochrome B and cryptochrome 2. *Nature*, **408**, 207-211.

McWatters, H.G., Bastow, R.M., Hall, A. and Millar, A.J. (2000) The ELF3 zeitnehmer regulates light signalling to the circadian clock. *Nature*, **408**, 716-720.

Meier, C., Bouquin, T., Nielsen, M.E., *et al.* (2001) Gibberellin response mutants identified by luciferase imaging. *Plant J.*, **25**, 509-519.

Menzel, G., Apel, K. and Melzer, S. (1996) Identification of two MADS box genes that are expressed in the apical meristem of the long-day plant *Sinapis alba* in transition to flowering. *Plant J.*, **9**, 399-408.

Michaels, S. and Amasino, R. (1999) FLOWERING LOCUS C encodes a novel MADS domain protein that acts as a repressor of flowering. *Plant Cell*, **11**, 949-956.

Michaels, S. and Amasino, R. (2000) Memories of winter: vernalization and the competence to flower. *Plant Cell Environ.*, **23**, 1145-1154.

Michaels, S. and Amasino, R. (2001) Loss of *FLOWERING LOCUS C* activity eliminates the late-flowering phenotype of *FRIGIDA* and autonomous-pathway mutations, but not responsiveness to vernalization. *Plant Cell*, **13**, 935-942.

Millar, A.J., Straume, M., Chory, J., Chua, N.H. and Kay, S.A. (1995) The regulation of circadian period by phototransduction pathways in *Arabidopsis*. *Science*, **267**, 1163-1166.

Mizukami, Y. and Ma, H. (1997) Determination of *Arabidopsis* floral meristem identity by *AGAMOUS*. *Plant Cell*, **9**, 393-408.

Mockler, T.C., Guo, H., Yang, H., Duong, H. and Lin, C. (1999) Antagonistic action of *Arabidopsis* cryptochromes and phytochrome B in the regulation of floral induction. *Development*, **126**, 2073-2082.

Napp-Zinn, K. (1979) On the genetical basis of vernalization requirement in *Arabidopsis thaliana* (L.) Heynh in *La Physiologie de la Floraison* (eds. P. Champagnat and R. Jaques), Coll. Int. CNRS, Paris, pp. 217-220.

Nelson, D.C., Lasswell, J., Rogg, L.E., Cohen, M.A. and Bartel, B. (2000) *FKF1*, a clock-controlled gene that regulates the transition to flowering in *Arabidopsis*. *Cell*, **101**, 331-340.

Onouchi, H., Igeno, M.I., Perilleux, C., Graves, K. and Coupland, G. (2000) Mutagenesis of plants overexpressing *CONSTANS* demonstrates novel interactions among *Arabidopsis* flowering-time genes. *Plant Cell*, **12**, 885-900.

Osborn, T.C., Kole, C., Parkin, I.A.P., *et al.* (1997) Comparison of flowering time genes in *Brassica rapa, B. napus* and *Arabidopsis thaliana*. *Genet. Soc. Am.*, **146**, 1123-1129.

Parcy, F., Nilsson, O., Busch, M.A., Lee, I. and Weigel, D. (1998) A genetic framework for floral patterning. *Nature*, **395**, 561-566.

Park, D.H., Somers, D.E., Kim, Y.S., *et al.* (1999) Control of circadian rhythms and photoperiodic flowering by the *Arabidopsis GIGANTEA* gene. *Science*, **285**, 1579-1582.

Poethig, R.S. (1997) Leaf morphogenesis in flowering plants. *Plant Cell*, **9**, 1077-1087.

Putterill, J., Robson, F., Lee, K., Simon, R. and Coupland, G. (1995) The *CONSTANS* gene of *Arabidopsis* promotes flowering and encodes a protein showing similarities to zinc finger transcription factors. *Cell*, **80**, 847-857.

Quail, P.H., Boylan, M.T., Parks, B.M., Short, T.W., Xu, Y. and Wagner, D. (1995) Phytochromes: photosensory perception and signal transduction. *Science*, **268**, 675-680.

Ratcliffe, O.J., Amaya, I., Vincent, C.A., Rothstein, S., Carpenter, R., Coen, E.S. and Bradley, D.J. (1998) A common mechanism controls the lifecycle and architecture of plants. *Development*, **125**, 1609-1615.

Ratcliffe, O.J., Bradley, D.J. and Coen, E.S. (1999) Separation of shoot and floral identity in *Arabidopsis*. *Development*, **126**, 1109-1120.

Redei, G.P. (1962) Supervital mutants in *Arabidopsis*. *Genetics*, **47**, 443-460.

Reed, J.W., Nagpal, P., Bastow, R.M., *et al.* (2000) Independent action of ELF3 and phyB to control hypocotyl elongation and flowering time. *Plant Physiol.*, **122**, 1149-1160.

Ruiz-Garcia, L., Madueno, F., Wilkinson, M., Haughn, G., Salinas, J. and Martinez-Zapater, J.M. (1997) Different roles of flowering-time genes in the activation of floral initiation genes in *Arabidopsis*. *Plant Cell*, **9**, 1921-1934.

Samach, A. and Coupland, G. (2000) Time measurement and the control of flowering in plants. *BioEssays*, **22**, 38-47.

Samach, A., Onouchi, H., Gold, S.E., *et al.* (2000) Distinct roles of CONSTANS target genes in reproductive development of *Arabidopsis*. *Science*, **288**, 1613-1616.

Sanda, S.L. and Amasino, R.M. (1996) Interaction of *FLC* and late-flowering mutations in *Arabidopsis thaliana*. *Mol. Gen. Genet.*, **251**, 69-74.

Schaffer, R., Ramsay, N., Alon, S., *et al.* (1998) The *late elongated hypocotyl* mutation of *Arabidopsis* disrupts circadian rhythms and the photoperiodic control of flowering. *Cell*, **93**, 1219-1229.

Schomburg, F.M., Patton, D.A., Meinke, D.W. and Amasino, R.M. (2001) *FPA*, a gene involved in floral induction in *Arabidopsis thaliana* encodes a protein containing RNA-recognition motifs. *Plant Cell*, **13**, 1427-1436.

Shannon, S. and Meeks-Wagner, D.R. (1991) A mutation in the *Arabidopsis TFL1* gene affects inflorescence meristem development. *Plant Cell*, **3**, 877-892.

Sheldon, C.C., Burn, J.E., Perez, P.P., *et al.* (1999) The FLF MADS box gene. A repressor of flowering in arabidopsis regulated by vernalization and methylation. *Plant Cell*, **11**, 445-458.

Sheldon, C.C., Rouse, D.T., Finnegan, E.J., Peacock, W.J. and Dennis, E.S. (2000) The molecular basis of vernalization: the central role of *FLOWERING LOCUS C (FLC)*. *Proc. Natl. Acad. Sci. USA*, **97**, 3753-3758.

Simon, R., Igeno, M.I. and Coupland, G. (1996) Activation of floral meristem identity genes in *Arabidopsis*. *Nature*, **384**, 59-62.

Simpson, G.G., Gendall, A.R. and Dean, C. (1999) When to switch to flowering. *Annu. Rev. Cell Dev. Biol.*, **99**, 519-550.

Somers, D.E., Devlin, P.F. and Kay, S.A. (1998a) Phytochromes and cryptochromes in the entrainment of the *Arabidopsis* circadian clock. *Science*, **282**, 1488-1490.

Somers, D.E., Webb, A.A.R., Pearson, M. and Kay, S.A. (1998b) The short-period mutant, toc-1, alters circadian clock regulation of multiple outputs throughout development in *Arabidopsis thaliana*. *Development*, **125**, 485-494.

Somers, D.E., Schultz, T.F., Milnamow, M. and Kay, S.A. (2000) *ZEITLUPE* encodes a novel clock-associated PAS protein from *Arabidopsis*. *Cell*, **101**, 319-329.

Strayer, C., Oyama, T., Schultz, T.F., *et al.* (2000) Cloning of the *Arabidopsis* clock gene *TOC1*, an autoregulatory response regulator homolog. *Science*, **289**, 768-771.

Suarez-Lopez, P., Wheatley, K., Robson, F., Onouchi, H., Valverde, F. and Coupland, G. (2001) CONSTANS mediates between the circadian clock and the control of flowering in *Arabidopsis*. *Nature*, **410**, 1116-1120.

Sugano, S., Andronis, C., Green, R.M., Wang, Z.Y. and Tobin, E.M. (1998) Protein kinase CK2 interacts with and phosphorylates the *Arabidopsis* circadian clock-associated 1 protein. *Proc. Natl. Acad. Sci. USA*, **95**, 11020-11025.

Sugano, S., Andronis, C., Ong, M.S., Green, R.M. and Tobin, E.M. (1999) The protein kinase CK2 is involved in regulation of circadian rhythms in *Arabidopsis*. *Proc. Natl. Acad. Sci. USA*, **96**, 12362-12366.

Swarup, K., Alonso-Blanco, C., Lynn, J.R., *et al.* (1999) Natural allelic variation identifies new genes in the *Arabidopsis* circadian system. *Plant J.*, **20**, 67-77.

Telfer, A., Bollman, K.M. and Poethig, R.S. (1997) Phase change and the regulation of trichome distribution in *Arabidopsis thaliana*. *Development*, **124**, 645-654.

Thomas, B. and Vince-Prue, D. (1997) *Photoperiodism in Plants*, Academic Press, San Diego.

Tilghman, S.M. (1999) The sins of the fathers and mothers: genomic imprinting in mammalian development. *Cell*, **96**, 185-193.

Tranquilli, G. and Dubcovsky, J. (2000) Epistatic interaction between vernalization genes *Vrn-Am1* and *Vrn-Am2* in *Triticum monococcum*. *J. Hered.*, **91**, 304-306.

Vetrilova, M. (1973) Genetic and physiological analysis of induced late mutants of *Arabidopsis thaliana* (L.) Heynh. *Biol. Plantarum*, **15**, 391-397.

Wang, Z. and Tobin, E.M. (1998) Constitutive expression of the *CIRCADIAN CLOCK ASSOCIATED 1* (*CCA1*) gene disrupts circadian rhythms and suppresses its own expression. *Cell*, **93**, 1207-1217.

Wang, Z.Y., Kenigsbuch, D., Sun, L., Harel, E., Ong, M.S. and Tobin, E.M. (1997) A Myb-related transcription factor is involved in the phytochrome regulation of an *Arabidopsis Lhcb* gene. *Plant Cell*, **9**, 491-507.

Weigel, D., Alvarez, J., Smyth, D.R., Yanofsky, M.F. and Meyerowitz, E.M. (1992) LEAFY controls floral meristem identity in *Arabidopsis*. *Cell*, **69**, 843-859.

Weigel, D. and Nilsson, O. (1995) A developmental switch sufficient for flower initiation in diverse plants. *Nature*, **377**, 495-500.

Weigel, D., Ahn, J.H., Blazquez, M.A., *et al.* (2000) Activation tagging in *Arabidopsis*. *Plant Physiol.*, **122**, 1003-1013.

Wellensiek, S.J. (1962) Dividing cells as the locus for vernalization. *Nature*, **195**, 307-308.

Weller, J.L., Murfet, I.C. and Reid, J.B. (1997a) Pea mutants with reduced sensitivity to far-red light define an important role for phytochrome A in day-length detection. *Plant Physiol.*, **114**, 1225-1236.

Weller, J.L., Reid, J.B., Taylor, S.A. and Murfet, I.C. (1997b) The genetic control of flowering in pea. *Trends Plant Sci.*, **2**, 412-418.

Wilson, R.N., Heckman, J.W. and Somerville, C.R. (1992) Gibberellin is required for flowering in *Arabidopsis thaliana* under short days. *Plant Physiol.*, **100**, 403-408.

Yanovsky, M.J., Izaguirre, M., Wagmaister, J.A., *et al.* (2000a) Phytochrome A resets the circadian clock and delays tuber formation under long days in potato. *Plant J.*, **23**, 223-232.

Yanovsky, M.J., Mazzella, M.A. and Casal, J.J. (2000b) A quadruple photoreceptor mutant still keeps track of time. *Curr. Biol.*, **10**, 1013-1015.

Zagotta, M.T., Hicks, K.A., Jacobs, C.I., Young, J.C., Hangarter, R.P. and Meeks-Wagner, D.R. (1996) The *Arabidopsis ELF3* gene regulates vegetative photomorphogenesis and the photoperiodic induction of flowering. *Plant J.*, **10**, 691-702.

Zeevaart, J.A.D. (1984) Photoperiodic induction, the floral stimulus and flower-promoting substances, in *Light and the Flowering Process* (eds. D. Vince-Prue, B. Thomas and K.E. Cockshull), Academic Press, London, pp. 137-141.

3 The genetic and molecular control of ovule development

David Chevalier, Patrick Sieber and Kay Schneitz

3.1 Introduction

Flowers are central to sexual reproduction in higher plants. Evolution led to a baffling multitude of colors, shapes and sizes of floral organs. This variety relates to their role in attracting pollinators, but also catches our attention when we admire the beauty that springs up everywhere. Yet it is the ovule, buried deep within the carpel and thus hidden from the curious eye, where the ultimate events come to pass. Within the ovule, the female aspects of the *Generationswechsel*, the alternation between the diploid sporophytic and the haploid gametophytic phase of the plant's lifecycle take place. It is the organ in which the egg cell develops, where double fertilization occurs, and in which the embryo develops during seed development.

The ovule thus represents the major female reproductive organ. Questions relating to its development and function have attracted the interest of botanists for several hundred years and many of these questions are still puzzling us today. For example, what regulates the switch from the sporophytic to the gametophytic phase? What ensures meiosis that proceeds correctly? What controls the identity of the ovule? What is the mechanism of early pattern formation within the ovule primordium and how is this process orchestrated with its outgrowth? What interactions occur between the sporophytic tissue and the developing female gametophyte?

Ovules are simply built structures. They consist essentially of three elements: at the top, the nucellus that harbors the multicellular haploid embryo sac or female gametophyte; a central chalaza, which is characterized by the integuments that initiate at its flanks; and at the bottom, the funiculus or stalk that connects the ovule to the carpel tissue (Esau, 1977). Despite the apparent overall simplicity in architecture, the ovules are composed of a number of clearly differentiated cell types. These features, combined with the ease with which sterile mutants can be detected in mutagenesis experiments, have allowed the identification of a large, and still growing, number of genes that affect various aspects of ovule development. On the basis of their functional analysis, several laboratories are contributing to a developing framework describing the genetic and molecular control of ovule development.

Botanists have investigated the morphology of ovule development and the ultrastructure of individual cell types in a large variety of plant species. This review, however, clearly reflects the fact that much of the recent molecular and genetic knowledge regarding ovule development is rooted in work using *Arabidopsis thaliana* as the model system. Nevertheless, important experiments were carried out with other species as well, for example tobacco or petunia, and are included in this overview. A series of reviews covering the topic has been published (Angenent and Colombo, 1996; Gasser *et al.*, 1998; Reiser and Fischer, 1993; Schneitz *et al.*, 1998b). Here, we focus on the sporophytic aspects of ovule development, such as placenta and ovule primordium formation, or integument morphogenesis. A number of recent reviews have summarized the various aspects of gametogenesis in great detail (Bhatt *et al.*, 2001; Drews *et al.*, 1998; Grossniklaus and Schneitz, 1998; Schneitz, 1999; Yang and Sundaresan, 2000).

3.2   Morphological aspects

This section describes the morphology of ovule development in *Arabidopsis thaliana*, which has been described in great detail (Bajon *et al.*, 1999; Christensen *et al.*, 1997; Mansfield *et al.*, 1991; Modrusan *et al.*, 1994; Robinson-Beers *et al.*, 1992; Schneitz *et al.*, 1995). It serves mainly to outline the individual steps that occur during ovule ontogeny. In principle these steps are shared among angiosperm ovules; some of the details of course vary from species to species. Embryo sac development in *Arabidopsis* follows the monosporic *Polygonum*-type, a pattern that is found in the majority of angiosperm species (Maheswari, 1950).

3.2.1   Carpel development

Ovules develop within carpels, thus, a brief summary of wild-type carpel development is given (Bowman *et al.*, 1999; Hill and Lord, 1989; Liu *et al.*, 2000; Okada *et al.*, 1989; Sessions and Zambryski, 1995). The syncarpous gynoecium of *Arabidopsis* consists of two congenitally fused carpels that form a single ovary (figure 3.1). A septum divides the ovary into two locules. Viewed from outside (abaxial) the two valves of the carpel are separated by the replum and topped with a short, solid style and stigmatic papillae. The medial margins of the two fused carpels give rise to the placenta, the ovules, the septum, the replum as well as stylar and stigmatic tissue. The four adaxial (inside the carpel) meristematic placentae are located at the junction between septum and inner carpel wall. They give rise to four rows of interdigitated ovules. Different species produce different numbers of ovules. *Arabidopsis* usually bears about 40–50 ovules per gynoecium. Thus, a total of several hundreds to one thousand or more ovules regularly develop within a single plant.

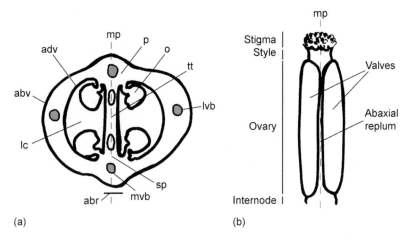

Figure 3.1 Schematic overview of an *Arabidopsis* gynoecium. (a) Horizontal section; (b) outside view. Abbreviations: mp, medial plane; p, placenta; o, ovule; tt, transmitting tract; lvb, lateral vascular bundle; sp, septum; mvb, medial vascular bundle; abr, abaxial replum; lc, locule; abv, abaxial valve; adv, adaxial valve.

3.2.2   Ovule primordium formation

The ovules are initiated by periclinal divisions at distinct positions in the sub-epidermal tissue of the placenta (Hill and Lord, 1994; Okada *et al.*, 1989). Further cell proliferation results in a growing bulge and eventually in extended finger-like protrusions (figure 3.2). Clonal analysis indicated that an *Arabidopsis* ovule for the most part consists of tissue derived from the L1 and L2 layers of the placenta (Jenik and Irish, 2000). Within this primordium there exists a regular filar arrangement of cells, suggesting that largely anticlinal cell divisions, relative to the surface of the growing ovule primordium, contribute to the outgrowth of the primordium. At the tip of the primordium, sporogenesis and gametogenesis take place within the nucellus. The integuments originate at the center or chalaza, and the funiculus develops from the bottom third of the primordium. Cell proliferation leads to an elongation of the bottom region. In addition, the vascular strand develops within this tissue.

3.2.3   Megasporogenesis

After the primordium has become distinct, the first sign of cellular differentiation consists in the formation of the megaspore mother cell (MMC), or megasporo-cyte, in the nucellus. One can consider the differentiation of the MMC as the first step of the transition from the sporophytic to the gametophytic phase of the plant's lifecycle. *Arabidopsis* ovules fall into the tenuinucellate class, in part owing to the small size of the nucellus (Dahlgren, 1927). The MMC becomes

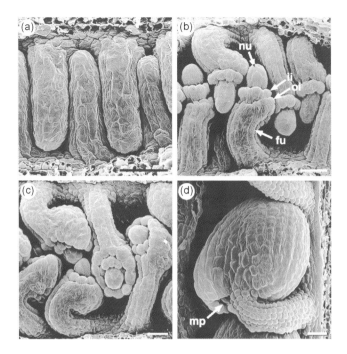

Figure 3.2 Scanning electron micrographs (SEMs) depicting ovule development in wild type at four stages. (a) Stage 2-I. The elongated ovule primordium. (b) Stage 2-III. The two integuments have initiated. (c) Stage 2-V. The integuments extend towards the apex of the nucellus. (d) Stage 4-I. Mature ovule. Abbreviations: fu, funiculus; ii, inner integument; mp, micropyle; nu, nucellus; oi, outer integument. Scale bars, 20 μm.

apparent as a large, hypodermal cell with an intrinsic polarity as evidenced by the enrichment of DNA-containing organelles at the proximal or chalazal pole (Bajon *et al.*, 1999; Webb and Gunning, 1990). This cell will directly undergo meiosis. In contrast, crassinucellate ovules generally feature a thick nucellus. In addition, a progenitor cell of the MMC, an archesporial cell, can be distinguished. The archespore enlarges, divides, and gives rise to the MMC and a parietal cell (Gifford and Foster, 1989). The parietal cell can undergo additional rounds of divisions. On the basis of such criteria, an archespore cannot be discriminated in *Arabidopsis*. Some reports, however, discriminate an archespore that then develops directly into the MMC (Yang and Sundaresan, 2000). Upon meiosis a tetrad of spores is formed. Only the chalazally oriented spore survives and will develop into the embryo sac. The other three spores undergo cell death. There is some variation on this theme since some plant species utilize a bisporic or tetrasporic pathway. In those instances, several or all meiotic products contribute to embryo sac development (Gifford and Foster, 1989; Maheswari, 1950).

3.2.4 Gametogenesis

Embryo sac development is characterized by an initially coenocytic phase, in which three rounds of nuclear divisions occur in the absence of cytokinesis, followed by cellularization. Morphologically, development begins with an enlargement of the surviving chalazal megaspore and its nucleus. At the same time, it assumes a tear-drop shape. This mononuclear embryo sac undergoes one round of nuclear division followed by the appearance of a large vacuole. The two nuclei have now become located to opposite poles. Two additional rounds of nuclear division occur, resulting in four nuclei at each pole. One nucleus of the proximal or chalazal pole becomes displaced to the distal pole and will be a constituent of the binuclear central cell. During cellularization the three nuclei closest to the distal or micropylar pole become integrated into two synergids and the egg cell, respectively. These three cells constitute the egg apparatus. One of the two synergids degenerates before fertilization. During the fertilization process, the pollen tube enters the embryo sac through this degenerated synergid. The other three proximal nuclei become part of the antipodal cells. The function of the antipodal cells in *Arabidopsis* is unknown. They degenerate at around the fertilization stage.

3.2.5 Integument development

Concomitantly with megasporogenesis and megagametogenesis, integument development occurs at the flank of the chalaza. Two integuments develop and thus *Arabidopsis* carries bitegmic ovules. *Arabidopsis* is regarded as an evolutionarily advanced species; nevertheless, bitegmy is considered the primitive condition in angiosperms as it is the prevailing type in the magnoliales (Gifford and Foster, 1989). Many other advanced dicotyledoneous species carry unitegmic ovules, i.e., ovules with only one integument.

Both integuments consist of two cell layers until just prior to fertilization. They eventually cover the nucellus, leaving a cleft—the micropyle—through which the pollen tube will enter during the fertilization process. The outer integument develops in an asymmetric fashion with less growth at the inner, or adaxial, side and more growth at the outer, or abaxial, side of the ovule. This leads to the characteristic curvature of the ovule as shown in figure 3.2. The degree of curvature varies significantly between species. It ranges from no curvature, the erect or orthotropous condition, to extreme curvature, the bend or anatropous condition in which the micropyle is located next to the funiculus.

3.3 Ovule identity

As outlined above, the ovules are borne on the placenta, which in turn is a part of the carpel. How much of ovule development is influenced by carpel tissue?

Tissue culture experiments in tobacco, using placental explants, have shown that ovules can develop independently from other parts of the carpel (Evans and Malmberg, 1989; Hicks and McHughen, 1974, 1977). Still, there is evidence from orchids that the combined action of ethylene and auxin triggers development of the ovary and thereby indirectly of the ovule (Zhang and O'Neill, 1993). The experiments with tobacco also indicated that there exists an early labile phase, when the primordium can be distinguished but prior to any visible sign of differentiation, in which the developing tissue is not yet committed to the ovular fate. In addition, they suggested that the commitment to ovular fate undergoes sequential restrictions from first gynoecial to second ovular fate (Evans and Malmberg, 1989).

What then controls the commitment to ovule fate? In tobacco two mutants, *Mgr3* and *Mgr9*, also show carpelloid outgrowths instead of ovules (Evans and Malmberg, 1989) but have not been characterized at the molecular level. As is the case for other floral organs (Coen and Meyerowitz, 1991; Ma, 1994; Weigel and Meyerowitz, 1994), it has been shown that genes encoding MADS-domain transcription factors regulate ovule identity (Angenent *et al.*, 1995; Colombo *et al.*, 1995). In petunia, simultaneous cosuppression of two MADS box genes, *FLORAL BINDING PROTEIN 7* (*FBP7*) and *FBP11* resulted in carpelloid tissue in place of ovules (Angenent *et al.*, 1995). Since the putative proteins of the two genes are 90% identical, it is likely that they act in a redundant fashion. Furthermore, ectopic expression of *FBP11* was sufficient for the formation of ectopic ovules in certain regions of the perianth (Colombo *et al.*, 1995). The identity of carpels is regulated by so-called C function genes (Coen and Meyerowitz, 1991; Weigel and Meyerowitz, 1994). Ectoptic expression of a *Brassica napus* C function gene, a homologue of the *AGAMOUS* (*AG*) gene of *Arabidopsis thaliana*, can lead to development of carpelloid tissue from the placenta in tobacco (Mandel *et al.*, 1992). This raised the possibility that *FBP7/11* may act by repressing C function activity in the ovule. However, the C function genes of petunia, *FBP6* and *pMADS3* (Angenent *et al.*, 1993; Tsuchimoto *et al.*, 1993), are also expressed during wild-type ovule development and seem not to be under the transcriptional control of *FBP7/FBP11* (Angenent *et al.*, 1995). It was therefore suggested that the balance of C function and *FBP7/FBP11* activities is critical (Angenent *et al.*, 1995).

The molecular basis of identity control in *Arabidopsis* is less well understood. A number of MADS box genes are expressed in the ovule (Ma *et al.*, 1991; Rounsley *et al.*, 1995; Savidge *et al.*, 1995) including the *SHATTERPROOF* (*SHP*) genes (Ferrándiz *et al.*, 2000; Liljegren *et al.*, 2000), previously known as *AGL1* and *AGL5* (Ma *et al.*, 1991), which are close homologues of *FBP7/11*. Functional analysis revealed that the *SHP* genes play a role in carpel valve margin differentiation (Ferrándiz *et al.*, 2000; Liljegren *et al.*, 2000). The possible function of these genes in ovule development remains to be determined. Carpelloid outgrowths can occur in *ap2-6* mutants (Modrusan *et al.*, 1994), indicating that *AP2* also has a role in ovule fate commitment. A second gene with a possible role in ovule fate regulation is *BELL1* (*BEL1*) (Modrusan *et al.*, 1994; Robinson-Beers *et al.*, 1992; Schneitz *et al.*, 1997). The *BEL1* locus encodes a putative homeodomain transcription factor (Reiser *et al.*, 1995). However, only a subregion of the ovule experiences a possible alteration fate. In *bel1* mutants the inner integument is not formed and instead of the outer integument a collar-like protrusion develops. This indicates that *BEL1* may control outer integument identity, or the identity of the entire chalaza (see also below). At very late stages the outgrowths can develop carpelloid features and express *AG* (Modrusan *et al.*, 1994; Ray *et al.*, 1994). In addition, *Arabidopsis* plants overexpressing a Brassica *AG* homologue carry ovules with a phenotype resembling ovules from *bel1* mutants (Ray *et al.*, 1994). It was suggested that *BEL1* exerts a cadastral function and prevents the misexpression of *AG* in the ovule (Modrusan *et al.*, 1994; Ray *et al.*, 1994). However, the situation is again more complex. *AG* is expressed in the ovule throughout its development and its RNA expression pattern overlaps with that of *BEL1* (Bowman *et al.*, 1991a; Reiser *et al.*, 1995), indicating that *BEL1* is not a transcriptional regulator of *AG*. It is still possible that the BEL1 and AG proteins interact as do other MADS-domain and homeodomain transcription factors (Gehring *et al.*, 1994; Shore and Sharrocks, 1995). Plants mutant for *AP2* and *AG* carry first-whorl sepal-carpel structures that bear ovules, indicating that neither gene is required for ovule development (Bowman *et al.*, 1991b). In contrast, further genetic analysis, using triple mutant combinations involving mutations in *BEL1*, *AG* and *AP2*, indicated that *AG* and *BEL1* genes are members of separate pathways that specify ovule identity in *Arabidopsis* (Western and Haughn, 1999).

3.4   Placenta formation

The placenta is specialized meristematic medial ridge tissue of the carpel from which ovules develop (figure 3.1). Recently, genes with a role in margin formation have been characterized in some detail. The *LEUNIG* (*LUG*) and *AINTEGUMENTA* (*ANT*) genes are important positive regulators of the development of margin tissue and thus of placenta and ovule formation (Liu *et al.*, 2000). Both genes are required repeatedly during plant development. *LUG* has multiple functions during floral development. During early floral primordium formation it acts as a negative regulator of *AG* expression in the first two whorls (Liu and Meyerowitz, 1995). Carpel and ovule development is impaired as well in *lug* mutants (Liu and Meyerowitz, 1995; Roe *et al.*, 1997b; Schneitz *et al.*, 1997). The carpel margins often do not fuse, the number of ovules is reduced and the shape of the integuments is altered. *ANT* is generally required for cell proliferation control in organ primordia (Elliott *et al.*, 1996; Klucher *et al.*, 1996; Krizek, 1999; Long and Barton, 1998; Mizukami and Fischer,

2000; Schneitz *et al.*, 1998a). In *ant* mutants, carpel fusion is sometimes aberrant and about half the regular number of ovules are formed. In addition, integument development is impaired (see below). Strikingly, in *lug ant* double mutants, carpels develop that show no sign of margin tissue as indicated by the absence of septum, placenta and ovules, replum, style and stigma (Liu *et al.*, 2000).

What is the biochemical function of the ANT and LUG proteins? The two genes have been cloned and sequence analysis indicated that both loci encode putative transcription factors (Conner and Liu, 2000; Elliott *et al.*, 1996; Klucher *et al.*, 1996). LUG localizes to the nucleus but lacks a known DNA-binding domain. It is, however, rich in glutamines and carries seven WD repeats. It shares its motif structure with known transcriptional co-repressors from yeast and *Drosophila* (Conner and Liu, 2000). ANT represents a putative transcription factor of the AP2 class (Elliott *et al.*, 1996; Klucher *et al.*, 1996; Weigel, 1995). It carries two AP2 domains that are known to bind DNA (Riechmann and Meyerowitz, 1998). Taken together, the data indicate that LUG and ANT act redundantly to promote carpel margin development through regulating downstream target genes at the transcriptional level.

What might these target genes be? *AG* and *CRABS CLAW* (*CRC*) are two excellent candidates. For instance, *AG* is expressed early in the medial ridge tissue and this expression is under a quantitative positive control of *LUG* (Liu *et al.*, 2000). *CRC* is the canonical member of the YABBY family of putative transcription factors (Bowman, 2000; Bowman and Smyth, 1999). The YABBY members are involved in polarity control and specify abaxial identity in lateral organs (Bowman, 2000; Siegfried *et al.*, 1999). *CRC*, in combination with a number of other factors, is required for the proper adaxial–abaxial polarity during carpel development (Eshed *et al.*, 1999). In the medial ridge tissue of the carpel, *CRC* is expressed in four patches adjacent to the placenta, and again this expression is under the positive control of *LUG* (Bowman and Smyth, 1999). Additional candidate genes include *SPATULA* (*SPT*), *TOUSLED* (*TSL*), *PERIANTHIA* (*PAN*), and *CUP-SHAPED COTYLEDON 2* (*CUC2*). In *spt* single mutants, defects in carpel margin ontogenesis are observed that are enhanced in *spt crc* double mutants (Alvarez and Smyth, 1999). The role of *TSL*, encoding a nuclear serine/threonine kinase (Roe *et al.*, 1993, 1997a), and *PAN*, encoding a putative bZIP protein (Chuang *et al.*, 1999; Running and Meyerowitz, 1996), is inferred from genetic analysis as well, since *lug tsl* and *tsl pan* double mutants show reduced marginal tissue development (Roe *et al.*, 1997b). The molecular nature of *CUC1* is not known, but *CUC2* encodes a putative NAC-type protein (Aida *et al.*, 1997). *CUC1* and *CUC2* are redundantly required for organ separation (Aida *et al.*, 1997, 1999; Ishida *et al.*, 2000). In *cuc1 cuc2* double mutants, septum formation is impaired and the number of ovules is greatly reduced. Furthermore, *CUC2* is expressed in the margin tissue, including the placenta (Ishida *et al.*, 2000).

3.5   The formation of the ovule primordium

The meristematic activity of the placenta ultimately leads to the formation of a row of bulges that will give rise to ovules. These will develop into distinct protrusions that consist of cells of uniform appearance and thus show little signs of cell differentiation. None the less, pattern formation—the spatial and temporal control of cell fate—must act at around this and subsequent stages. This quickly becomes evident as differentiation proceeds along two axes of polarity. The nucellus, chalaza and funiculus become morphologically recognizable along the longitudinal, or proximal–distal (P-D), axis (figure 3.2b). There is also an adaxial–abaxial polarity that becomes evident slightly later than the P-D axis and is oriented roughly normal to the P-D axis. Its earliest morphological sign is the asymmetric development of the outer integument (figure 3.2b). It becomes initiated at the abaxial side and will only eventually grow around the entire circumference of the primordium. Further asymmetric development, with more abaxial than adaxial growth, in part leads to the curvature of the upper part of the ovule (figure 3.2c). In addition, the funiculus develops asymmetrically along the adaxial–abaxial axis as well (figure 3.2b,c), resulting in its final bent shape.

Both axes of polarity become evident quite early during development, and with respect to the P-D axis it was reasoned that the patterning mechanism is likely to act during the primordium outgrowth phase (Schneitz *et al.*, 1995). Thus, important questions relate to the mechanism underlying the outgrowth and the patterning of the primordium, and how these two processes are interconnected.

3.5.1   Controlling the outgrowth

So far, no single mutant completely devoid of ovules has been reported, raising the possibility that several genes act redundantly to control this process. The current evidence supports this notion. In addition, the process may be under control of the plant growth hormone auxin, since treatment of developing wild-type gynoecia with a polar auxin transport inhibitor leads to a loss of ovules (Nemhauser *et al.*, 2000). It is not known what factors mediate this response to auxin.

With respect to the genetic control, several different loci, including *ANT*, *HUELLENLOS* (*HLL*), *SHORT INTEGUMENTS 2* (*SIN2*), and *CUC1* and *CUC2*, have to date been implicated to play a role in the outgrowth of the ovule primordium (Broadhvest *et al.*, 2000; Elliott *et al.*, 1996; Ishida *et al.*, 2000; Schneitz *et al.*, 1998a). In *ant* and *sin2* mutants, the number of ovules is reduced and the individual ovules are separated by a greater distance than normal. In *hll* and *ant* mutants the P-D axis is reduced owing to shorter funiculi. In all three mutants, integument development is affected as well (see below). *HLL* also appears to be required for the control of cell death in the ovule since in *hll* mutants cell degeneration occurs regularly in this tissue. There are also defects in gametogenesis: in *sin2* mutants a MMC is not formed, while in *ant* and *hll* mutants a block usually happens at or before the formation of the tetrad.

Analysis of double mutants revealed a redundant role for these three genes in primordium outgrowth (Broadhvest *et al.*, 2000; Schneitz *et al.*, 1998a). The *ant hll* double mutants carry ovules that show a variable but clearly greater reduction in their P-D extension than either single mutant (Schneitz *et al.*, 1998a). The nucellus is regularly formed, as indicated by the presence of a MMC that often develops further to a tetrad or early embryo sac stage. The nucellus, even at a late stage, is attached directly to the placenta in about 30% of the ovules, but more often, in about 70% of the instances, is situated on top of a small cell cluster. Using the *BEL1* expression as a marker for the chalaza (Reiser *et al.*, 1995), it was shown that this cell cluster has chalazal identity. The phenotypes of *sin2 ant* double mutants did not differ from those of *ant* mutants, indicating that *ant* is epistatic to *sin2* with respect to ovule development (Broadhvest *et al.*, 2000). The *hll sin2* double mutants, however, exhibited an ovule phenotype similar to but even stronger than *ant hll* double mutants (Broadhvest *et al.*, 2000). In *hll sin2* double mutants, ovules are nearly eliminated. Pistils bear only rudimentary ovules that have arrested development following only very limited growth. In addition, cell death was observed in the entire primordium.

Based on the evidence presented above, *ANT*, *HLL* and *SIN2* all seem to have positive roles in ovule primordium outgrowth. How do they relate to each other? *HLL* appears to be a central factor, as indicated by the synergism of *ant hll* and *hll sin2* double mutants. The observation that *ant* is epistatic to *sin2* allows the interpretation that *ANT* and *SIN2* act in the same pathway and that *ANT* most likely acts before *SIN2* (Broadhvest *et al.*, 2000). This is supported by the finding that the *ANT* mRNA expression pattern is unaltered in the *sin2* mutant (Broadhvest *et al.*, 2000). Additional work is needed to understand this aspect in more detail.

The function of *CUC1* and *CUC2* in ovule primordium outgrowth is inferred from the phenotype of the *cuc1 cuc2* double mutant and the mRNA expression pattern of *CUC2* (Ishida *et al.*, 2000). As detailed above, *cuc1 cuc2* double mutants have a reduced number of ovules. This could be due to defects in placenta differentiation. However, the ovules that are formed show a reduced funiculus. Furthermore, *CUC2* is expressed in a region flanking the budding ovule primordium, suggesting that *CUC2* indirectly affects primordium outgrowth. It will be interesting to investigate how *CUC1* and *CUC2* function relates to the activities of *ANT*, *HLL* and *SIN2*.

3.5.2   *Pattern formation and the regulation of growth*

It has been speculated that the nucellus, chalaza and funiculus are distinct pattern elements that are established during the early outgrowth of the primordium

(Schneitz *et al.*, 1995). One basis for this idea was the finding that many mutations affecting ovule development resulted in early aberrant development of only a part of the ovule, mostly the chalaza (Gasser *et al.*, 1998; Schneitz *et al.*, 1998b). The chalaza-specific expression pattern of *BEL1* provided the first molecular evidence for this idea. The analysis of the *ant hll* double mutant yielded additional indirect evidence (Schneitz *et al.*, 1998a), since one could draw the conclusion that nucellus development can be initiated independently of the presence of a chalaza and funiculus. Thus, it was realized that the nucellus can be formed as an autonomous element, although later nucellus development may depend on other ovule tissues (Schneitz *et al.*, 1998a).

The molecular basis of pattern formation in early ovule development is not understood. None the less, the study of four genes, *BEL1*, *NOZZLE* (*NZZ*), *INNER NO OUTER* (*INO*) and *ANT*, has provided more insight into the patterning mechanism and how it relates to ovule primordium outgrowth. *BEL1* seems to be required for the development of the chalaza and the integuments, and *INO* appears to promote adaxial–abaxial patterning in the chalaza and the outer integument. What keeps their activities in check? Genetic analysis identified *NZZ* as a negative regulator of the 'cell proliferation' gene *ANT* and the patterning genes *BEL1* and *INO*. In addition, *NZZ* seems to share an overlapping function with *BEL1* in providing chalaza identity. The present evidence is in accordance with the view that *NZZ* may be part of a mechanism coupling aspects of patterning and primordium outgrowth.

3.5.3   The proximal–distal polarity

The homeobox gene *BEL1* is an excellent candidate for a role in ovule patterning. As mentioned above, in *bel1* irregular protrusions develop where normally an outer integument is found (Modrusan *et al.*, 1994; Robinson-Beers *et al.*, 1992; Schneitz *et al.*, 1997). Furthermore, after an initially ubiquitous distribution, its mRNA becomes restricted to a central domain, roughly corresponding to the prospective chalaza, before signs of integument morphogenesis are recognizable (Reiser *et al.*, 1995). On the basis of the nature of the putative BEL1 protein, the *bel1* mutant phenotype and its expression pattern, it was suggested that *BEL1* is responsible for chalazal development and either part of the P-D mechanism or an early downstream target of it (Reiser *et al.*, 1995).

A second good candidate is *NZZ* (Schiefthaler *et al.*, 1999), also known as *SPOROCYTELESS* (*SPL*) (Yang *et al.*, 1999). In *nzz* mutants the nucellus fails to develop properly, a MMC is usually not formed, the integuments are variably shorter, and the funiculus exhibits hyperplasia. In the single *spl* mutant that was isolated, the integument and funicular aspects of the *nzz* phenotype were not observed. In contrast, the epidermis of the nucellar-like tissue at the tip underwent some very late extra growth in *spl* but not in the three *nzz* mutants. In addition, *NZZ/SPL* is required for anther development. The putative NZZ protein is of unknown biochemical function, but computer analysis suggested sequence features expected for transcription factors. In accordance with this notion, it localizes to the nucleus (Yang *et al.*, 1999). At the mRNA level, *NZZ* is ubiquitously expressed in the ovule primordium. Prior to integument initiation its level is somewhat reduced but it is still present throughout the ovule. Elevated levels of expression can be seen in the developing integuments (Schiefthaler *et al.*, 1999).

What is the cause of the nucellar aspect of the *nzz* phenotype? Two proposals have been put forward. In the first model *NZZ* was assigned a primary function in MMC differentiation, and the reduction of the nucellar tissue was considered a secondary effect. It was postulated that the MMC and the surrounding nucellus interact. Thus, nucellar development depends on the correct formation of the MMC formation (Yang *et al.*, 1999). The second model suggested a primary role for *NZZ* in early ovule development, including nucellus formation (Schiefthaler *et al.*, 1999). Results obtained from recent experiments favor the latter view but leave open a role for *NZZ* in MMC differentiation (Balasubramanian and Schneitz, 2000).

An intriguing aspect of these experiments was the finding that *ant* is essentially epistatic to *nzz* (figure 3.3). The explanation for this epistasis is complex. The results suggest that *NZZ* is a general negative regulator of *ANT* during nucellus and funiculus development. Furthermore, *ANT* seems to act prior to *NZZ* in early integument ontogenesis. The notion that *NZZ* acts as a repressor of *ANT* in the ovule is corroborated by the phenotype of plants ectopically expressing *ANT* (Krizek, 1999). Those plants carry ovules that show a partial *nzz* phenocopy.

It is not yet clear what the molecular basis is for the repression of *ANT* by *NZZ*. There is evidence that the mechanism may involve transcriptional control (Balasubramanian and Schneitz, 2000). In wild type, *ANT* is initially expressed throughout the ovule but quickly becomes excluded from the nucellus and is found in the chalaza, the developing integuments, and the upper part of the funiculus (Elliott *et al.*, 1996). In *nzz* mutants, *ANT* expression fails to be excluded from the distal region of the primordium (Balasubramanian and Schneitz, 2000). This is particularly well seen in *nzz ant* (and *nzz ino*, see below) double mutants, which clearly form a nucellus (S. Balasubramanian and K. Schneitz, unpublished observations). Ectopic expression of *ANT* alone, however, is not sufficient to lead to a reduced or absent nucellus (Krizek, 1999; Mizukami and Fischer, 2000). There are additional complications. Transcriptional control of *ANT* by *NZZ* could not be shown for the funiculus (Balasubramanian and Schneitz, 2000). Furthermore, since *ANT* and *NZZ* show extensive spatial and temporal co-expression in the ovule primordium, the repressor mechanism must involve additional factors. In addition, posttranscriptional aspects cannot be excluded and may play a role as well. It will be important to demonstrate that the distal *ANT* misexpression is a direct consequence of the absence of *NZZ* function.

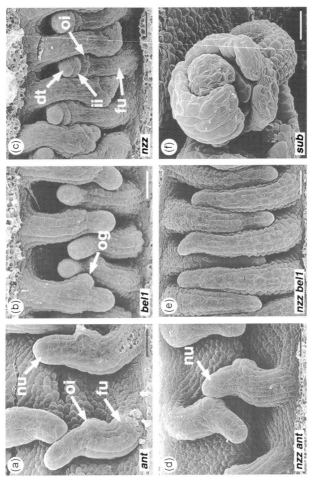

Figure 3.3 SEMs of ovule of different single and double mutants. Stages: (a–e) about stage 2III/V: (f) stage 3-1. (a) *ant-72F5*. (b) *bel1-1460*. (c) *nzz-2*. Instead of a nucellus a small patch of tissue is seen. Note the already elongated funiculus. (d) *nzz-2 ant-72F5*. (e) *nzz-2 bel1-1460*. Only elongated structures develop. (f) *sub-1*. No regular integument development. Abbreviations: dt, distal tissue in place of nucellus; fu, funiculus; ii, inner integument; nu, nucellus; og, outgrowth; oi, outer integument. Scale bars, 20 μm.

With respect to integument development, the situation appears less intricate. The epistasis of *ant* over *nzz* raises the possibility that *ANT* is a positive regulator of *NZZ* during early integument development. Indeed, the elevated levels of *NZZ* transcription in the integuments are not detected in *ant* mutants, although a basal level is still present (Balasubramanian and Schneitz, 2000). The result indicates that *ANT* is an early positive regulator of *NZZ* transcription during integument initiation.

Additional evidence for an early role of *NZZ* in patterning the primordium comes from the analysis of *nzz bel1* double mutants (figure 3.3). In plants lacking regular *NZZ* and *BEL1* function, tuberous outgrowths replace the ovules. On the basis of morphology, these structures consist of largely funicular cells topped by *nzz*-like distal tissue. One interpretation of this phenotype suggests that the chalaza is replaced by funicular tissue. If true, it indicates that *BEL1* and *NZZ* share overlapping functions in specifying the chalaza. The mechanistic basis of the interactions is presently unknown. Experiments showed that both genes are still expressed in the central region of ovules from both single mutants (Balasubramanian and Schneitz, 2000) and the double mutant (S. Balasubramanian and K. Schneitz, unpublished observation). Surprisingly, however, *BEL1* expression is not excluded from the distal tip region in *nzz* mutants. This indicates that *NZZ* represses, directly or indirectly, *BEL1* in the distal region of the ovule primordium. However, similar reservations as described for the interactions between *NZZ* and *ANT* apply.

3.5.4   The adaxial–abaxial axis

What of the adaxial–abaxial axis of polarity? So far two genes, *SUPERMAN* (*SUP*), also known as *FLO10* (Schultz *et al.*, 1991), and *INO*, have been implicated in adaxial–abaxial patterning in the ovule. Besides other defects in floral development (Bowman *et al.*, 1992; Sakai *et al.*, 1995; Schultz *et al.*, 1991), *sup* mutants exhibit symmetrical, rather than the usual asymmetric, integument development owing to increased adaxial cell proliferation (Gaiser *et al.*, 1995). *SUP* encodes a putative transcription factor with a single zinc-finger domain (Sakai *et al.*, 1995). In *ino* mutants, the outer integument fails to initiate correctly and thus ovules develop that carry only an inner integument (Baker *et al.*, 1997; Gaiser *et al.*, 1995; Schneitz *et al.*, 1997; Villanueva *et al.*, 1999). In addition, a protuberance emerges on the adaxial side of the ovule and undergoes limited expansion. The mutant phenotype is compatible with *INO* being a primary determinant of adaxial–abaxial polarity within the chalaza (Villanueva *et al.*, 1999). The *INO* gene was cloned (Villanueva *et al.*, 1999) and shown to represent a member of the YABBY gene family (Bowman, 2000; Bowman and Smyth, 1999), substantiating the notion that *INO* controls adaxial–abaxial polarity in the ovule. *INO* mRNA was detected only in the outer integument on the abaxial side of the ovule primordium. In *sup* mutants, *INO* expression extends across both sides of the outer integument and into the funiculus (Villanueva *et al.*, 1999). It is likely that the *sup* mutant ovule phenotype is due to the ectopic expression of *INO*. This would result in the abaxialization of the adaxial side and thus lead to symmetrical growth of the outer integument (Villanueva *et al.*, 1999). Since *SUP* expression is observed only in a small patch on the adaxial side of the funiculus (Sakai *et al.*, 1995), it is likely that *SUP* negatively regulates *INO* through a cell nonautonomous mechanism (Villanueva *et al.*, 1999). Recently, it was found that *INO* expression is restricted to the abaxial cell layer of the outer integument (Balasubramanian and Schneitz, 2000). This raises the possibility that *INO* is also required to determine the polarity within the two cell-layered outer integument.

A genetic analysis suggested surprisingly complex interactions between *NZZ* and *INO* (Balasubramanian and Schneitz, 2000). It turned out that *ino* is essentially epistatic to *nzz* as well (Balasubramanian and Schneitz, 2000). Similarly to the scenario with *ANT*, see above, this suggests that *NZZ* negatively regulates *INO* during nucellus and funiculus development and that *INO* positively controls *NZZ* activity during early outer integument development. The latter interaction appears to occur at the transcriptional level since the elevated levels of *NZZ* expression are missing in the outer but not the inner integument in *ino* mutants. Thus, INO is a positive transcriptional regulator of *NZZ*. Again, it remains to be shown whether INO directly binds to enhancer elements in the *NZZ* promoter. The mechanistic basis for the negative regulation of *INO* by *NZZ* during nucellus and funiculus development, as suggested by the genetic analysis, is not understood. In contrast to what might have been expected, the *INO* mRNA expression pattern is not altered in a very obvious way as assayed by *in situ* hybridization experiments using sectioned material. The overall spatial expression of *INO* looks relatively normal in *nzz* mutants. Nevertheless, it seems that *INO* expression comes on precociously and possibly shifted distally by a few cells. Additional experiments are certainly required to corroborate this notion. Regardless of the outcome of those experiments, it is already clear that *INO* mRNA is not expressed in the prospective nucellus and funiculus in *nzz* mutants. Thus, it is very likely that *INO* influences the development of those tissues in a cell nonautonomous fashion in a *nzz* mutant and possibly also in wild type.

3.6   Integument morphogenesis

Integument development in *Arabidopsis* is characterized by several features. First, the integuments are of epidermal origin. Second, the inner integument usually initiates slightly earlier than the outer integument. Third, the outer integument undergoes asymmetric growth. This aspect has been discussed above. Fourth, growth occurs by cell proliferation and cell elongation. Particularly the distal half of the outer integument consists of files of elongated, cuboidal cells (figure 3.2). Fifth, for most of prefertilization development both integuments are composed of two cell layers. What regulates these intricate aspects of cell proliferation, cell size, and cell shape during integument morphogenesis?

It was recently shown that the plant growth hormone ethylene plays a role in the growth of the integuments. For example, impairing ethylene production in the ovules of tobacco, by inhibiting the expression of an ovule-specific gene coding for the ethylene-forming enzyme 1-aminocyclopropane-1-carboxylate oxidase (ACC oxidase or ACO), resulted in ovules with a block in integument development (De Martinis and Mariani, 1999). Furthermore, the *ETHYLENE RESPONSE 2* (*ETR2*) gene, encoding an ethylene receptor, is expressed in ovules (Hua *et al.*, 1998; Sakai *et al.*, 1998).

3.6.1 Early integument development

So far, genetic analysis has revealed that integument morphogenesis is a complex process. The finding that many of the corresponding double mutants show additive phenotypes indicates that integument morphogenesis consists of numerous independently regulated steps. Mutations in some genes, however, lead to more drastic, and early, consequences than mutations in other loci.

For example, in *ant* and *hll* mutants, growth of both integuments is not correctly initiated, with the defect in *hll* mutants occurring slightly later than in *ant* mutants (Elliott *et al.*, 1996; Klucher *et al.*, 1996; Schneitz *et al.*, 1997, 1998a). Thus, *ant* or *hll* mutants bear ovules without integuments. A more subtle effect is seen in *nzz* mutants. Here the timing of integument initiation is affected (Balasubramanian and Schneitz, 2000). As discussed above, *INO* is an essential factor for outer integument development. *ino* mutants fail to form an outer integument, suggesting that inner and outer integument initiation are not dependent on each other. This may not be entirely true. In the *aberrant testa shape* (*ats*) mutant (Léon-Kloosterziel *et al.*, 1994), both integuments initiate but are situated more closely to each other than in wild type and eventually two integument cell layers degenerate, giving the impression of a single, fused integument (K. Schneitz, unpublished observation). Plants that lack *ATS* and *INO* wild-type function show ovules in which development of both integuments is drastically impaired (Baker *et al.*, 1997). In addition, a mutant lacking only the inner integument has not yet been described in the literature.

3.6.2 Later aspects of integument ontogenesis

Once the integuments have begun development, cell division and cell elongation have to be maintained. A number of genes with a role in later aspects of integument morphogenesis have been identified (table 3.1). *SHORT INTEGUMENT 1* (*SIN1*) and *SIN2* seem to be involved in the control these two aspects (Broadhvest *et al.*, 2000; Lang *et al.*, 1994; Robinson-Beers *et al.*, 1992). Both genes are

Table 3.1 Characteristics of the known *Arabidopsis* mutants affected in integument morphogenesis

Name	Integument phenotype[a]	Function	Other aspect	Protein	Reference
bag	No recognizable integuments, balloon-shaped cells		Flower; size		Schneitz et al. (1997)
cuc1 cuc2	Short integuments		Flower	CUC2 NAC protein	Aida et al. (1997), Ishida et al. (2000),
lal	Reduced oi Protrusion ii				Schneitz et al. (1997)
lug	Protrusion ii		Flower	Glutamine-rich, WD repeats	Conner and Liu (2000), Liu and Meyerowitz (1995), Roe et al. (1997b), Schneitz et al. (1997)
mog	Balloon-shaped cells		Flower		Schneitz et al. (1997)
nzz	Variably shorter integuments	Putative transcriptional regulator	Ovule, anther	Novel protein	Balasubramanian and Schneitz (2000), Schiefthaler et al. (1999)
sin1	Short integuments	Cell elongation	Flowering time, maternal effect		Robinson-Beers et al. (1992), Lang et al. (1994)
sin2	Short integuments	Cell division	Carpel, sepal		Broadhvest et al. (2000)
sub	Distal protrusions oi Arrested oi	Cell proliferation	Flower; size	Putative receptor kinase	Schneitz et al. (1997), Chevalier and Schneitz, unpublished
sup	Symmetrical growth oi	Adaxial/abaxial patterning	Flower	Zinc finger protein	Bowman et al. (1992), Sakai et al. (1995) Schultz et al. (1991)
tso1	No layer organization Irregular shaped cells	Orientation of cell division	Flower, fasciated	Cysteine rich protein	Hauser et al. (1998), Hauser et al. (2000), Liu et al. (1997), Song et al. (2000)
tsl	Protrusion ii		Flower, leaves, Flowering time	Nuclear ser/thr kinase	Roe et al. (1993), Roe et al. (1997a,b)

[a] Abbreviations: ii, inner integument; oi, outer integument.

repeatedly required during reproductive development (Broadhvest *et al.*, 2000; Ray A. *et al.*, 1996; Ray S. *et al.*, 1996) and in both mutants the integuments are reduced and fail to cover the nucellus. However, the functions of the two genes differ: *SIN1* appears to be responsible for cell elongation and *SIN2* seems to control cell division. The *sin1 sin2* double mutants exhibit an additive effect, further indicating that *SIN1* and *SIN2* act in parallel processes (Broadhvest *et al.*, 2000).

The *SIN1* and *SIN2* genes have not yet been cloned. Other loci, however, including *CUC2*, *TSO1*, *TOUSLED* (*TSL*) and *STRUBBELIG* (*SUB*), have been characterized at the molecular level. In *cuc1 cuc2* double mutants, the integuments variably fail to develop following integument initiation (Ishida *et al.*, 2000). Within the ovule, *CUC2* is expressed in a stripe between the nucellus and the chalaza, suggesting an indirect effect of *CUC2* on integument development. Mutations in *TSO1* affect floral development, including integument ontogenesis. Strong *tso1* alleles cause the growth of callus-like tissues in place of floral organs (Liu *et al.*, 1997). It was proposed that *TSO1* affects cell division since *tso1* mutants exhibited nuclear defects and partially formed cell walls. In contrast, weak *tso1* alleles interfere little with the development of the flowers but show aberrant ovule integument morphogenesis (Hauser *et al.*, 1998). The weak *tso1* mutations lead to disturbances of the shapes and alignments of integument cells, a defect that becomes more severe as development progresses. The analysis of these aberrations led to the hypothesis that *TSO1* is part of a mechanism that regulates directional cell expansion or its coordination among adjacent cells. The *TSO1* locus has been cloned (Hauser *et al.*, 2000; Song *et al.*, 2000) and further analysis indicated that the TSO1 protein localizes to the nucleus. The protein carries two cysteine-rich repeats (CXCs), also called TCRs (*TSO1* cysteine-rich repeats). A number of genes, found in animals and plants, encode such double repeats. In addition, these cysteine-rich domains share homology with the single CXC domain of the *Drosophila* Polycomb group protein Enhancer of zeste (E(z)) (Jones and Gelbart, 1993), its plant homologues CURLY LEAF (CLF) (Goodrich *et al.*, 1997) and MEDEA (MEA) (Grossniklaus *et al.*, 1998; Kiyosue *et al.*, 1999), and also with the *Drosophila* kinesin-like protein KLP3A (Hauser *et al.*, 2000; Williams *et al.*, 1995, 1997). One function of the Polycomb-group genes in *Drosophila* is to negatively regulate the segment-identity genes through modifying the chromatin structure (Pirrotta, 1997, 1998). Interestingly, *CLF* negatively regulates *AG* expression in vegetative tissues (Goodrich *et al.*, 1997). On the other hand, the KLP3A protein associates with the cytoskeletal spindles and is required for spindle assembly and cytokinesis (Williams *et al.*, 1995, 1997). Considering the differences in functions of CXC domain proteins described above, what is the biochemical function of TSO1? At present the answer is not known. However, *in vitro* experiments using CPP1 from soybean, a homologue of TSO1, have shown that CPP1 binds to promoter sequences of the *leghemoglobin c3* (*Gmlbc3*) gene with a domain containing both CXC repeats (Cvitanich *et al.*, 2000). In addition, experiments with transgenic plants indicated that CPP1 acts as a repressor of the *Gmlbc3* gene (Cvitanich *et al.*, 2000). Interestingly, *TSO1* shows synergistic interactions with *LUG* (Hauser *et al.*, 1998), a gene encoding a putative transcriptional co-repressor (Conner and Liu, 2000). Ovules of *lug* mutants show a recessed outer and protruding inner integument (Roe *et al.*, 1997b; Schneitz *et al.*, 1997). The ovules of *tso1 lug* double mutants exhibit highly reduced integuments with variable morphology. This raises the possibility that *LUG* and *TSO1* may act in similar processes during ovule development (Hauser *et al.*, 1998).

The *tsl* mutants show a pleiotropic phenotype affecting leaf morphology, flowering time, and a stochastic decrease in the number of sepals, petals and stamens (Roe *et al.*, 1993, 1997b). In addition, there are defects in the development of apical and internal tissues of the gynoecium (see above). The ovules of *tsl* mutants resemble somewhat ovules from *lug* mutants. They also show a protruding inner integument and ovules of *lug tsl* double mutants did not show any distinct deviation from the single mutant phenotype, indicating that *TSL* and *LUG* may play similar roles during ovule development (Roe *et al.*, 1997b). On the basis of the genetic and molecular analysis, it was suggested that *TSL* plays a role in promoting cell division within the meristem (Roe *et al.*, 1997b). *TSL* encodes a nuclear serine/threonine protein kinase that requires self-assocation for its protein kinase activity (Roe *et al.*, 1993, 1997a). Additional potential insight into the function of TSL comes from studies using mammalian cell culture systems. Two human homologues of *TSL* have been isolated, the so-called Tousled-like kinases (*TLK1* and *TLK2*) (Silljé *et al.*, 1999). Interestingly, both TLKs are cell cycle-regulated kinases with maximal activity during S phase. This regulation appears to be achieved through differential phosphorylation. Further studies indicated that TLK activity is tightly linked to ongoing DNA replication. The authors proposed that the TLKs function in transcription and/or chromatin remodelling (Silljé *et al.*, 1999), since chromatin assembly is tightly linked to DNA replication (Krude, 1999) and chromatin structure is important for the establishment and maintenance of gene expression patterns (Hagstrom and Schedl, 1997). It remains to be shown whether *Arabidopsis* TSL kinase activity is cell cycle regulated as well.

In addition to *TSL*, the *SUB* gene (Schneitz *et al.*, 1997) highlights the importance of signal transduction during integument morphogenesis. Mutations in *SUB* lead to a pleiotropic phenotype affecting the entire plant. With respect to ovule development, either a reduction in integument size or the formation of outgrowths at the distal tip of the outer integument can be observed (figure 3.3). This indicates that *SUB* controls cell proliferation during integument morphogenesis. Recent studies showed that *SUB* encodes a putative transmembrane receptor kinase (D. Chevalier and K. Schneitz, unpublished results), indicating that *SUB* may be involved in coordinating the cell division patterns between cells of the integuments.

3.7   Summary and outlook

Until recently, very little was known about the molecular basis of ovule development in any plant species. This has clearly changed. A large number of genes with a role in ovule development have been identified by several laboratories. This review has attempted to outline the present framework that rapidly emerged from their genetic and molecular analysis in just a few years. Despite these efforts, however, it has become clear that mutational saturation of this developmental process was not achieved. A number of additional factors will certainly be identified through second-site mutagenesis experiments in mutant backgrounds, activation tagging (Weigel *et al.*, 2000), enhancer trap screens (Sundaresan *et al.*, 1995), yeast two-hybrid screens (Chien *et al.*, 1991), and reverse genetics-based approaches. Furthermore, there is an urgent need for additional marker genes with region- and cell type-specific expression patterns to augment the phenotypic analysis based on morphological criteria. Many genes with a function in ovule development will be required for other aspects of the plant's lifecycle as well and therefore may not be identified in screens based on ovule mutant phenotypes. Thus, it is important to devise tools that allow the isolation and analysis of mutations in those genes based on experiments involving ovule mutant phenotypes. With some effort, such tools can be made available to the community.

Experiments making use of these methods will lead to a better understanding of several questions, such as what controls the specification of the ovule, or how genes such as *NZZ* regulate cell proliferation and pattern formation. Work on *TSL* and *SUB* highlights the importance of signaling pathways during integument morphogenesis. This particular field will certainly advance quickly in the future, and it will be interesting to find out, for example, how hormonal action, mediated by growth regulators such as ethylene, feeds into this signaling network. Another unresolved issue deals with embryo sac development and its dependence on regular integument morphogenesis. Part of the answer will also be provided by the cloning and molecular characterization of some of the female gametophytic genes that have been identified in recent years by mutational analysis (Christensen *et al.*, 1998; Drews *et al.*, 1998; Grini *et al.*, 1999; Howden *et al.*, 1998; Moore *et al.*, 1997; Sheridan and Huang, 1997; Shimizu and Okada, 2000). No doubt, these are exciting times for people interested in ovules and the future looks even brighter.

Note added in proof

While this review was in press, Heisler *et al.* (2001) reported the cloning of the *SPT* gene and Takada *et al.* (2001) published the identification of the *CUC1* gene.

Takada, S., Hibara, K., Ishida, T. and Tasaka, M. (2001) The *CUP-SHAPED COTYLEDON1* gene of *Arabidopsis* regulates shoot apical meristem formation. *Development*, **128**, 1127-1135.

Heisler, M.G., Atkinson, A., Bylstra, Y.H., Walsh, R. and Smyth, D. (2001) *SPATULA*, a gene that controls development of carpel margin tissues in *Arabidopsis*, encodes a bHLH protein. *Development*, **128**, 1089-1098.

References

Aida, M., Ishida, T., Fukaki, H., Fujisawa, H. and Tasaka, M. (1997) Genes involved in organ separation in *Arabidopsis*: an analysis of the *cup-shaped cotyledon* mutant. *Plant Cell*, **9**, 841-857.

Aida, M., Ishida, T. and Tasaka, M. (1999) Shoot apical meristem and cotyledon formation during *Arabidopsis* embryogenesis: interaction among the *CUP-SHAPED COTYLEDON* and *SHOOT MERISTEMLESS* genes. *Development*, **126**, 1563-1570.

Alvarez, J. and Smyth, D.R. (1999) *CRABS CLAW* and *SPATULA*, two *Arabidopsis* genes that control carpel development in parallel with *AGAMOUS*. *Development*, **126**, 2377-2386.

Angenent, G.C. and Colombo, L. (1996) Molecular control of ovule development. *Trends Plant Sci.*, **1**, 228-232.

Angenent, G.C., Franken, J., Busscher, M., Colombo, L. and van Tunen, A.J. (1993) Petal and stamen formation in petunia is regulated by the homeotic gene *fbp1*. *Plant J.*, **3**, 101-112.

Angenent, G.C., Franken, J., Busscher, M., *et al.* (1995) A novel class of MADS box genes is involved in ovule development in Petunia. *Plant Cell*, **7**, 1569-1582.

Bajon, C., Horlow, C., Motamayor, J.C., Sauvanet, A. and Robert, D. (1999) Megasporogenesis in *Arabidopsis thaliana* L.: an ultrastructural study. *Sexual Plant Reproduction*, **12**, 99-109.

Baker, S.C., Robinson-Beers, K., Villanueva, J.M., Gaiser, J.C. and Gasser, C.S. (1997) Interactions among genes regulating ovule development in *Arabidopsis thaliana*. *Genetics*, **145**, 1109-1124.

Balasubramanian, S. and Schneitz, K. (2000) *NOZZLE* regulates proximal–distal pattern formation, cell proliferation and early sporogenesis in *Arabidopsis thaliana*. *Development*, **127**, 4227-4238.

Bhatt, A.M., Canales, C. and Dickinson, H.G. (2001) Meiosis: the means to 1N. *Trends Plant Sci.*, **6**, 114-121.

Bowman, J.L. (2000) The YABBY gene family and abaxial cell fate. *Curr. Opin. Plant Biol.*, **3**, 17-22.

Bowman, J.L. and Smyth, D.R. (1999) *CRABS CLAW*, a gene that regulates carpel and nectary development in *Arabidopsis*, encodes a novel protein with zinc finger and helix-loop-helix domains. *Development*, **126**, 2387-2396.

Bowman, J.L., Drews, G.N. and Meyerowitz, E.M. (1991a) Expression of the *Arabidopsis* floral homeotic gene *AGAMOUS* is restricted to specific cell types late in flower development. *Plant Cell*, **3**, 749-758.

Bowman, J.L., Smyth, D.R. and Meyerowitz, E.M. (1991b) Genetic interactions among floral homeotic genes of *Arabidopsis*. *Development*, **112**, 1-20.

Bowman, J.L., Sakai, H., Jack, T., Weigel, D., Mayer, U. and Meyerowitz, E.M. (1992) *SUPERMAN*, a regulator of floral homeotic genes in *Arabidopsis*. *Development*, **114**, 599-615.

Bowman, J.L., Baum, S.F., Eshed, Y., Putterill, J. and Alvarez, J. (1999) Molecular genetics of gynoecium development in *Arabidopsis*. *Curr. Topi. Dev. Biol.*, **45**, 155-205.

Broadhvest, J., Baker, S.C. and Gasser, C.S. (2000) *SHORT INTEGUMENTS 2* promotes growth during *Arabidopsis* reproductive development. *Genetics*, **155**, 899-907.

Chien, C.T., Bartel, P.L., Sternglanz, R. and Fields, S. (1991) The two-hybrid system: a method to identify and clone genes for proteins that interact with a protein of interest. *Proc. Natl. Acad. Sci. USA*, **88**, 9578-9582.

Christensen, C.A., King, E.J., Jordan, J.R. and Drews, G.N. (1997) Megagametogenesis in *Arabidopsis* wild type and the *Gf* mutant. *Sexual Plant Reproduction*, **10**, 49-64.

Christensen, C.A., Subramanian, S. and Drews, G.N. (1998) Identification of gametophytic mutations affecting female gametophyte development in *Arabidopsis*. *Dev. Biol.*, **202**, 136-151.

Chuang, C.-F., Running, M.P., Williams, R.W. and Meyerowitz, E.M. (1999) The *PERIANTHA* gene encodes a bZIP protein involved in the determination of floral organ number in *Arabidopsis thaliana*. *Genes Dev.*, **13**, 334-344.

Coen, E.S. and Meyerowitz, E.M. (1991) The war of the whorls: genetic interactions controlling flower development. *Nature*, **353**, 31-37.

Colombo, L., Franken, J., Koetje, E., *et al.* (1995) The petunia MADS box gene *FBP11* determines ovule identity. *Plant Cell*, **7**, 1859-1868.

Conner, J. and Liu, Z. (2000) *LEUNIG*, a putative transcriptional corepressor that regulates *AGAMOUS* expression during flower development. *Proc. Natl. Acad. Sci. USA*, **97**, 12902-12907.

Cvitanich, C., Pallisgaard, N., Nielsen, K.A., *et al.* (2000) CPP1, a DNA-binding protein involved in the expression of a soybean *leghemoglobin c3* gene. *Proc. Natl. Acad. Sci. USA*, **97**, 8163-8168.

Dahlgren, K.V.O. (1927) Die Morphologie des Nuzellus mit besonderer Berücksichtigung der deckzellosen Typen. *Jahrb. Wissens. Botanik*, **67**, 347-426.

De Martinis, D. and Mariani, C. (1999) Silencing gene expression of the ethylene-forming enzyme results in a reversible inhibition of ovule development in transgenic tobacco plants. *Plant Cell*, **11**, 1061-1072.

Drews, G.N., Lee, D. and Christensen, C.A. (1998) Genetic analysis of female gametophyte development and function. *Plant Cell*, **10**, 5-17.

Elliott, R.C., Betzner, A.S., Huttner, E., *et al.* (1996) *AINTEGUMENTA*, an *APETALA2*-like gene of *Arabidopsis* with pleiotropic roles in ovule development and floral organ growth. *Plant Cell*, **8**, 155-168.

Esau, K. (1977) *Anatomy of Seed Plants*, Wiley, New York.

Eshed, Y., Baum, S.F. and Bowman, J.L. (1999) Distinct mechanisms promote polarity establishment in carpels of *Arabidopsis*. *Cell*, **99**, 199-209.

Evans, P.T. and Malmberg, R.L. (1989) Alternative pathways of tobacco placental development: time of commitment and analysis of a mutant. *Dev. Biol.*, **136**, 273-283.

Ferrándiz, C., Liljegren, S.J. and Yanofsky, M.F. (2000) Negative regulation of the *SHATTERPROOF* genes by *FRUITFULL* during *Arabidopsis* fruit development. *Science*, **289**, 436-438.

Gaiser, J.C., Robinson-Beers, K. and Gasser, C.S. (1995) The *Arabidopsis SUPERMAN* gene mediates asymmetric growth of the outer integument of ovules. *Plant Cell*, **7**, 333-345.

Gasser, C.S., Broadhvest, J. and Hauser, B.A. (1998) Genetic analysis of ovule development. *Ann. Rev. Plant Physiol. Plant Mol. Biol.*, **49**, 1-24.

Gehring, W., Affolter, M. and Bürglin, T. (1994) Homeodomain proteins. *Annu. Rev. Biochem.*, **63**, 487-526.

Gifford, E.M. and Foster, A.S. (1989) *Morphology and Evolution of Vascular Plants*, W.H. Freeman, New York.

Goodrich, J., Puangsomlee, P., Martin, M., Long, D., Meyerowitz, E.M. and Coupland, G. (1997) A polycomb-group gene regulates homeotic gene expression in *Arabidopsis*. *Nature*, **386**, 44-51.

Grini, P.E., Schnittger, A., Schwarz, H., *et al.* (1999) Isolation of ethyl methanesulfonate-induced gametophytic mutants in *Arabidopsis thaliana* by a segregation distortion assay using the multimarker chromosome 1. *Genetics*, **151**, 849-863.

Grossniklaus, U. and Schneitz, K. (1998) The molecular and genetic basis of ovule and megagametophyte development. *Semin. Cell Dev. Biol.*, **9**, 227-238.

Grossniklaus, U., Vielle-Calzada, J.-P., Hoeppner, M.A. and Gagliano, W.B. (1998) Maternal control of embryogenesis by *MEDEA*, a *Polycomb* group gene in *Arabidopsis*. *Science*, **280**, 446-450.

Hagstrom, K. and Schedl, P. (1997) Remembrance of things of the past: maintaining gene expression patterns with altered chromatin. *Curr. Opin. Genet. Dev.*, **7**, 814-821.

Hauser, B.A., Villanueva, J.M. and Gasser, C.S. (1998) Arabidopsis *TSO1* regulates directional processes in cells during floral organogenesis. *Genetics*, **150**, 411-423.

Hauser, B.A., He, J.Q., Park, S.O. and Gasser, C.S. (2000) TSO1 is a novel protein that modulates cytokinesis and cell expansion in *Arabidopsis*. *Development*, **127**, 2219-2226.

Hicks, G.S. and McHughen, A. (1974) Altered morphogenesis of placental tissues of tobacco *in vitro*: stigmatoid and carpelloid outgrowths. *Planta*, **121**, 193-196.

Hicks, G.S. and McHughen, A. (1977) Ovule development *in vitro* from isolated tobacco placental tissue. *Plant Sci. Lett.*, **8**, 141-145.

Hill, J.L. and Lord, D.R. (1994) Wild-type flower development: gynoecial initiation. In *Arabidopsis: An Atlas of Morphology and Development* (ed. J. Bowman), Springer Verlag, New York, pp. 158-159.

Hill, J.P. and Lord, E.M. (1989) Floral development in *Arabidopsis thaliana*: a comparison of the wild type and the homeotic pistillata mutant. *Can. J. Botany*, **67**, 2922-2936.

Howden, R., Park, S.K., Moore, J.M., Orme, J., Grossniklaus, U. and Twell, D. (1998) Selection of T-DNA-tagged male and female gametophytic mutants by segregation distortion in Arabidopsis. *Genetics*, **149**, 621-631.

Hua, J., Sakai, H., Nourizadeh, S., *et al.* (1998) EIN4 and ERS2 are members of the putative ethylene receptor gene family in Arabidopsis. *Plant Cell*, **10**, 1321-1332.

Ishida, T., Aida, M., Takada, S. and Tasaka, M. (2000) Involvement of *CUP-SHAPED COTYLEDON* genes in gynoecium and ovule development in *Arabidopsis thaliana*. *Plant Cell Physiol.*, **41**, 60-67.

Jenik, P.D. and Irish, V.F. (2000) Regulation of cell proliferation patterns by homeotic genes during *Arabidopsis* floral development. *Development*, **127**, 1267-1276.

Jones, R.S. and Gelbart, W.M. (1993) The *Drosophila* Polycomb-group gene *Enhancer of zeste* contains a region with sequence similarity to *trithorax*. *Mol. Cell. Biol.*, **10**, 6357-6366.

Kiyosue, T., Ohad, N., Yadegari, R., *et al.* (1999) Control of fertilization-independent endosperm development by the *MEDEA* polycomb gene in *Arabidopsis*. *Proc. Natl. Acad. Sci. USA*, **96**, 4186-4191.

Klucher, K.M., Chow, H., Reiser, L. and Fischer, R.L. (1996) The *AINTEGUMENTA* gene of *Arabidopsis* required for ovule and female gametophyte development is related to the floral homeotic gene *APETALA2*. *Plant Cell*, **8**, 137-153.

Krizek, B.A. (1999) Ectopic expression of *AINTEGUMENTA* in *Arabidopsis* plants results in increased growth of floral organs. *Dev. Genet.*, **25**, 224-236.

Krude, T. (1999) Chromatin replication: finding the right connection. *Curr. Biol.*, **9**, R394-R396.

Lang, J.D., Ray, S. and Ray, A. (1994) *sin1*, a mutation affecting female fertility in *Arabidopsis*, interacts with *mod1*, its recessive modifier. *Genetics*, **137**, 1101-1110.

Léon-Kloosterziel, K.M., Keijzer, C.J. and Koornneef, M. (1994) A seed shape mutant of *Arabidopsis* that is affected in integument development. *Plant Cell*, **6**, 385-392.

Liljegren, S.J., Ditta, G.S., Eshed, Y., Savidge, B., Bowman, J.L. and Yanofsky, M.F. (2000) *SHATTERPROOF* MADS-box genes control seed dispersal in *Arabidopsis*. *Nature*, **404**, 766-770.

Liu, Z. and Meyerowitz, E.M. (1995) *LEUNIG* regulates *AGAMOUS* expression in *Arabidopsis* flowers. *Development*, **121**, 975-991.

Liu, Z., Running, M. and Meyerowitz, E.M. (1997) *TSO1* functions in cell division during *Arabidopsis* flower development. *Development*, **124**, 665-672.

Liu, Z., Franks, R.G. and Klink, V.P. (2000) Regulation of gynoecium marginal tissueformation by *LEUNIG* and *AINTEGUMENTA*. *Plant Cell*, **12**, 1879-1892.

Long, J.A. and Barton, M.K. (1998) The development of apical embryonic pattern in *Arabidopsis*. *Development*, **125**, 3027-3035.

Ma, H. (1994) The unfolding drama of flower development: recent results from genetic and molecular analyses. *Genes Dev.*, **8**, 745-756.

Ma, H., Yanofsky, M.F. and Meyerowitz, E.M. (1991) AGL1-ALG6, an *Arabidopsis* gene family with similarity to floral homeotic and transcription factor genes. *Genes Dev.*, **5**, 484-495.

Maheswari, P. (1950) *An Introduction to the Embryology of Angiosperms*, McGraw-Hill, New York.

Mandel, M.A., Bowman, J.L., Kempin, S.A., Ma, H., Meyerowitz, E.M. and Yanofsky, M.F. (1992) Manipulation of flower structure in transgenic tobacco. *Cell*, **71**, 133-143.

Mansfield, S.G., Briarty, L.G. and Erni, S. (1991) Early embryogenesis in *Arabidopsis thaliana*. I. The mature embryo sac. *Can. J. Botany*, **69**, 447-460.

Mizukami, Y. and Fischer, R.L. (2000) Plant organ size control: *AINTEGUMENTA* regulates growth and cell numbers during organogenesis. *Proc. Natl. Acad. Sci. USA*, **97**, 942-947.

Modrusan, Z., Reiser, L., Feldmann, K.A., Fischer, R.L. and Haughn, G.W. (1994) Homeotic transformation of ovules into carpel-like structures in *Arabidopsis*. *Plant Cell*, **6**, 333-349.

Moore, J.M., Vielle Calzada, J.-P., Gagliano, W. and Grossniklaus, U. (1997) Genetic characterization of *hadad*, a mutant disrupting female gametogenesis in *Arabidopsis thaliana*. *Cold Spring Harbor Symp. Quant. Biol.*, **62**, 35-47.

Nemhauser, J.L., Feldman, L.J. and Zambryski, P.C. (2000) Auxin and *ETTIN* in *Arabidopsis* gynoecium morphogenesis. *Development*, **127**, 3877-3888.

Okada, K., Komaki, M.K. and Shimura, Y. (1989) Mutational analysis of pistil structure and development of *Arabidopsis thaliana*. *Cell Differ. Dev.*, **28**, 27-38.

Pirrotta, V. (1997) PcG complexes and chromatin silencing. *Curr. Opin. Genet. Dev.*, **7**, 249-258.

Pirrotta, V. (1998) Polycombing the genome: PcG, trxG, and chromatin silencing. *Cell*, **93**, 333-336.

Ray, A., Robinson-Beers, K., Ray, S., *et al.* (1994) *Arabidopsis* floral homeotic gene BELL (*BEL1*) controls ovule development through negative regulation of AGAMOUS gene (*AG*). *Proc. Natl. Acad. Sci. USA*, **91**, 5761-5765.

Ray, A., Lang, J.D., Golden, T. and Ray, S. (1996) *SHORT INTEGUMENT* (*SIN1*), a gene required for ovule development in *Arabidopsis*, also controls flowering time. *Development*, **122**, 2631-2638.

Ray, S., Golden, T. and Ray, A. (1996) Maternal effects of the *short integument* mutation on embryo development in *Arabidopsis*. *Dev. Biol.*, **180**, 365-369.

Reiser, L. and Fischer, R.L. (1993) The ovule and the embryo sac. *Plant Cell*, **5**, 1291-1301.

Reiser, L., Modrusan, Z.L.M., Samach, A., Ohad, N., Haughn, G.W. and Fischer, R.L. (1995) The *BELL1* gene encodes a homeodomain protein involved in pattern formation in the *Arabidopsis* ovule primordium. *Cell*, **83**, 735-742.

Riechmann, J.L. and Meyerowitz, E.M. (1998) The AP2/EREBP familiy of plant transcription factors. *Biol. Chem.*, **379**, 633-646.

Robinson-Beers, K., Pruitt, R.E. and Gasser, C.S. (1992) Ovule development in wild-type *Arabidopsis* and two female-sterile mutants. *Plant Cell*, **4**, 1237-1249.

Roe, J.L., Sessions, R.A., Feldmann, K.A. and Zambryski, P.C. (1993) The *Tousled* gene in *A. thaliana* encodes a protein kinase homolog that is required for leaf and flower development. *Cell*, **75**, 939-950.

Roe, J.L., Durfee, T., Zupan, J.R., Repetti, P., McLean, B.G. and Zambryski, P.C. (1997a) TOUSLED is a nuclear serine/threonine protein kinase that requires a coiled-coil region for oligomerization and catalytic activity. *J. Biol. Chem.*, **272**, 5838-5845.

Roe, J.L., Nemhauser, J.L. and Zambryski, P.C. (1997b) *TOUSLED* participates in apical tissue formation during gynoecium development in *Arabidopsis*. *Plant Cell*, **9**, 335-353.

Rounsley, S.D., Ditta, G.S. and Yanofsky, M.F. (1995) Diverse roles for MADS box genes in *Arabidopsis* development. *Plant Cell*, **7**, 1259-1269.

Running, M.P. and Meyerowitz, E.M. (1996) Mutations in the *PERIANTHIA* gene of *Arabidopsis* specifically alter floral organ number and initiation pattern. *Development*, **122**, 1261-1269.

Sakai, H., Medrano, L.J. and Meyerowitz, E.M. (1995) Role of *SUPERMAN* in maintaining *Arabidopsis* floral whorl boundaries. *Nature*, **378**, 199-203.

Sakai, H., Hua, J., Chen, Q.G., *et al.* (1998) *ETR2* is an *ETR1*-like gene involved in ethylene signaling in *Arabidopsis*. *Proc. Natl. Acad. Sci. USA*, **95**, 5812-5817.

Savidge, B., Rounsley, S.D. and Yanofsky, M. (1995) Temporal relationship between the transcription of two *Arabidopsis* MADS box genes and the floral organ identity genes. *Plant Cell*, **7**, 721-733.

Schiefthaler, U., Balasubramanian, S., Sieber, P., Chevalier, D., Wisman, E. and Schneitz, K. (1999) Molecular analysis of *NOZZLE*, a gene involved in pattern formation and early sporogenesis during sex organ development in *Arabidopsis thaliana*. *Proc. Natl. Acad. Sci. USA*, **96**, 11664-11669.

Schneitz, K. (1999) The molecular and genetic control of ovule development. *Curr. Opin. Plant Biol.*, **2**, 13-17.

Schneitz, K., Hülskamp, M. and Pruitt, R.E. (1995) Wild-type ovule development in *Arabidopsis thaliana*: a light microscope study of cleared whole-mount tissue. *Plant J.*, **7**, 731-749.

Schneitz, K., Hülskamp, M., Kopczak, S.D. and Pruitt, R.E. (1997) Dissection of sexual organ ontogenesis: a genetic analysis of ovule development in *Arabidopsis thaliana*. *Development*, **124**, 1367-1376.

Schneitz, K., Baker, S.C., Gasser, C.S. and Redweik, A. (1998a) Pattern formation and growth during floral organogenesis: *HUELLENLOS* and *AINTEGUMENTA* are required for the formation of the proximal region of the ovule primordium in *Arabidopsis thaliana*. *Development*, **125**, 2555-2563.

Schneitz, K., Balasubramanian, S. and Schiefthaler, U. (1998b) Organogenesis in plants: the molecular and genetic control of ovule development. *Trends Plant Sci.*, **3**, 468-472.

Schultz, E.A., Pickett, F.B. and Haughn, G.W. (1991) The *FLO10* gene product regulates the expression domain of homeotic genes *AP3* and *PI* in *Arabidopsis* flowers. *Plant Cell*, **3**, 1221-1237.

Sessions, R.A. and Zambryski, P.C. (1995) *Arabidopsis* gynoecium structure in the wild type and in *ettin* mutants. *Development*, **121**, 1519-1532.

Sheridan, W.F. and Huang, B.-Q. (1997) Nuclear behavior is defective in the maize (*Zea mays* L.) *lethal ovule2* female gametophyte. *Plant J.*, **11**, 1029-1041.

Shimizu, K.K. and Okada, K. (2000) Attractive and repulsive interactions between female and male gametophytes in *Arabidopsis* pollen tube guidance. *Development*, **127**, 4511-4518.

Shore, P. and Sharrocks, A.D. (1995) The MADS-box family of transcription factors. *Eur. J. Biochem.*, **229**, 1-13.

Siegfried, K.R., Eshed, Y., Baum, S.F., Otsuga, D., Drews, G.N. and Bowman, J.L. (1999) Members of the *YABBY* gene family specify abaxial cell fate in *Arabidopsis*. *Development*, **126**, 4117-4128.

Silljé, H.H., Takahashi, K., Tanaka, K., Van Houwe, G. and Nigg, E.A. (1999) Mammalian homologues of the plant *Tousled* gene code for cell-cycle-regulated kinases with maximal activities linked to ongoing DNA replication. *EMBO J.*, **18**, 5691-5702.

Song, J.-Y., Leung, T., Ehler, L.K., Wang, C. and Liu, Z. (2000) Regulation of meristem organization and cell division by *TSO1*, an *Arabidopsis* gene with cysteine-rich repeats. *Development*, **127**, 2207-2217.

Sundaresan, V., Springer, P., Volpe, T., *et al.* (1995) Patterns of gene action in plant development revealed by enhancer trap and gene trap transposable elements. *Genes Dev.*, **9**, 1797-1810.

Tsuchimoto, S., van der Krol, A.R. and Chua, N.-H. (1993) Ectopic expression of pMADS3 in transgenic petunia phenocopies the petunia blind mutant. *Plant Cell*, **5**, 843-853.

Villanueva, J.M., Broadhvest, J., Hauser, B.A., Meister, R.J., Schneitz, K. and Gasser, C.S. (1999) *INNER NO OUTER* regulates abaxial/adaxial patterning in *Arabidopsis* ovules. *Genes Dev.*, **13**, 3160-3169.

Webb, M.C. and Gunning, B.E.S. (1990) Embryo sac development in *Arabidopsis thaliana*. I. Megasporogenesis, including the microtubular cytoskeleton. *Sexual Plant Reprod.*, **3**, 244-256.

Weigel, D. (1995) The APETALA2 domain is related to a novel type of DNA binding domain. *Plant Cell*, **7**, 388-389.

Weigel, D. and Meyerowitz, E.M. (1994) The abcs of floral homeotic genes. *Cell*, **78**, 203-209.

Weigel, D., Ahn, J.H., Blazquez, M.A., *et al.* (2000) Activation tagging in *Arabidopsis*. *Plant Physiol.*, **122**, 1003-1013.

Western, T.L. and Haughn, G.W. (1999) *BELL1* and *AGAMOUS* genes promote ovule identity in *Arabidopsis thaliana*. *Plant J.*, **18**, 329-336.

Williams, B., Riedy, M., Williams, E., Gatti, M. and Goldberg, M. (1995) The *Drosophila* kinesin-like protein KLP3A is a midbody component required for central spindle assembly and initiation of cytokinesis. *J. Cell Biol.*, **129**, 709-723.

Williams, B.C., Dernburg, A.F., Puro, J., Nokkala, S. and Goldberg, M.L. (1997) The *Drosophila* kinesin-like protein KLP3A is required for proper behavior of male and female pronuclei at fertilization. *Development*, **124**, 2365-2376.

Yang, W.-C. and Sundaresan, V. (2000) Genetics of gametophyte biogenesis in *Arabidopsis*. *Curr. Opin. Plant Biol.*, **3**, 53-57.

Yang, W.-C., Ye, D., Xu, J. and Sundaresan, V. (1999) The *SPOROCYTELESS* gene of *Arabidopsis* is required for initiation of sporogenesis and encodes a novel nuclear protein. *Genes Dev.*, **13**, 2108-2117.

Zhang, X.S. and O'Neill, S.D. (1993) Ovary and gametophyte development are coordinately regulated by auxin and ethylene following pollination. *Plant Cell*, **5**, 403-418.

4   The developmental biology of pollen

David Twell

4.1   Introduction

The pollen grains of flowering plants harbour the haploid male partners in sexual reproduction. Their biological role is to deliver the two sperm cells, via the pollen tube, to the embryo sac to effect double fertilization. The study of pollen development provides access to a number of unique cellular events and processes in plants that have evolved to specify male gametogenesis and plant fertility. Apart from its intrinsic importance for sexual reproduction, pollen development can also be considered a microcosm of cellular development, providing an attractive system in which to dissect the regulation of the fundamental cellular processes of cell division, cell fate determination, gene expression and cellular differentiation.

Pollen development and function have been the subjects of intense investigation for over a century and there exists an impressive literature concerning gamete development, transport and fertilization in a wide range of species. The scope of this review concerns the cell biology and the molecular and genetic regulation of pollen development in flowering plants. Current knowledge is presented in two major parts. In the first part important sporophytic–gametophytic interactions that influence pollen structure and function are discussed, together with landmark developmental events. These include the development of microspore polarity; asymmetric cell division; determination of vegetative and generative cell fate; intracellular migration of the generative cell, and assembly of the male germ unit. In essence, the discussion reflects the ontogeny of a single cell from microspore inception to pollen release, with emphasis on gametophytic control and on recent progress in using *Arabidopsis* as a model. Detailed discussion of the progamic phase is beyond the scope of this review, but there are excellent reviews that discuss recent progress in pollen–stigma interactions, pollen tube growth, guidance and fertilization (Taylor and Hepler, 1997; Malho, 1998; Franklin-Tong, 1999; Pruitt, 1999; Wilhelmi and Preuss, 1999; Lord, 2000; Palanivelu and Preuss, 2000). In the second part, the repertoire and regulation of gametophytically expressed genes is reviewed, focusing on those genes for which evidence of function has been demonstrated. Finally, this section provides a discussion of recent advances in molecular genetic analyses, which provide powerful new tools for the dissection of pollen development and function.

Cell biology and differentiation

4.2 Microsporogenesis

The general ontogeny of microsporogenesis is outlined here, together with some recent progress in genetic analysis of this process. In the young anther, groups of archesporial cells divide periclinally to form an outer, primary parietal layer and an inner sporogenous layer. The primary parietal layer undergoes periclinal and anticlinal divisions, giving rise to several concentric layers that differentiate into the endothecium, middle layer(s) and the innermost tapetum. The primary sporogenous layer gives rise to the microsporocytes or meiocytes directly, or after further limited mitotic divisions.

Prior to meiosis, the meiocytes, tapetum and wall layers are all interconnected by plasmodesmata forming a syncytium. Early during meiotic prophase, the connections between cell wall layers are severed. At this stage, individual meiocytes are surrounded by a thick callose wall, but larger cytomictic channels extend between the meiocytes. Syncytium formation between meiocytes may be important in maintaining meiotic synchrony within individual locules.

During meiosis, cytoplasmic reorganization results in the degradation of most of the meiocyte RNA and ribosomes, and dedifferentiation of plastids and mitochondria. These changes are likely to be important in erasing diplophase informational macromolecules, providing a 'neutral' cellular environment in which to establish the gametophytic programme (Dickinson, 1994). Finally, postmeiotic cleavage of the meiocyte occurs according to two different developmental programmes. In the majority of dicot species cytokinesis of the meiocyte occurs simultaneously after completion of meiosis I and II, in others cleavage occurs sequentially, first to form cellular dyads that undergo cleavage to form tetrads. Completion of meiotic cytokinesis results in the deposition of callose around individual microspores and their symplastic isolation from each other and from the diploid parent.

Although mechanisms regulating the production of meiocytes have remained obscure, an exciting recent development is the identification of a regulatory gene required for the initiation of male and female sporogenesis, termed *SPOROCYTELESS* (*SPL*), (or *NOZZLE*) in *Arabidopsis* (Schiefthaler *et al.*, 1999; Yang *et al.*, 1999). In *spl*, mutants, archesporial cells divide to form sporogenous cells and primary parietal cells, but further development of both cell layers is arrested. *SPL* encodes a nuclear protein with limited similarities to MADS box transcription factors. In anthers, *SPL* expression is restricted to sporogenous cells and meiocytes, suggesting a primary role in formation of the microsporocytes. Defects in the anther wall layers are therefore indirect, which further suggests that development of the anther walls and tapetum are dependent upon signals from the microsporocytes (Yang *et al.*, 1999). A more extensive number of mutants specifically affecting meiosis have been described in *Arabidopsis* and maize that affect meiotic chromosome pairing, segregation and cell division, some of which are specific to male meiosis (see Yang and Sundaresan, 2000). Male-specific mutants have also been described that block meiotic cytokinesis: *tes* (Spielman *et al.*, 1997) and *stud* (Hülskamp *et al.*, 1997). These mutants provide evidence for specialized events associated with simultaneous cytokinesis and have interesting implications for microspore polarity (see section 4.6.5).

4.3 Microgametogenesis

The completion of meiosis of the microsporocytes marks the initiation of a unique pathway of cellular differentiation—microgametogenesis—that leads, through a simple cell lineage, to the formation of the mature pollen grains

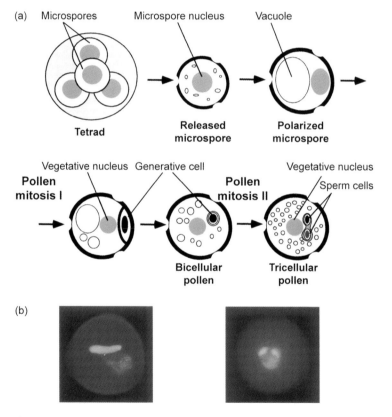

Figure 4.1 Schematic diagram of microgametogenesis in *Arabidopsis*. (a) Major cytological events are illustrated based on light and electron microscopy. (b) Mature pollen of bicellular tobacco pollen (left) and tricellular *Arabidopsis* pollen (right) stained with DAPI to reveal nuclei.

(figure 4.1a). Although pollen shows remarkable diversity in form, development appears to be regular throughout the angiosperms. All known species follow one of two general pathways that are distinguished on the basis of the timing of generative cell division. In the large majority, pollen is shed in a bicellular condition, and the generative cell divides to form the two sperm cells within the pollen tube; in tricellular pollen species, in contrast, the generative cell divides before pollen is released from the anther (figure 4.1b).

4.4 Sporophytic–gametophytic interactions

Microsporogenesis and microgametogenesis take place within the anther loculus, or pollen sac, and depend on complex interactions between sporophytic and gametophytic cells. The vital role of sporophytic anther tissues is clear from the large numbers of sporophytic mutations that disrupt pollen development and cause male sterility. Male sterile mutants have been described in more than 100 species and a significant number of these result in aberrant tapetal cell development (Kaul, 1988; van der Meer *et al.*, 1992; Chaudhury, 1993; Aarts *et al.*, 1997; Taylor *et al.*, 1998; Sanders *et al.*, 1999). The vital role of the tapetum in microspore development has been directly demonstrated in experiments in which tapetal cells were genetically ablated by tapetal-specific expression of a fungal ribonuclease (Mariani *et al.*, 1990).

The tapetum is a transient cell layer that is usually present only during microspore development, before it undergoes programmed cell death around the time of pollen mitosis I (reviewed in Pacini, 1990, 1997). The tapetum actively secretes β(1-3) glucanase, which is required for callose dissolution and microspore release and, in cooperation with other anther wall layers, controls the supply of amino acids, lipids and carbohydrates required for pollen nutrition. The tapetum also controls the synthesis and deposition of secondary metabolites required for elaboration of the unique pollen wall, UV protection, pollinator attraction and pollen fertility. Finally, the tapetum cells undergo a carefully orchestrated polar degeneration, depositing their cell contents within the exine cavities to form the pollen coat (Murgia *et al.*, 1991). In the following sections (4.4.1 to 4.4.5), the impact of these key sporophytic–gametophytic interactions on microspore and pollen structure and function is discussed.

4.4.1 The unique pollen wall

One of the earliest signs of microspore differentiation, initiated during the tetrad stage, is the synthesis of the outer pollen wall or exine. The exine layer, which is often elaborately patterned and sculpted, is the most striking structural feature of pollen grains (figure 4.2a) and is composed of sporopollenin, one of the most highly resistant biopolymers known. The exine forms an incomplete wall layer that is highly reduced or absent in regions that define the positions of the

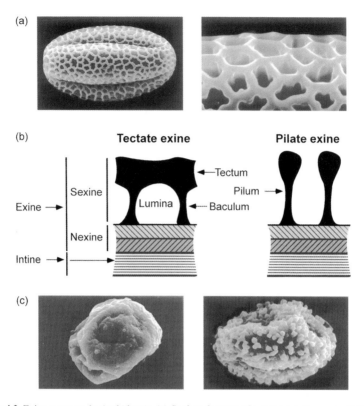

Figure 4.2 Exine structure in *Arabidopsis*. (a) Surface features of a mature pollen grain (left) and at higher magnification (right) viewed by scanning electron microscopy. (b) Diagram of wall layers in species with tectate and pilate exine structures (adapted from Heslop-Harrison, 1971). (c) *Arabidopsis* pollen wall mutants *pow1* (left) and *pow2* (right).

germination apertures. Aperture number, position and exine ornamentation are useful taxonomic characters, and have allowed reconstruction of past vegetation from the fossil record. In addition to the obvious mechanical protective function of the exine, exine sculpturing plays a role in attachment to pollinators and adhesion to the stigmatic surfaces. In wind-pollinated species, particularly grasses, pollen grains often lack elaborate surface patterning and appear smooth. Beneath the exine, which is defined by the outer sexine and one or two basal nexine layers, a second major wall layer surrounds the pollen grain, termed the intine (figure 4.2b). While the exine is composed of sporopollenin, a polymer containing fatty acids and phenylpropanoids, the intine is largely composed of pectin and cellulose (for reviews see Erdtman, 1952; Blackmore and Barnes, 1990; Scott, 1994).

The origin and genetic control of the synthesis of the two pollen wall layers differ. Exine synthesis and patterning are under sporophytic control, through the contribution of both the inherited early microspore cytoplasm and the tapetum (Heslop-Harrison, 1971). Exine synthesis is initiated with the formation of the primexine layer at the tetrad stage. Patterning is thought to be established rapidly, following meiotic cleavage, and is evident before microspore release from the tetrad. When meiosis is disturbed, microspore nuclei are formed with unequal amounts of genomic material, and a normal exine layer is formed even in enucleated cells. Therefore, agents responsible for the patterning are sporophytically controlled and must be present in the cytoplasm before meiosis (Heslop-Harrison, 1971). Although the precise mechanisms of exine synthesis and patterning remain obscure, cytoplasmic elements have been identified that could be involved in localized deposition responsible for patterning. These include small protein-coated vesicles, the microtubule cytoskeleton and negative regulation of the formation of exine pattern via association with localized endoplasmic reticulum (ER) (Scott, 1994). After initiation of patterning, the exine is elaborated through sporopollenin deposits, released by the tapetal cells onto the surface of the microspores. This leads to thickening of the exine layer, and of baculae and the tectum in species with patterned exines (figure 4.2b).

In contrast, intine wall synthesis is initiated during the free microspore stage before completion of exine growth and microspore vacuolation (Knox and Heslop-Harrison, 1970; Knox, 1971), and is thought to be largely under gametophytic control. Microspore-expressed or microspore-specific genes such as the *Brassica Bcp1* (Theerakulpisut *et al.*, 1991), *Arabidopsis APG* (Roberts *et al.*, 1991) and tobacco *NTM19* (Oldenhof *et al.*, 1996) genes, which encode secretory proteins, may play an important structural role in the microspore wall. This could be important for structural stability of the wall during morphogenesis, as the microspore changes shape from a lobed structure to more rounded, expanded spore.

4.4.2 Exine pattern mutants

The isolation of mutants that are defective in pattern provides a direct route to identify genes involved in exine patterning, and to evaluate the role of this pattern in pollen–pistil interactions. To date, only a few mutants have been isolated that show defects in exine structure, which often lead to pollen abortion and male sterility. Paxson-Sowders *et al.* (1997) described a mutant in *Arabidopsis*, *dex1* that shows rare invaginations of the plasma membrane at the tetrad stage, which are associated with irregular deposition of sporopollenin and microspore degeneration. In *dex1*, the tapetum appears normal, and a callose wall and primexine are present. Therefore, it was concluded that primexine and the callose wall do not play a direct role in pattern formation, but the plasma membrane is likely to be important (Paxson-Sowders *et al.*, 1997). The *Arabidopsis* male sterile (*ms2*) mutant forms inviable pollen grains and does not show signs of exine synthesis (Aarts *et al.*, 1997). *ms2* is defective in the expression of a tapetum-specific gene encoding a putative fatty acyl reductase and may play a role in the reduction of fatty acids to fatty alcohols, which could act as precursors in sporopollenin synthesis (Aarts *et al.*, 1997). Recently, a tapetal-specific chalcone synthase-like gene (*NSCHSLK*) from *Nicotiana sylvestris* has been isolated that could also be involved in exine biosynthesis (Atanassov *et al.*, 1998). In two other *Arabidopsis* male sterile mutants, *ms9* and *ms12*, the tapetum degenerates at an early stage and almost no tapetally derived sporopollenin is added to the microspores (Taylor *et al.*, 1998). Such mutants could act specifically, but equally could encode proteins with general cellular functions in the tapetum.

Jackson *et al.* (2000), have recently described a sporophytically determined spineless pollen wall mutation in *Haplopappus gracilis* that renders the otherwise functional pollen grains completely unrecognizable as Compositae. The mutation appears to affect one specific structural element of the exine. Whereas the endexine layer was found similar to the wild type, the organization of ektexine units was disturbed and mutant pollen lacked the caveae and foot layer characteristic of normal pollen. Pollen morphogenesis screens in *Arabidopsis* have led to the isolation of several sporophytic *pollen wall* (*pow*) mutants that show defective exine patterning (figure 4.2c). Such mutants are expected to provide access to molecular components involved in exine synthesis and patterning.

Recent experiments in *Arabidopsis* provide compelling evidence that the exine may play an important role in adhesion of the pollen grain to the stigmatic surfaces. Zinkl *et al.* (1999) showed that when the wild-type pollen coat was removed with cyclohexane, or in mutant pollen lacking a pollen coat (*cer* mutants), pollen grains remained adhered to the pistil. Pollen adhesion was shown to be taxon-specific or species-specific and to cause imprints on the papillar cells (Zinkle *et al.*, 1999). Further evidence for the role of exine in pollen adhesion is suggested by a cited mutant, *lap1*, that has defects in adhesion and a disturbed exine structure (Zinkl and Preuss, 2000)

4.4.3   Phenylpropanoids

The tapetum is also a source of enzymes and products of secondary metabolism deposited within the pollen wall. These include phenylpropanoids such as the tetrahydroxy chalcone forming the major yellow coloured pigment in pollen of *Petunia* and *Tulipa* lines (de Vlaming and Kho, 1976). Flavonoids may function both in UV protection and in pollinator attraction and are largely under sporophytic control. For example, the petunia mutation, *an4*, affects UDPglucose:flavonoid-3-*O*-glucosyltransferase activity, and anthers lack anthocyanins (de Vlaming *et al.*, 1984; Gerats *et al.*, 1985). The petunia *Po* mutation, in the chalcone isomerase *chiA* gene, produces yellow or greenish pollen as a result of the accumulation of yellow chalcones mixed with anthocyanins (van Tunen *et al.*, 1991). Although such mutants show normal pollen development and fertility, other classes of phenylpropanoids are known to play an important role in pollen fertility in (reviewed in Taylor and Hepler, 1997). Maize and petunia mutants deficient in chalcone synthase (CHS) expression in anthers, produce flavonol-deficient pollen that is defective in germination and pollen tube growth (Taylor and Jorgensen, 1992; van der Meer *et al.*, 1992; Napoli *et al.*, 1999). This defect can be rescued by the application of classes of aglycone flavonols, such as kaempferol and quercitin (Mo *et al.*, 1992; Vogt *et al.*, 1995; Napoli *et al.*, 1999), which are thought to play a signalling role in pollen tube growth.

Potential targets of flavonol signalling in pollen could be genes involved in pollen germination and tube growth. To identify such genes, Guyon *et al.*, (2000) used a subtracted library of conditional male fertile (CHS-deficient) petunia supplied with flavonols to identify 22 petunia germinating pollen clones (Guyon *et al.*, 2000). All but two were expressed exclusively in pollen, and transcript levels of all clones increased in response to kaempferol. Strongly induced transcripts encoded proteins with putative regulatory or signalling functions in pollen tube growth. These included a leucine-rich repeat protein (S/D4), a LIM-domain protein (S/D1), a putative Zn^{2+} finger protein (D14) with a haem-binding site and a predicted translation product (S/D3) with features in common with a neuropeptide that regulates guidance and growth in the tips of extending axons.

Flavonol synthesis appears to involve both sporophytic and gametophytic control. In addition to CHS, phenylalanine ammonia-lyase (PAL), which catalyses the first committed step in the phenylpropanoid pathway, is an important regulatory point (Bate *et al.*, 1994). PAL enzyme was localized to the tapetal cells, and PAL activity in immature anthers was positively correlated with pollen fertility in cytoplasmic male sterile *Brassica oleracea* L. (Kishitani *et al.*, 1993). Furthermore, tapetum-specific sense and antisense expression of a sweet potato PAL cDNA in tobacco resulted in abnormal flavonol-deficient pollen and partial male sterility (Matsuda *et al.*, 1996). A putative anther-specific transcriptional activator (NtmybAS1) of PAL has recently been identified in tobacco (Sweetman and Twell, genbank accession AF198498). NtmybAS1 is encoded by an anther-specific myb-related gene, which is coordinately expressed with *PAL* in the tapetum, and can transactivate native *PAL* promoters in transgenic tobacco (W.-C. Yang, S.A. Amirsadeghi and D. Twell, unpublished).

In contrast, gametophytic transcription of genes encoding two enzymes involved in later steps of flavonol synthesis, flavanone 3-hydroxylase and flavanol synthase, was demonstrated in microspores and immature bicellular pollen of potato (van Eldik *et al.*, 1997). In species such as maize and petunia, in which the primary role of pollen flavonols appears to be in pollen germination and tube growth, flavonols exist *in planta* as both the aglycone and glycosyl conjugates. Recently, a gametophytically expressed pollen-specific flavonol 3-*O*-galactosyltransferase from petunia that could control the formation of pollen-specific glycosylated flavonols has been identified (Miller *et al.*, 1999).

Flavonols may also play some role in pollen maturation, since specific flavonols accelerate the maturation and germination of tobacco pollen cultured *in vitro* (Ylstra *et al.*, 1992). However, the flavonol-deficient maize and petunia *chs* mutants reveal that flavonols are not vital for determinative events such as asymmetric division, generative cell morphogenesis and division. Furthermore, the complete absence of flavonoids in the *Arabidopsis tt4* (*chs*) mutant does not affect pollen fertility (Burbulis *et al.*, 1996), so the flavonol requirements can also differ between species.

4.4.4   The pollen coat

The exine of the mature pollen grain is often covered by a lipid-rich coat, which has a several potential functions. These include UV protection, pollinator attraction, pollen–pollen and pollen–stigma adhesion, pollen hydration and pollen–pistil signalling (for recent reviews see Pacini, 1997; Dickinson *et al.*, 2000). According to Pacini (1997), the pollen coat may consist of simple pollen coatings (pollenkitt) rich in lipids and pigments, formed from the complete degeneration of the tapetum, or more complex coatings (tryphine) resulting from partial degeneration of the tapetum, to form a dense matrix including intact organelles.

The tryphine of Brassicaeae may comprise 10–15% of the mass of the pollen grain and consists of layers containing lipid, protein, glycoconjugates and pigments such as carotinoids and flavanoids. Lipids are thought to maintain the fluidity of the pollen coat and comprise nonpolar esters of medium- and long-chain fatty acids. Very long-chain lipids are also present that are essential for pollen hydration on the stigma. For example, mutations in the *Arabidopsis CER1* and *CER6* genes, which encode enzymes involved in the synthesis of very long-chain lipids on vegetative surfaces, and in the pollen coat, cause sterility (Preuss *et al.*, 1993; Aarts *et al.*, 1995; Hülskamp *et al.*, 1995).

Formation of the pollen coat is thought to involve two abundant lipid-rich organelles in the tapetum cells, the tapetosomes possessing oleosins and triacylglycerols (TAGs), and the elaioplasts having unique polypeptides and neutral esters (Ting *et al.*, 1998). In a recent detailed study, lipids and proteins of the tapetosomes and elaioplasts were shown to co-accumulate, but appear to be selectively degraded or retained during transfer to the pollen surface in *B. campestris* (Ting *et al.*, 1998).

Sporophytically expressed oleosin-like proteins, derived from the tapetosomes, also form a major component of the *Brassica* pollen coat (Ross and Murphy, 1996; Ruiter, 1997). In seeds, oleosins regulate oil body size and prevent lipid aggregation; therefore pollen coat oleosins may stabilize lipids in the pollen coat during dehydration and rehydration, and/or promote adhesion of the coat to the exine or stigma. The recent analysis of an *Arabidopsis* mutation in an oleosin domain protein (GRP17) provides conclusive evidence for the role of pollen coat oleosins in pollen–stigma interactions (Mayfield and Preuss, 2000). GRP17 (M_r 49 000) represents the most abundant pollen coat protein in *Arabidopsis* and contains a conserved oleosin domain and a highly charged domain. In a T-DNA insertion mutant in *GRP17*, removing the oleosin domain, pollen hydration was delayed on the pistil, but not *in vitro*. Therefore, GRP17 is an important pollen coat protein that affects the efficiency of pollen hydration. Grp17-1 mutants otherwise showed normal fertility, which could suggest functional redundancy with other oleosins present in the pollen coat (Mayfield and Preuss, 2000).

Several other pollen coat proteins such as aquaporins, water stress-inducible proteins and dehydrins may also play a role in hydration/dehydration steps. Pollen coat proteins also include enzymes, such as esterases and acid phosphatases (for review see Dickinson *et al.*, 2000). Although the pollen coat is reduced or absent in anemophilous species, proteins are present. For example, the predominant protein on the surface of maize pollen was identified as a tapetally expressed endoxylanase that could function in remodelling of pollen or stigmatic cell walls on germination (Bih *et al.*, 1999).

The pollen coat of *Brassica* also contains families of small highly charged, cysteine-rich proteins, termed pollen coat proteins (PCP), that function in adhesion and recognition. PCP-A1, with high affinity to S-locus glycoproteins (SLGs), although unlinked to the S locus, was isolated from *Brassica oleracea* (Doughty *et al.*, 1998). PCP-A1 transcripts accumulate specifically in pollen at the late binucleate stage, rather than in the tapetum, which previously was taken to be the principal source of the pollen coat. Using an optical biosensor immobilized with S gene family proteins, two SLR1-binding proteins, members of the PCP class A family (SLR1-BP1 and SLR1-BP2), were identified in pollen coat extracts of *Brassica campestris* (Takayama *et al.*, 2000a). Similarly to PCP-A1, the SLR1-BP gene was gametophytically expressed in maturing pollen; thus some pollen coat proteins clearly have a gametophytic origin. This raises the interesting possibility that phenotypically 'sporophytic' self-incompatibility (SI) may in fact be 'gametophytic' in terms of gene expression (Dickinson *et al.*, 2000). Indeed, Schopfer *et al.* (1999) reported the identification of the male determinant of SI as the *S-LOCUS CYSTEINE-RICH* (*SCR*) gene, which is quite similar to the PCPs. SCR is predicted to be located in the pollen coat, which was previously shown to be the location of the male determinant of SI in *Brassica oleracea* (Stephenson *et al.*, 1997). *SCR* was shown to be expressed gametophytically in microspores and sporophytically in the tapetal cells, although the physiological relevance of expression at both sites remains to be determined (Schopfer *et al.*, 1999; Takayama *et al.*, 2000b). Several excellent reviews provide further discussion of recent research on the pollen coat and cell-signalling in the SI response (Dickinson *et al.*, 2000; Heizmann *et al.*, 2000; Nasrallah, 2000).

4.4.5   Microspore nutrition and metabolism

Details of the dynamics of microspore nutrition remain largely unknown, although recent studies of genes involved in sugar and amino acid uptake and metabolism suggest both sporophytic and gametophytic regulation. In some species, such as tobacco, biphasic cycling of starch synthesis and degradation occurs such that starch is first synthesized in amyloplasts during microspore development, but is metabolized before pollen mitosis I (PMI). Starch synthesis is then often reinitiated during early–mid pollen development, and is metabolized during the final stages of maturation before dehydration (tobacco, *Brassica*). There is, however, considerable variation in patterns of storage and metabolism between different pollen species. In maize, abundant starch-filled amyloplasts persist in the mature pollen grains and are utilized during early pollen tube growth (Bedinger, 1992). Enzymes involved in starch synthesis (ADP-glucose pyrophophorylase) are activated in microspores in maize (Olmedilla *et al.*, 1991), and there is direct evidence for gametophytic control of starch synthesis from the phenotypic segregation of *waxy* mutations. More recently, reduced amylose synthesis was observed in transgenic rice pollen harbouring an antisense *Waxy* gene (Terada *et al.*, 2000). In other species, such as *Brassica*, few starch grains are present and lipid is accumulated and stored in dense ER membrane profiles. These are metabolized during rapid pollen tube growth, which involves extensive membrane synthesis. In a detailed study of lipid metabolism, developmental regulation of acyl carrier protein and other enzymes of lipid metabolism was demonstrated, suggesting that microspores, and in particular early bicellular pollen, synthesize and store large amount of lipid (Evans *et al.*, 1992).

The accumulation of sugars and amino acids, in addition to serving a storage and nutritional role, also represent compatible solutes, and may serve as osmoprotectants for cellular membranes and proteins during pollen dehydration. Proline represents the most abundant amino acid in pollen of petunia, tomato and many grass species (Schwacke *et al.*, 1999). Physiological evidence for the role of proline as an osmoprotectant is suggested from the positive correlation between proline content and pollen survival at high temperatures in sorghum (Zhang and Croes, 1983; Lansac *et al.*, 1996). In other species, such as spinach and wheat, glycine betaine accumulates and could play a similar role (see Schwacke *et al.*, 1999).

Mature and germinating pollen import external proline, so that sporophytic tissues are the most likely source of precursors of proline synthesis (Schwacke *et al.*, 1999). However, *tomPRO1* and *tomPRO2* mRNAs encoding a key enzyme of proline synthesis (δ-pyrroline-5-carboxylate synthetase) were not increased in developing anthers (Fujita *et al.*, 1998). Therefore, proline accumulation is proposed to occur via increased uptake activity through the gametophytic expression of an identified pollen-specific proline transporter (LeProT1) during pollen maturation (Schwacke *et al.*, 1999). Further metabolism of proline in pollen is thought to occur by dehydrogenation by proline dehydrogenase (ProDH), and the elevated expression of the tomato ProDH gene in pollen is consistent with an increased accumulation of proline in maturing pollen. Other potential transporters include a pollen-specific amino acid permease, NsAAP1, from *Nicotiana sylvestris* (Lalanne *et al.*, 1997). mRNA of *NsAAP1* increased in abundance during the first hour of germination and tube growth, suggesting a potential role in amino acid transport during pollen tube growth.

The control of lysine synthesis in anthers may be under both sporophytic and gametophytic control. Lysine synthesis in higher plants starts with the condensation of L-aspartate-β-semialdehyde (L-ASA) and pyruvate into dihydrodipicolinic acid. The enzyme that catalyses this step, dihydrodipicolinate synthase (DHDPS) is thought to have regulatory control in lysine synthesis. The *Arabidopsis* DHDPS promoter was expressed in vegetative tissues, and in anthers expression was found in the tapetum of young anthers and maturing pollen (Vauterin *et al.*, 1999).

The *Arabidopsis AtSTP2* gene encodes a high-affinity, low-specificity monosaccharide carrier that can transport hexoses and pentoses (Truernit *et al.*, 1999). *AtSTP2* mRNA and AtSTP2 protein first appear during callose degradation and microspore release from the tetrads. However, *AtSTP2* mRNA and AtSTP2 protein are no longer detected after the mitotic divisions, including mature or germinating pollen. The putative role of AtSTP2 may therefore be in the uptake of glucose units resulting from callose degradation during microspore release (Truernit *et al.*, 1999).

The high sucrose content of mature pollen has also been linked to desiccation tolerance and is suggested to be a key factor in preserving membranes in dry pollen (Hoekstra *et al.*, 1989). Gametophytically regulated sugar uptake occurs during pollen maturation and tube growth. *In vitro* germination assays showed that petunia pollen can utilize both sucrose and the monosaccharides glucose or fructose as carbon sources (Ylstra *et al.*, 1998). The *Petunia* pollen-specific cDNA *pmt1* encodes a putative monosaccharide transporter with a typical late pollen expression profile, and so could function in sugar uptake during late pollen development and during tube growth (Ylstra *et al.*, 1998). Furthermore, when pollen was cultured in a medium containing 2% sucrose there was a rapid and complete conversion of sucrose into equimolar amounts of glucose and fructose. This indicates the presence of wall-bound invertase activity and uptake of sugars in the form of monosaccharides by the growing pollen tube (Ylstra *et al.*, 1998). More recently, the *Arabidopsis* AtSUC1 protein has been characterized as a plasma membrane H^+-sucrose symporter (Stadler *et al.*, 1999). AtSUC1 was only detected in activated pollen and growing pollen tubes, suggesting that it may be involved in sucrose import for pollen tube growth, and/or in the cell-specific modulation of water potential (Stadler *et al.*, 1999).

An important concept that has emerged from research into anther culture and microspore/pollen embryogenesis is that nutrient levels can play an important role in redirecting microspore and pollen cell fate. From the mid-microspore stage until pollen maturity, the functions of the anther locule environment can be substituted *in vitro* with simple defined media containing essential inorganic nutrients, glutamine, sucrose and amino acids (Tanaka and Ito, 1980, 1981; Benito Moreno *et al.*, 1988; Tupy *et al.*, 1991). Sucrose and glutamine in particular were shown to maintain gametophytic development and therefore could be important sporophytically derived nutrient signals to the developing microspores and pollen.

In summary, the sporophytic anther tissues influence pollen development through effects on nutrition, osmotic and UV protection, and surface features, enabling effective pollen–pistil interactions. However, it is also apparent that male gametophytic development is to a large extent dependent upon the intrinsic gametophytic gene expression programme (section 4.8). This degree of autonomy presumably reflects the need to establish an independent genetic programme for the specialized development, structure and function of the pollen grain. Key cellular processes that specify microgametogenesis, from microspore release to pollen shed, are examined in the following discussion.

4.5 Microspore development

4.5.1 Microspore isolation

Microspore isolation after meiosis leads to the independent development of individual spores, with some variation in the rate of development, since spores show asynchrony. This may permit the haploid genomes to function independently of each other and of the diploid parent. In the majority of species, individual microspores are released from the tetrad by the action of callase, thought to be secreted from the tapetum (Steiglitz, 1977). However, microspores and pollen grains may remain attached in tetrads (e.g. *Caluna*), or in higher-order, polyad arrangements (e.g. *Acacia*), which are associated with a reduction in callose deposition at the tetrad stage (Blackmore and Crane, 1988). In massulate orchid species, microspore development and pollen mitosis I are highly synchronized within each massula, since intersporal connections are maintained between microspores (Heslop-Harrison, 1968).

Callose appears to function as a structural polymer enabling microspore isolation and facilitating separation, since defective or delayed callose dissolution in *petunia* male sterile mutants leads to failure of microspore separation (Izhar and Frankel, 1971). Callose may also play a role in development of the primary pollen wall, since premature callose dissolution leads to microspore collapse and defects in exine patterning (Worrall *et al.*, 1992). In the *Arabidopsis quartet* mutants, *qrt1* and *qrt2*, the microspores remain attached in a tetrad (Preuss *et al.*, 1994). However, failure to separate is not associated with defects in callose synthesis or degradation but with persistence of pectic polysaccharides surrounding the callose wall (Rhee and Somerville, 1998). Therefore, in addition to callose dissolution, degradation of pectic components surrounding the meiocytes and tetrads appears to be an important step in microspore release. *QRT1* and *QRT2* appear to function in this cell type-specific removal of pectin (Rhee and Somerville, 1998).

4.5.2 Vacuole morphogenesis, microspore growth and polarity

A conspicuous feature of microspore development is the carefully orchestrated process of vacuole morphogenesis (figure 4.1). Vacuole growth is likely to be important in achieving rapid expansion of the microspores, which increase in volume by a factor of 2 to 3. In newly released microspores, vacuoles are few and small, but these rapidly increase in size and number. During mid-microspore development, vacuoles enlarge and fuse to create several larger vacuoles, which subsequently fuse to produce a single vacuole occupying the majority of the cell volume (Owen and Makaroff, 1995).

In association with vacuole growth, the microspore nucleus is displaced to an eccentric position against the microspore wall. Vacuole growth and morphogenesis may therefore play a role in establishing cellular polarity through nuclear displacement. Ultrastructural analyses in *Brassica* and *Arabidopis* have revealed that nuclear migration begins before the single large vacuole is formed (Hause *et al.*, 1991; Owen and Makaroff, 1995), which suggest that nuclear movement may operate independently of vacuole growth. Even so, the vacuole is expected to provide stability in maintaining the eccentric position of the microspore nucleus. In *Brassica napus* and *Arabidopsis*, the vacuole is pressed firmly against the microspore nucleus, causing an indentation at the vegetative face (Hause *et al.*, 1991). In this regard, disturbances in vacuole morphogenesis could influence nuclear migration and/or positioning, which in turn could affect division asymmetry and cell fate (section 4.6). However, in some orchid species, such as *Phalaenopsis*, in which microspores remain attached as tetrads, vacuoles do not develop, excluding a role for vacuoles in nuclear positioning in this system (Brown and Lemmon, 1991a).

4.6 Asymmetric division, cell fate and polarity

Asymmetric division of the microspore at pollen mitosis I (PMI), by definition, signifies the termination of microspore development and the initiation of pollen development. PMI is an intrinsically asymmetric division, after which daughter cells are immediately different (Twell *et al.*, 1998). The large vegetative cell (VC) shows dispersed nuclear chromatin and constitutes the bulk of the pollen cytoplasm. In contrast, the small generative cell (GC) shows condensed nuclear chromatin and contains relatively few organelles and stored metabolites. Whereas the vegetative cell exits the cell cycle, the generative cell completes a further mitotic division to form the two sperm cells (figure 4.1). This dimorphism is also reflected in the expression of specific genes. For example, the tomato pollen-specific *LAT52* promoter drives expression specifically in the vegetative cell after pollen mitosis I, and provides a gametophytic cell fate marker (figure 4.3; Eady *et al.*, 1995).

Manipulation of division asymmetry using microtubule inhibitors, or by centrifugation, has demonstrated that division asymmetry at PMI has a vital role in cell fate determination (Tanaka and Ito, 1980, 1981; Terasaka and Niitsu, 1987; Zaki and Dickinson, 1990, 1991; Eady *et al.*, 1995; Touraev *et al.*, 1995; Zonia *et al.*, 1999). For example, induction of symmetrical division in tobacco produces two apparently identical cells that resemble vegetative cells and express the VC-specific *lat52-gus* marker (figure 4.3b; Eady *et al.*, 1995). Blocking division at PMI produced 'uninucleate pollen grains' that still expressed *lat52-gus* (figure 4.3c). Together, these data demonstrate that asymmetric division is required for generative cell differentiation and that vegetative cell-specific gene activation can be uncoupled from nuclear division and cytokinesis. These studies also led to the hypothesis that vegetative cell genes are controlled by gametophytic transcription factors, which normally reach a threshold of activity at pollen mitosis I (figure 4.4). The activation of such factors, which are proposed to act as vegetative cell fate determinants, must also occur independently of nuclear division and cytokinesis (Eady *et al.*, 1995).

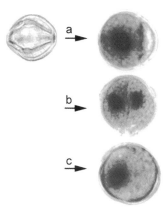

Figure 4.3 Manipulation of division asymmetry and cell fate in tobacco microspores. Uninucleate microspores (top left image) matured *in vitro* undergo asymmetric division (a) to produce a large vegetative cell expressing the nuclear-localized *LAT52-gus/nia* marker and a smaller generative cell in which *LAT52* is repressed. (b) Maturation in the presence of low levels of colchicine induces symmetric divisions with *LAT52* expression in both daughter cells. (c) In the presence of high levels of colchicine, division is blocked but *LAT52* is still activated.

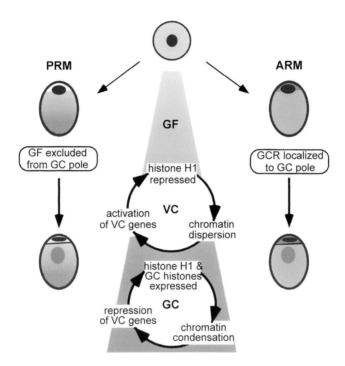

Figure 4.4 General models of pollen cell-fate determination. In both models, vegetative cell-specific gene expression is activated via the accumulation of gametophytic transcription factors (GF). In the passive repression model (PRM) vegetative cell-specific genes are repressed in the generative cell as a result of exclusion of GF from the generative cell pole. In the active repression model (ARM) localization of a generative cell repressor (GCR) to the generative cell pole blocks expression. Furthermore, attenuation of histone H1 levels in the vegetative cell is proposed to allow chromatin dispersion and vegetative cell-specific gene expression via a positive feedback loop. Conversely, maintenance of histone H1 and expression of histone variants in the generative cell would block chromatin dispersion and activation of vegetative cell-specific genes.

4.6.1 Asymmetric cytokinesis—sealing cell fate

Asymmetric cytokinesis following PMI effectively seals the fate of the smaller generative cell. Given the specialized requirements of this event, it is not surprising that cytokinesis at PMI differs from that in somatic cells. First, no preprophase band of microtubules marks the future division plane, and second, although a typical centrifugally expanding cell plate develops in the internuclear zone, a unique curved cell plate is formed to enclose the eccentric generative nucleus. Curved profiles of phragmoplast microtubules guide the growth of the cell plate at its margins during the second phase (Brown and Lemmon, 1991a, 1992, 1994; Teraska and Niitsu, 1990, 1995). Therefore, the cell plate at PMI is formed in a two-phase process involving dynamic reorganization of phragmoplast microtubules. Gametophytic mutants that fail to complete cytokinesis at PMI (*tio*), or that disturb cell plate position and growth (*gem1*), may reveal molecular components that control this unique division (see section 4.6.5).

4.6.2   Microspore and pollen embryogenesis

An alternative pathway of development can be induced in cultured microspores and bicellular pollen leading to embryogenesis (Touraev *et al.*, 1997). In *Brassica* this developmental switch is induced by high-temperature stress (Keller and Armstrong, 1979), whereas in tobacco it is induced by a sucrose and/or nitrogen starvation treatment (Touraev *et al.*, 1996). Although embryogenic development is marked by a symmetrical first cell division (Zaki and Dickinson, 1990, 1991), division symmetry does not appear to be causally related to embryogenic induction. For example, in the absence of stress, markers of gametophytic development are expressed in cultured tobacco microspores even after induction of symmetric divisions (Eady *et al.*, 1995; Touraev *et al.*, 1995). Whereas generative cell development is strictly dependent upon asymmetric division, external conditions (i.e. the absence of stress) appear more important than division symmetry for the maintenance of vegetative cell fate. The switch to embryogenic development might therefore involve the destruction or inactivation of gametophytic transcription factors through nutrient signalling. Simultaneous signalling to the cell cycle machinery may also be involved since the induction of pollen embryogenesis leads to derepression of the vegetative cell cycle (Zarsky *et al.*, 1992). Current effort is focused on identifying genes that are potential targets for stress-induced signalling during the embryogenic induction of microspores (Touraev *et al.*, 1997). In this regard, Kyo *et al.* (2000) recently purified three phosphoproteins from tobacco that appeared specifically during embryogenic induction of microspores by glutamine starvation. These NtEPs show moderate homology with type-1 copper-binding glycoproteins and with an early nodulin. The expression of NtEPc was correlated with dedifferentiation, but not with pollen development or cell division, providing an early marker for embryogenic induction.

4.6.3   Models of cell-fate determination

Two general models have been proposed to account for differential gene expression resulting from asymmetric division at PMI (Eady *et al.*, 1995). Both assume that vegetative cell gene activation is the default pathway resulting from the accumulation of a gametophytic factor, and provide alternative mechanisms to explain how vegetative cell-specific genes are repressed in the generative cell. In the passive repression model a gametophytic factor is simply excluded from the generative cell pole. However, in the active repression model a hypothetical generative cell repressor, at the generative cell-pole, acts to block vegetative cell-specific gene activation (figure 4.4).

These models must also take into account the condensed generative cell chromatin compared with that of the vegetative cell (figure 4.1b). Chromatin condensation is likely to limit the access of transcription factors and might be directly involved in the initiation and/or maintenance of transcriptional repression. Differential chromatin condensation is first observed during late anaphase–early telophase before the chromosomes are enclosed within the daughter nuclei (Terasaka and Tanaka, 1974), suggesting cytoplasmic polarity immediately before division. Unequal chromatin behaviour within the 'generative cell domain' could therefore involve a generative cell pole-localized 'condensation factor' that may be equivalent to the hypothetical generative cell repressor (figure 4.4).

Proteins known to regulate chromatin packaging and gene expression include histones or histone-associated factors. The identification of generative cell-specific core histones gH2B and gH3 in lily (Ueda and Tanaka, 1994, 1995a,b), and the vegetative cell-specific decline in histone H1 levels during pollen maturation, support this hypothesis (Tanaka, 1997; Tanaka et al., 1998). However, species-specific patterns of histone H1 regulation may operate, since in tobacco histone H1 levels appear to be maintained in the vegetative nucleus (Oakeley et al., 1997). This study also showed a specific decrease in the degree of cytosine methylation in the generative nucleus compared with the vegetative nucleus, which could also play a role in differential gene expression and chromatin dispersion.

4.6.4 Determination of microspore polarity

Given the critical role of asymmetric division in the control of pollen cell fate, how is microspore polarity achieved? The proposed role of the polarity system is twofold: first, to ensure nuclear migration; and second, to ensure that the correct spindle axis is established, perpendicular to the microspore wall (Twell et al., 1998). The polar axis of the microspore is defined by the proximal–distal axis of the microspore in the tetrad (Erdtman, 1952). The direction of nuclear migration can be either along this axis towards the inner (proximal) or outer (distal) walls of the tetrad, or perpendicular to the polar axis, towards a radial wall (Maheshwari, 1950; Park et al., 1998). So, although the direction of polarity can differ between species, nuclear migration appears to be predetermined, towards a fixed site. In Arabidopsis microspores, the polar axis is aligned with the long axis of the three germination pores in the radial walls—parallel to the proximal–distal axis. Nuclear migration is always towards a radial wall and the plane of division is roughly aligned with the polar axis (Park et al., 1998). Therefore, hypothetical polarity determinants controlling nuclear migration could be asymmetrically distributed as a gradient between the radial walls, or located in the microspore wall, in the plasma membrane, or in a cortical domain at the future generative cell pole.

The localization of a microspore polarity determinant could be an early event generated during microsporogenesis and/or it could be intrinsically programmed by early gametophytic gene expression. Early determination is attractive, as in most dicot species meiotic cytokinesis is simultaneous resulting in 'quadri-partitioning' of the postmeiotic cytoplasm into four equal 'spore domains' (Brown and Lemmon, 1991b). Intrinsic polarity could be generated within these domains during meiotic phragmoplast development, which involves specialized microtubule and microfilament arrays (Van Lammeren et al., 1989; Brown and Lemmon, 1991b). Such systems could deliver polarity determinants to membrane or wall locations. Evidence for early cytoplasmic polarity is provided by observations that show that future sites of apertures are defined by localized ER associated with the plasma membrane. These polar ER configurations appear to be dependent upon spindle orientation at meiosis, since disruption of meiotic spindles prevents aperture formation or leads to apertures at incorrect positions (Heslop-Harrison, 1971).

The cytoskeleton is also likely to play a central role in determining both nuclear migration and the division plane. First, the microspore nuclear migration is microtubule (Mt)-dependent (Terasaka and Niitsu, 1990; Eady et al., 1995). Evidence for a polar system that could direct nuclear migration comes from studies of the orchid Phalaenopsis, in which a specialized generative pole microtubule system is assembled at the future generative pole (Brown and Lemmon, 1991a). Similar polarized Mt systems have not been observed in other species, although Mt configurations do show dynamic changes associated with nuclear migration (Teraska and Niitsu, 1990; Hause et al., 1991; Brown and Lemmon, 1994; Gervais et al., 1994). In a recent study in tobacco, an asymmetric Mt cage with a tether at one end was observed enclosing the microspore nucleus (Zonia et al., 1999). The tether appears to contact first the radial wall and then the pole, according to the direction of nuclear migration. This suggests that independent Mt-dependent mechanisms may operate to control nuclear migration in different species. However, maintainence of nuclear position appears to involve both microtubules (Teraska and Niitsu, 1990; Tanaka, 1997) and microfilaments (Gervais et al., 1994; Zonia et al., 1999). An actin net surrounds the nucleus at the GC pole in tobacco (Zonia et al., 1999), and individual Mts have also been observed 'tethering' the microspore nucleus to the plasma membrane (Hause et al., 1991; Brown and Lemmon, 1992).

4.6.5 Mutants affecting microspore polarity, division and cell fate

Genetic screens in Arabidopsis have led to the isolation of a range of mutations that affect pollen cell divisions at PMI (Chen and McCormick, 1996; Howden et al., 1998; Park et al., 1998; Twell and Howden, 1998a,b; Twell, 1999). The analysis of these mutants confirms the requirement of asymmetric division for generative cell differentiation, and provides evidence in support of a model of polarity determination by asymmetrically localized factors.

Mutations in the *STUD* and *TETRASPORE* genes act sporophytically to prevent the formation of intersporal callose walls after meiosis II (Hülskamp *et al.*, 1997; Spielman *et al.*, 1997). This creates 'coenocytic' microspores composed of four nuclei within a common cytoplasm. Up to four sperm cell-pairs have been observed in mature *stud* and *tes* pollen grains, demonstrating that all four nuclei can undergo nuclear migration and asymmetric division. The *stud* and *tes* mutants demonstrate that meiotic cytokinesis and cytoplasmic isolation are not prerequisites for microspores to achieve polarity, and suggest that polarity can operate independently within each spore domain.

The gametophytic mutant *sidecar pollen* (*scp*) specifically affects microspore division symmetry (Chen and McCormick, 1996). In plants carrying *scp*, pollen is either aborted, or shed in a divided condition with an extra vegetative-like cell (figure 4.5). Mutant *scp* microspores first undergo a premature symmetrical division at the microspore stage, followed by one of the daughter cells dividing asymmetrically (figure 4.5). This suggests that polarity is not expressed at the time when *scp* microspores divide, but polarity is not defective, since it is re-established in one daughter cell. This supports the asymmetric localization of a polarity determinant, before division in *scp*. *SCP* may encode a repressor of cell division normally active until after the microspore achieves polarity, or *scp* may act as a gain-of-function mutation that promotes premature entry into mitosis (Chen and McCormick, 1996). *scp* also provides evidence for the importance of coordinating the mitotic cell cycle with the expression of polarity to ensure correct asymmetric division at PMI.

The gametophytic mutant, *gemini pollen 1*, displays a range of pollen phenotypes characterized by equal, unequal and partial divisions at PMI (Park *et al.*, 1998). In contrast to *scp*, symmetrical divisions in *gem1* do not occur precociously and neither daughter cell polarizes or completes a further asymmetric division (figure 4.5). Cell fate analysis in *gem1* revealed a quantitative relationship between vegetative cell fate and division symmetry. This supports the role of cell size or the nuclear/cytoplasmic ratio as a factor in determination of cell fate. *gem1* may result directly from a lesion in a localized polarity determinant or indirectly in genes required for the expression of polarity. In this regard, nuclear migration appears to be incomplete or unstable immediately before PMI. A further interesting phenotype in the *gem1* mutant is that nuclear division is often uncoupled from cytokinesis, producing enucleate cytoplasmic compartments (Park *et al.*, 1998; Park and Twell, 2001). These 'uncoupling' phenotypes demonstrate that, although cytokinesis and karyokinesis are normally tightly linked, these processes can operate independently. In this context, *GEM1* may function in the spatial coordination of nuclear and cytoplasmic division.

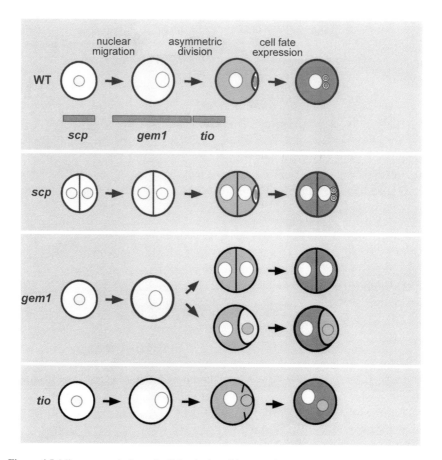

Figure 4.5 Microspore polarity and cell fate in the wild type and *scp*, *gem1* and *tio* mutants. Premature division and polarization of only one daughter cell in *scp* suggests that polarity is asymmetrically determined before microspore division in *scp*. In *gem1* equal and unequal divisions occur at the correct time at PMI and result from unstable nuclear migration and/or cytokinesis. Smaller daughter cells in unequally divided microspores do not polarize or divide further and show intermediate cell fate. Intermediate cell fate may be caused by aberrant segregation of gametophytic factors (see figure 4.4). In *tio* microspores polarity is normal, but cytokinesis at PMI is defective. Resultant binucleate daughter cells express the default vegetative cell fate, but one nucleus remains relatively condensed, suggesting differential inheritance of nuclear factors. Cytoplasmic shading indicates the intensity of *LAT52-gus* (vegetative cell fate) expression. Nuclear shading indicates the degree of chromatin condensation (generative cell fate).

In *two-in-one pollen* (*tio*) (previously termed *solo pollen*; Twell and Howden, 1998a), which acts gametophytically, mature pollen contains two nuclei resulting from failure of cytokinesis at PMI (figure 4.5). *tio* spores undergo vegetative cell maturation and express the *lat52-gus* marker, supporting the hypothesis that vegetative cell fate is the default programme (Twell *et al.*, 1998). However, one nucleus remains relatively condensed, suggesting differential inheritance of nuclear factors promoting and/or limiting chromatin dispersion. *tio* pollen mutants may result from lesions in gametophytically expressed structural or regulatory components of the cytokinetic apparatus.

4.7 Generative and sperm cell development

The generative cell (GC) divides within the vegetative cell (VC) cytoplasm and is dependent on the vegetative cell cytoplasm for its support and nutrition (reviewed in Chaboud and Perez, 1992; Palevitz and Tiezzi, 1992). Direct evidence for this dependence was provided by targeted cell ablation in which the vegetative cell was ablated by the expression of the cytotoxic diphtheria toxin A chain (DTA), under the control of the vegetative cell-specific *LAT52* promoter (Twell, 1995). Expression of *LAT52-DTA*, led to rapid loss of membrane integrity and death of the VC; however, the GC remained viable for up to 24 h. Furthermore, the GC failed to move into the degenerating VC, demonstrating the dependence of migration on the presence of an intact VC.

4.7.1 Generative cell migration

Generative cell migration is a uniquely specialized cell–cell cooperation that creates a 'cell within a cell' structure, which has evolved to enable internalization and transport of the male gametes within the pollen tube. Generative cell migration follows asymmetric partitioning of the microspore cytoplasm at PMI by the hemispherical callose wall. This callose wall is subsequently degraded, presumably involving targeted secretion of $\beta(1\text{-}3)$-glucanases, allowing the generative cell to detach from the intine and move inwards. The cytoskeleton is also implicated in generative cell migration, since Mt arrays are strongly associated with the internal membrane of the GC and the GC face of the vegetative nucleus immediately after PMI (Zonia *et al.*, 1999).

Recent results suggest that a mutational approach may be a valuable tool to identify cellular components involved in this unique intracellular migration. The gametophytic mutant *limpet pollen* (*lip*) blocks generative cell migration after pollen mitosis I and generative or sperm cells remain at the periphery (Howden *et al.*, 1998). It is significant that division of the GC to form the sperm cells is commonly completed in *lip*, demonstrating that PMII is not strictly dependent on the position of the GC in the vegetative cytoplasm. Since peripheral generative cells appear to be separated from the vegetative cell by persistent wall material, it is conceivable that *LIP* regulates the delivery of $\beta(1\text{-}3)$ glucanases required for degradation of the transient callose wall. Alternatively, *LIP* may encode a factor involved in the cytoplasmic reorganization required for cell migration.

4.7.2   Generative cell morphogenesis

After detachment from the pollen wall, the GC rounds up to become spherical. Subsequently, the generative cell undergoes further morphogenesis to form an elongated, lenticular or spindle shape (reviewed in Palevitz and Tiezzi, 1992). Generative cell morphogenesis is associated with the reorganization of Mt arrays within the cortical GC cytoplasm. Following generative cell division, the newly formed sperm cells also show similar elongated profiles, associated with basket-like arrays of Mts aligned with their long axes. The importance of internal Mt bundles in maintaining GC shape is suggested by the fact that Mts in elongated GCs are rapidly disorganized by Mt inhibitors *in situ*, leading to a spherical shape (Sanger and Jackson, 1971; Heslop-Harrison *et al.*, 1988). However, Mt arrays are not sufficient to maintain cell shape after isolation of GCs. Therefore, unknown interactions between the VC cytoplasm and GC surface may be involved in signalling Mt organization in the generative and sperm cells. Although initially controversial, the weight of evidence has demonstrated the absence of F-actin in the generative and sperm cells, excluding an internal role in GC morphogenesis (Palevitz and Liu, 1992).

Mitochondria and sometimes plastids are segregated into the GC at PMI. Detailed studies of the developmental fate of plastid and mitochondrial DNA in the GC provide evidence for differential regulation of DNA synthesis and degradation in mitochondria and plastids between species. In a recent study, four groups of species showed patterns of organelle DNA degradation that were correlated with different modes of cytoplasmic inheritance of organelles (discussed in Nagata *et al.*, 1999).

4.7.3   Generative cell division

Generative cell division takes place suspended within a membrane-bounded compartment of the vegetative cytoplasm. DNA replication takes place soon after GC detachment, whereas the vegetative nucleus remains arrested in G_1 (Zarsky *et al.*, 1992). Recent evidence for the strict regulation of the cell cycle in generative and sperm cells has come from studies of the DNA content of sperm cells in *Arabidopsis* (Friedman, 1999). The DNA content of *Arabidopsis* sperm was approximately 1.5C, suggesting that DNA replication is incomplete in mature pollen grains. S phase was sustained throughout progamic development and not completed until after fertilization. The observed variation in the regulation of generative and sperm cell cycle in different species could therefore be an important determinant of interspecific incompatibility (Friedman, 1999). Although the regulatory factors that control the generative cell division cycle are unknown, RT-PCR studies of isolated maize sperm cells demonstrated the expression of the cyclin-dependent kinase, A1 cyclin, and histone H3 (Sauter *et al.*, 1998). Two other maize B group cyclins, however, were not expressed in sperm, suggesting cell type-specific regulation of cyclins in sperm.

Structural studies demonstrate the presence of Mts in the GC mitotic apparatus, but F-actin is absent from the spindle, or from the phragmoplast that forms in some species at pollen mitosis II (see Palevitz and Tiezzi, 1992). This may reflect the key role of actin in the structural maturation of the cell plate in somatic cells. Given the apparent simple callosic cell plate at PMII, the lack of structural maturation may preclude the requirement for actin in the cell plate. In other species the generative cell divides by constriction rather than by conventional centrifugal phragmoplast growth. Both observations suggest that specialized mechanisms of cytokinesis operate during GC division.

A further important aspect of GC division is that it is asymmetric in some species. Sperm cell dimorphism, based upon cell and/or nuclear morphology and organelle content, has given rise to the concept of the 'polarized fertilization-unit' (reviewed in Mogensen, 1992). Thus, the sperm cells are proposed to be predetermined to fuse either with the egg nucleus or with the two polar nuclei in the embryo sac. Although there have been a number of reports describing species with dimorphic sperm, preferential fertilization, correlated with sperm cell dimorphism, has been demonstrated in only two species, *Zea mays* (Roman, 1948) and *Plumbago zeylanica* (Russell, 1985). In *Zea mays*, the sperm cell carrying an extra set of supernumerary B chromosomes preferentially fertilizes the eggs cell. In *Plumbago*, the sperm cells are dimorphic with regard to size and organelle content. The larger plastid-rich sperm cell preferentially fuses with the egg cell, and the plastid-free, mitochondria-rich sperm cell with the central cell (Russell, 1985). Current efforts are directed towards identifying asymmetrically expressed sperm cell genes that may be involved in preferential gamete recognition events.

The selection of mutants specifically affected in generative cell division provides a direct approach to identify genes regulating microgametogenesis. Male-specific gametophytic mutants have been isolated in maize, *gaMS-1* (Sari-Gorla *et al.*, 1996), and *Arabidopsis*, *mad1-3* (Grini *et al.*, 1999), that are arrested at the bicellular stage. However, pleiotropic aberrant pollen phenotypes suggest that failure of GC division results indirectly from vegetative cell defects. In contrast, morphological screening in *Arabiodpsis* led to the isolation of highly penetrant *duo pollen* (*duo*) mutants that produce mature pollen in a bicellular condition, containing a single vegetative nucleus and a generative-like cell (Twell and Howden, 1998a,b). Vegetative cell development appears to be otherwise normal and pollen is able to germinate normally *in vitro* (A. Durbarry and D. Twell, unpublished). *DUO* genes may therefore act specifically in the generative cell. The selection of such mutations suggests that specialized gametophytic factors, and/or regulatory mechanisms, have evolved to specify GC division in *Arabidopsis*.

4.7.4   The male germ unit

A key structural assemblage formed within the male gametophyte is known as
the male germ unit (MGU), which consists of a physical association between
the gametic cells (generative cell or sperm cell pair) and the vegetative nucleus
(Dumas *et al.*, 1984). The MGU contains all the nuclear and cytoplasmic
DNA of heredity, and is transported within the pollen tube to the embryo sac.
MGUs have been observed in mature tricellular pollen of a number of dicot
species. In mature bicellular pollen systems, and in tricellular pollen of monocot
species, such as maize, the MGU is not clearly assembled in mature pollen, but
assembles early during pollen tube development (reviewed in Mogensen, 1992).
Despite apparent differences in the timing of MGU assembly, the existence of
angiosperm MGUs is now widely accepted (Dumas *et al.*, 1998). Although yet
to be demonstrated, the proposed function of the male germ unit is to ensure the
coordinated delivery of the two sperm cells, and synchrony of the two sperm
cell fusion events with egg and central cell (Dumas *et al.*, 1984, 1998).

According to current models of gamete transport, generative and sperm cells,
which are coated in myosin, are proposed to be transported along cortical F-actin
bundles in the pollen tube cytoplasm (Heslop-Harrison, 1989; Heslop-Harrison
and Heslop-Harrison, 1989). Isolated sperm cells can be transported along the
actin cytoskeleton *in vitro* or when injected into algal cells (Russell, 1996).
Therefore, sperm cell transport may involve an actomyosin-like motility system
similar to that described in animal cells. Further evidence for the dependence
of MGU movement on ATP-dependent actomyosin interactions is provided by
inhibition of MGU transport by cytochalasin B treatment (Heslop-Harrison and
Heslop-Harrison, 1989).

4.7.4.1   Male germ unit morphogenesis
Since the discovery of the MGU, most studies have been devoted to the
ultrastructural characterization, but not to the process of assembly. MGU assembly during pollen germination has been described in species with bicellular
pollen (Hu and Yu, 1988; Rougier *et al.*, 1991). However, the nature of sperm
cell associations in a tricellular pollen species has been only briefly described
in *Brassica* (Dumas *et al.*, 1985; Charzinska *et al.*, 1989). A key question is
whether associations between the three components exist early during bicellular
development, or whether they are established later.

In *Arabidopsis*, which possesses tricellular pollen, the MGU develops during
pollen maturation within the anther and comprises the associated vegetative
nucleus and sperm cell pair positioned centrally within the cytoplasm (figure 4.6;
E. Lalanne and D. Twell, unpublished). Screening for nuclear morphogenesis
mutants in *Arabidopsis* has led to the isolation of two distinct classes of mutants
that affect the integrity or the positioning of the MGU respectively: *germ unit
malformed* (*gum*) and *male germ unit displaced* (*mud*) (figure 4.6). Mutants in

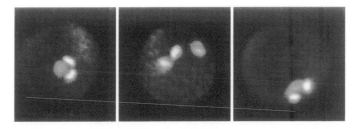

Figure 4.6 Mutants affecting male germ unit morphogenesis in *Arabidopsis*. Optical sections through mature DAPI-stained pollen of wild type (left) and *gum1* (*germ unit malformed*) and *mud1* (*male germ unit displaced*) mutants.

both classes act gametophytically and are fully penetrant such that heterozygous mutants produce 50% aberrant pollen. Both mutations specifically reduce pollen transmission and, when homozygous, show no sporophytic phenotypes, suggesting that *GUM* and *MUD* act specifically in the male gametophyte. The reduced male transmission of *mud* and *gum* mutants supports the hypothesis that the integrity and positioning of the MGU are important for pollen competitive ability.

Molecular biology and genetics

4.8   Male gametophytic gene expression

There is now overwhelming evidence that pollen development requires postmeiotic gene expression from the haploid genome. Early evidence was largely derived from genetic studies, including rare gametophytic mutants, deviations from Mendelian segregation ratios and gametophytic segregation of pollen isozymes (reviewed in Ottaviano and Mulcahy, 1989). The most direct and conclusive evidence for postmeiotic transcription has come from studies of the expression of specific genes and their associated regulatory sequences. A number of reviews chart progress made during the past decade (Bedinger, 1992; Mascarenhas, 1990, 1992; McCormick, 1991; Twell *et al.*, 1994; Sari-Gorla and Pe, 1999).

What has emerged is that a strikingly large proportion of the transcriptome appears to be gametophytically expressed. In the region of 20 000 unique mRNAs are estimated to be present in the pollen grain at maturity, and at least 10% of these appear to represent pollen-specific transcripts. From a handful of isolated pollen-expressed genes in 1990, this has expanded to over 50 distinct classes comprising approximately 150 unique genes (table 4.1). Here, current knowledge of the repertoire of gametophytically expressed genes is reviewed,

Table 4.1 Compilation of cloned microspore and pollen-expressed sequences

Function group	Species	Gene[1,2]	Specificity[3]	Reference
PECTIN METABOLISM				
Pectate lyase	Le	LAT56, LAT59	P, A	Wing *et al.* (1989)
	Aa	Amb a I/Amb a II*	P	Rafnar *et al.* (1991)/ Rogers *et al.* (1991)
	Nt	TP10,G10	**P**	Rogers *et al.* (1992)
	Zm	Zm58	**P**	Turcich *et al.* (1993)
	Nt	Nt59	P	Kulikauskas and McCormick (1997)
	At	At59	P	Kulikauskas and McCormick (1997)
Polygalacturonase	Oo	P2	P	Brown and Crouch (1990)
	Zm	PGI	P	Niogret *et al.* (1991)
	Zm	3C12	P	Allen and Lonsdale (1992, 1993)
	Zm	ZmPG	P	Barakate *et al.* (1993)
	Bn	Sta-44	P	Robert *et al.* (1993)
	Gh	G9	P	John and Peterson (1994)
	Nt	NPG1	P	Tebbut *et al.* (1994)
	Ms	P73	P	Qui and Erickson (1996)
	Sg	SgPG1, 2, 3, 4	**P**	Futamura *et al.* (2000)
Pectin esterase	Bn	Bp19	**P**	Albani *et al.* (1991)
	Ph	PPE1	P	Mu *et al.* (1994b)
	Zm	ZmC5	P	Wakeley *et al.* (1998)
	Sg	SgPME1	**P**	Futamura *et al.* (2000)
CYTOSKELETON				
Actin	Nt	TAC25	P	Thangavelu *et al.* (1993)
	At	ACT4/ACT12	P	Huang *et al.* (1996a)
	At	ACT1, ACT3	P	An *et al.* (1996)
	At	ACT11	P, V	Huang *et al.* (1997)
Profilin	Bv	Bet v II*	**P**	Valenta *et al.* (1991)
	Zm	**ZmPRO1, 2, 3**	P	Staiger *et al.* (1993)
	Pp	PpPFN	P	Valenta *et al.* (1994)
	Nt	NtPro	P	Mittermann *et al.* (1995)
	At	PFN4	P	Christensen *et al.* (1996)
	At	AthPRF4	P	Huang *et al.* (1996b)
	Zm	ZmPRO4	P, V	Gibbon *et al.* (1998)
	Le	LePro 1	P	Yu *et al.* (1998)
	Ha	Hel a 2*	P	Asturias *et al.* (1998)
	Ma	Mer a 1*	P	Vallverdu *et al.* (1998)
ADF-like	Ll	LMP131A	A	Kim *et al.* (1993)
	Bn	BMP1	A	Kim *et al.* (1993)
	Zm	ZmABP1, ZmABP2	P	Lopez *et al.* (1996)
α-Tubulin	At	TUA1	M, P	Carpenter *et al.* (1992)

Table 4.1 (continued)

Function group	Species	Gene[1,2]	Specificity[3]	Reference
β-Tubulin	Zm	TUB3, 4, 5	P, V	Rogers et al. (1993)
	Zm	TUB, 1, 4, 6, 7, 8	P, V	Villemur et al. (1994)
PROTEIN KINASES				
Receptor-like kinase	Pi	**PRK1**	P, O	Mu et al. (1994a); Lee et al. (1997)
	Le	LePRK1, LePRK2	**P**	Muschietti et al. (1998)
	At	RKF1	P, St	Takahashi et al. (1998)
CDPK	Zm	**CDPK**	P	Estruch et al. (1994)
CCaMK	Nt	TCCaMK-1/2	A	Liu et al. (1998)
	Ll	CCaMK	PMC, T, M	Poovaiah et al. (1999)
MAP kinase	Nt	NTF4	P	Wilson et al. (1997)
MAPK kinase kinase	At	AtMAP3Kgamma	P, V	Jouannic et al. (1999)
GSK3/shaggy protein kinase	Ph	PSK6	P	Tichtinsky et al. (1998)
	At	ASK-beta/theta		Tichtinsky et al. (1998)
	Nt	NSK6	P	Tichtinsky et al. (1998)
SIGNAL TRANSDUCTION				
Heterotrimeric G protein	Ll	41KDa	P	Ma et al. (1999)
Human RTP protein	Ha	sf21	P, Pi	Krauter-Canham et al. (1997)
Leucine-rich repeat protein	Ha	SF17	P	Reddy et al. (1995)
	Ph	S/D4	**P**	Guyon et al. (2000)
Disease resistance protein	Bv	Bet v I*	P	Breiteneder et al. (1989) Swoboda et al. (1995)
Rab GTPase	At	AtRAB2	P, V	Moore et al. (1997)
Rho GTPases	At	**Rop1At**	**P**	Li et al. (1998)
	At	Rop3At, Rop3At	P, V	Li et al. (1998)
Pollen coat proteins	Bo	PCP-A1	**P**	Doughty et al. (1998)
	Bc	SLR1-BP	**P**	Takayama et al. (2000a)
	Bcs	**SCR**	P, T	Takayama et al. (2000b)
TRANSCRIPTIONAL REGULATION				
LIM domain protein	Ha	HaPLIM	**P**	Baltz et al. (1992b)
	Nt	NtPLIM1a,b	**P**	Sweetman et al. (2000)
	Ph	S/D1	**P**	Guyon et al. (2000)
MADS domain protein	Zm	ZmMADS2	**P**	Heuer et al. (2000)
	Am	DEFH125	**P**	Zachgo et al. (1997)
	At	AGL18	P, Es, En	Alvarez-Buylla et al. (2000)
Zn finger protein	Ph	D14	**P**	Guyon et al. (2000)
MYB domain protein	Nt	NtMYBAS1/2	P, T	Yang et al. (2001)

Table 4.1 (continued)

Function group	Species	Gene[1,2]	Specificity[3]	Reference
CELL CYCLE				
Cyclin-dependent kinase (CDK)	Zm	Cdc2ZmA/B	Sc, V	Sauter *et al.* (1998)
Cyclin	Zm	Zeama:CycA1;1	Sc, V	Sauter *et al.* (1998)
HISTONES	Os	Histone H3	P, V	Raghavan (1989)
	Zm	Histone H3	Sc, V	Sauter *et al.* (1998)
	Li	gcH2A, gcH3	Gc/Sc	Xu *et al.* (1999b)
	Ll	gH2A, gH3	Gc/Sc	Ueda *et al.* (2000)
DNA REPAIR	Ll	ERCC1	Gc, V	Xu *et al.* (1998)
RNA METABOLISM				
Poly(A) binding protein	At	PAB5	P, O	Belostotsky and Meagher (1993)
	At	PAB2	P, V	Palanivelu *et al.* (2000)
La antigen-like (RNA binding protein)	Bn	Bn1	**P**	Smykal *et al.* (2000a)
S-RNase	Na	S2-RNase A/B	M, P, Pi	Dodds *et al.* (1993)
PROTEIN SYNTHESIS				
Initiation factor eIF-4A	Nt	neIF4A-8	**P**	Brander and Kuhlemeier (1995)
Initiation factor eIF-4E	Nt	NteIF4E	M, P*	J.P. Combe and D. Twell, unpub.
Initiation factor eIF-iso4E	Nt	NteIFiso4E	M, P*	J.P. Combe *et al.* unpub.
AA SYNTHESIS/ TRANSPORT				
Dihydrodipicolinate (lysine) synthase	At	dhdps	P, T, V	Vauterin *et al.* (1999)
Proline transporter	Le At	LePROT1	**P**	Schwacke *et al.* (1999)
Amino acid permease	Ns	NsAap1	**P**	Lalanne *et al.* (1997)
SUGAR TRANSPORT				
Monosaccharide transporter	Ph	Pmt1	**P**	Ylstra *et al.* (1998)
	At	AtSTP2	M	Truernit *et al.* (1999)
Sucrose transporter	At	AtSUC1	P, A, F	Stadler *et al.* (1999)
ENZYMES				
Alcohol dehydrogenase	At	**ADH1**	P, V	Chang and Meyerowitz (1986)
	Zm	**ADH1**	P, V	Chen *et al.* (1987)
	Np	**ADH1**	P, V	Rousselin *et al.* (1990)
	Ph	**ADH1**	P, V	Gregerson *et al.* (1991)
	Nt	**ADH1**	P	Bucher *et al.* (1995)

Table 4.1 (continued)

Function group	Species	Gene[1,2]	Specificity[3]	Reference
Aldehyde dehydrogenase	Nt	TobAldh2A	P, V	Op den Camp and Kuhlemeier (1987)
Pyruvate decarboxylase b-subunit mitochondrial	Nt	PDC2	P	Bucher et al. (1995)
ATPase	Ns	nsatp2.3	**P**	Lalanne et al. (1998)
	Ns	nsatp2.1/2.2	P, V	Lalanne et al. (1998)
Flavonol 3-O-galactosyltransferase	Ph	F3GalTase	**P**	Miller et al. (1999)
Starch synthase	Zm	ADP-GPP	M, P	Olmedilla et al. (1991)
VARIOUS				
Expansin-like	Lp	Lol p I*	P	Perez et al. (1990); Griffith et al. (1991)
	Lp	Lol p I/Lol p Ib*	P	Singh et al. (1991)
	Zm	Zea m I	P	Broadwater et al. (1993)
	Os	Ory s I*	A	Xu et al. (1995b)
Heat shock protein	Nt	NtHSP18P	P, V	Smykal et al. (2000b)
Phospholipid transfer protein	Bn	E2	M, T	Foster et al. (1992)
STRUCTURAL (UNKNOWN)				
Extensin-like	Zm	Pex1	**P**	Rubinstein et al. (1995a)
GlySer-rich/ amyloplast protein	Ll	LIM14	M, A	Mousavi et al. (1999)
Sperm surface protein	Ll	LGC1	Gc, Sc	Xu et al. (1999a)
Desiccation-induced protein	Ll	LLA23	**P**	Huang et al. (2000)
Ascorbate oxidase-like	Tp	Tpc44	P	Stinson et al. (1987)
	Le	LAT51	P	Ursin et al. (1989)
	Bn	Bp10	**P**	Albani et al. (1992)
	Nt	NTP303	**P**	Weterings et al. (1992)
	Sb	MSb2.1	P	Pe et al. (1994)
	Sb	Mpt1	P	Sari-Gorla and Pe (1999)
Kunitz trypsin Inhibitor-like	Le	**LAT52**	P, A	Twell et al. (1989)
	Zm	Zm13	**P**	Hanson et al. (1989)
	Sb	MSb8	P	Pe et al. (1994)
	Oe	Ole eI*	P	Villalba et al. (1994)
	Os	PS1	P	Zou et al. (1994)
	Nt	NTP52	P	Honys et al. (2000)
?	Bn	Bp4	M, A	Albani et al. (1990)
?	Bc	**Bcp1**	P, T	Theerakulpisut et al. (1991)
?	Pp	Poa pIX*	P	Silvanovich et al. (1991)
?	Pp	Poa p IX*	P	Olsen et al. (1991)
?	Oo	P1	**P**	Mascarenhas (1992)
?	At	APG	M, T	Roberts et al. (1993)

Table 4.1 (continued)

Function group	Species	Gene[1,2]	Specificity[3]	Reference
?	Tp	Tpc70	**P**	Turcich *et al.* (1994)
?	Nt	NTM19	M	Oldenhof *et al.* (1996)
?	Bn	Bnm1	**P**	Treacy *et al.* (1997)
?	At	ATA18, ATA21	P	Rubinelli *et al.* (1998)

[1]Asterisked genes, identified as pollen allergens.

[2]Genes underlined: *in vitro* evidence for protein function: Genes in **bold**: *in planta* evidence for protein function.

[3]Specificity of expression is represented by abbreviations as follows: A = anther; Es = embryo sac; En = endosperm; F = funiculus; Gc = generative cell; M = microspore; O = ovary; Pi = pistil; P = pollen; PMC = pollen mother cell; T = tapetum; Sc = sperm cells; St = stamen; V = vegetative tissues. **P** = strong evidence for pollen-specific expression.

[4]Key to species abbreviations: Aa = *Ambrosia artemisifolia*; Am = *Antirrhinum majus*; At = *Arabidopsis thaliana*; Bv = *Betula verrucosa*; Bc = *Brassica campestris*; Bn = *Brassica napus*; Bo = *Brassica oleracea*; Gh = *Gossypium hirsuta*; Ha = *Helianthus anuus*; Le = *Lycopersicon esculentum*; Ll = *Lilium longiflorum*; Lp = *Lolium perenne*; Ms = *Medicago sativa*; Ma = *Mercurialis annua*; Na = *Nicotiana alata*; Np = *Nicotiana plumbaginifolia*; Ns = *Nicotiana sylvestris*; Nt = *Nicotiana tabacum*; Oo = *Oenothera organensis*; Oe = *Olea europea*; Os = *Oryza sativa*; Ph = *Petunia hybrida*; Pi = *Petunia inflata*; Pp = *Poa pratense*; Sg = *Salix gilgiana*; Sb = *Sorghum bicolor*; Tp = *Tradescantia paludosa*; Zm = *Zea mays*.

? Refers to genes of unknown function which otherwise do not have close homologues in other species or organisms.

focusing on those classes for which evidence of function and/or activity has been demonstrated, and those that are preferentially or specifically expressed, implying their role in processes specialized for pollen development and function.

4.8.1 Developmental phases of expression

Microgametogenesis may be considered most simply to comprise two phases, microspore development, terminating with the completion of PMI, and pollen development, culminating in the production of mature tricellular pollen. The unique processes that occur during each of these phases demand the expression of unique combinations of genes. Although far from complete, there is now some detail of the division of labour between these two phases in terms of gene expression and the functional significance of a growing number of these genes.

The progression through pollen mitosis I is associated with a dramatically increased level of transcription and a major shift in the translatable mRNA populations following PMI (Vergné and Dumas, 1988; Bedinger and Edgerton, 1990; Mandaron *et al.*, 1990; Schrauwen *et al.*, 1990). Based on the analysis of the accumulation profiles of a number of pollen-expressed and pollen-specific mRNAs, two broad classes of genes are recognized, the early and the late pollen-expressed genes, that may be considered different regulatory or function groups (Mascarenhas, 1990). Early group transcripts are first expressed in microspores, but decrease in abundance before pollen maturation, whereas the late group are expressed after PMI and accumulate until maturity.

A further level of control operates to regulate the cell-specific expression of late pollen genes between the vegetative and generative cells. Evidence for the presence of specific transcripts within the vegetative cytoplasm was provided by *in situ* hybridization studies (Hanson *et al.*, 1989; Ursin *et al.*, 1989; Brown and Crouch, 1990; Reijnen *et al.*, 1991; Theerakulpisut *et al.*, 1991). Conclusive evidence for vegetative cell-specific regulation of late pollen genes was obtained by linking the promoter of the tomato *LAT52* promoter to a nuclear-targeted GUS fusion protein gene, which revealed that the *LAT52* promoter was activated specifically in the vegetative nucleus (Twell, 1992). Based on their abundance and isolation from whole pollen or anther cDNA libraries, most other late pollen-expressed transcripts are likely to be vegetative-cell expressed. Evidence for specific expression in gametic cells is discussed in section 4.8.6.

4.8.2   Transcription in developing microspores

Analysis of actin mRNA in maize revealed expression in young microspores, an increase in abundance after PMI, and a decline before maturity (Stinson *et al.*, 1987). Similar profiles have been reported for other gametophytically expressed genes, such as granule-bound starch synthase (Olmedilla *et al.*, 1991) and *Adh1* (Gregerson *et al.*, 1991). Several anther-specific cDNA clones isolated from *B. napus* showed overlapping patterns of accumulation appearing first in buds containing uninucleate microspores and decreasing in mature pollen grains (Albani *et al.*, 1990, 1991; Roberts *et al.*, 1991; Scott *et al.*, 1991a,b; Foster *et al.*, 1992). Other early groups are concerned with basic cellular functions required to support the microspore such as translation initiation factors (eIF-4A, eIF4E) and mitochondrial functions (b-ATPase) (table 4.1). Interestingly, these mRNAs show a biphasic pattern that peaks first in the early microspore and subsequently in early pollen, similar to the activity of the *APG* promoter (Twell *et al.*, 1993; J. Sweetman and D. Twell, unpublished).

A common feature of microspore-expressed genes is their sporophytic expression in tapetum, such as *BCP1* from *Brassica campestris* (Theerakulpisut *et al.*, 1991; Xu *et al.*, 1993) and *APG* from *Arabidopsis* (Roberts *et al.*, 1993). *BCP1* expression in both sporophyte and gametophyte has been shown to play an essential role in microspore development (Xu *et al.*, 1995a). *APG* encodes a proline-rich protein with a secretory signal, suggesting a wall location, and its promoter is first activated in the developing microspore prior to microspore nuclear migration (Roberts *et al.*, 1993; Twell *et al.*, 1993).

To date, there is convincing evidence for only one strictly microspore-specific gene, the tobacco *NTM19* gene, which encodes a small secretory protein that could be a structural or regulatory component of the microspore wall (Oldenhof *et al.*, 1996). *NTM19* transcripts appear after microspore release and are rapidly degraded before PMI. The *NTM19* promoter was shown to direct microspore-specific expression of GUS, and microspore-specific cell ablation, when used to drive expression of a fungal ribonuclease in tobacco (Custers *et al.*, 1997). This apparent paucity of evidence could reflect the particular molecular screening approaches used. However, microspores appear to be relatively undifferentiated compared with pollen grains and therefore presumably require fewer specialized cell functions.

4.8.3 Transcription in developing pollen

The late pollen genes represent a much larger group of gametophytically expressed sequences. Most of these probably function during the progamic phase as a result of their accumulation and storage in the mature dehydrated pollen. Stinson *et al.* (1987) first reported the isolation of pollen-specific cDNA clones from *Tradescantia paludosa*. Transcripts were first detectable soon after PMI and increased in abundance until pollen shed. This characteristic late pattern of transcript accumulation after PMI was subsequently reported for a number of pollen-specific genes from diverse species (Twell *et al.* 1989; Ursin *et al.*, 1989; Brown and Crouch, 1990; Rogers *et al.*, 1992; Weterings *et al.*, 1992). Exceptions such as the pollen-specific P6 transcript from *Oenothera* (Brown and Crouch, 1990), and the S/D10 transcript from petunia (Guyon *et al.*, 2000), which appear only in the final stages of pollen maturation, indicate additional complexity in patterns of regulation. Indeed, recent analysis of several late pollen genes from tobacco, including those encoding LIM domain proteins, polygalacturonase, pectate lyase and a myb-like protein (NtMybAS1), also revealed low level expression in microspores (Sweetman *et al.*, 2000; Yang *et al.*, 2001; and unpublished findings). This suggests that the late gametophytic gene expression programme is not an absolute switch set up after PMI. These game-tophytic genes are in fact activated in the unicellular microspore and are strongly enhanced after pollen mitosis I, presumably in response to gametophytic factors activated in the vegetative cell (Eady *et al.*, 1995; section 4.6.3).

Promoter studies have allowed the developmental regulation of a number of pollen-specific genes to be established. The repertoire of pollen-expressed and pollen-specific promoters has previously been reviewed in detail (Twell, 1994). For example, the tomato *LAT52* and *LAT59* promoters were coordinately activated in developing pollen grains, in close association with PMI (Twell *et al.*, 1990; Eady *et al.*, 1994), similarly to the promoter of the homologous maize gene, *Zm13* (Guerrero *et al.*, 1990). These and a number of other studies provide evidence for a developmental switch in transcriptional activity soon after pollen mitosis I. Promoter mutagenesis studies have further demonstrated that discrete regions of the *LAT52* promoter were active at different times during pollen maturation (Bate and Twell, 1998). This provides a composite structure that has evolved to allow rapid activation of *LAT52* promoter activity after PMI and maintain high levels throughout maturation.

4.8.4 Pollen-specific regulatory elements

With regard to the organization of regulatory elements, which has been defined in a number of late pollen-specific promoters, some general features have emerged. First, *cis*-regulatory elements sufficient to direct transcription in pollen are always located relatively close to the transcription start site, within 70 to 150 bp. Second, multiple upstream promoter elements increase transcriptional activity of these pollen-specific core promoter regions. Some of these elements are insufficient to activate pollen expression, while others also possess pollen-specific enhancer activity and are able to activate minimal promoter sequences. Third, none of these regulatory elements enhances expression in sporophytic tissues, suggesting a positive regulatory mechanism involving the pollen-specific expression or activation of transcription factors.

The most detailed understanding of pollen-specific regulatory elements has emerged from studies of three coordinately regulated *LAT* gene promoters from tomato (Twell *et al.*, 1991; Eyal *et al.*, 1995; Bate and Twell, 1998). The sequences involved in transcriptional activation within upstream domains of *LAT52* include the PBII motif (TGTGGTT), which functions within a pollen-specific enhancer element. The identical *LAT52* PBI element in the pollen-specific proximal region also occurs within the *LAT56* promoter as part of a region of extended sequence identity, termed the 52/56 box (TGTGGTTATATA), that has a quantitative role in both promoters (Twell *et al.*, 1991). A second shared regulatory element is the 56/59 box (GAAA/TTTGA), which occurs at similar positions in the *LAT56* (-97 bp) and *LAT59* (-108 bp) promoters. Mutational analysis showed that the 56/59 box is functional in both promoters, and that only the GTGA core motif is critical for activity (Twell *et al.*, 1991). Interestingly, only the GTGA motif within the 56/59 box is conserved in sequence and position between the homologous *LAT56* and tobacco *G10* promoters. This conserved GTGA core motif was demonstrated to function as a quantitative element within the *G10* promoter (Rogers *et al.*, 2001).

More detailed linker substitution analysis identified 30 bp proximal promoter regions of *LAT52* and *LAT59* that are essential for their expression in pollen (Eyal *et al.*, 1995; Bate and Twell, 1998). These regions conferred pollen specificity when fused to the heterologous cauliflower mosaic virus 35S promoter. Adjacent upstream elements, the 52/56 box in *LAT52* and the 56/59 box in *LAT59*, are involved in modulating the level of expression in pollen. These analyses suggest novel regulatory elements in the pollen-specific core region, and sequences conserved between *LAT52* and *LAT59* including an AGAAA motif at position -71 (Bate and Twell, 1998).

Substitution mutagenesis of the maize *Zm13* promoter revealed a similar structure to the homologous *LAT52* promoter. Within the core pollen-specific region, -119 to -37, a proximal region from -84 to -53 conferred pollen specificity. An upstream region from -107 to -102 (Q-element) could increase the expression of the proximal region, but was unable to activate pollen expression alone (Hamilton *et al.*, 1992, 1998). Despite the similar organization, sequences of the Q-element and quantitative elements of *LAT52* were not conserved; however, within the pollen-specificity region the AGAAA element was conserved and appears to function in both promoters. Other detailed promoter studies revealed a novel proximal regulatory element (AAATGA) in the pollen-specific tobacco *Ntp303* promoter (Weterings *et al.*, 1995), which is conserved in the *Brassica napus* homologue *Bp10* (Albani *et al.*, 1992).

The activity of common DNA sequences within the promoters of coordinately regulated late pollen genes strongly implies the involvement of shared transcription factors. Evidence in support of this comes from promoter competition experiments, in which co-transfected 30 bp pollen-specific promoter fragments were able to titrate out factors required for activation of heterologous promoters. This suggests that *LAT52* and *LAT59* promoters share regulatory factors that bind to their pollen-specific 30 bp core elements (Eyal *et al.*, 1995).

A recent analysis of unrelated promoters that are preferentially expressed in pollen provides evidence that shared regulatory elements are used by genes in diverse function groups. Promoter regions responsible for preferential expression in anthers and pollen were defined for three coordinately regulated nuclear-encoded mitochondrial complex I (nC1) genes from *Arabidopsis* (Zabaleta *et al.*, 1998). In two of these promoters the $-200/-100$ regions, which contained sequences similar to the core PB motif and/or the 52/56 box, were both necessary and sufficient to confer pollen-/anther-specific expression. Thus, nC1 promoters contain a regulatory module specifying developmental regulation in pollen, which appears to involve *cis*-elements used by other pollen-specific genes (Zabaleta *et al.*, 1998).

4.8.5 Posttranscriptional regulation

In most angiosperms, germination and pollen tube growth occur rapidly, within 2–60 minutes of pollination, requiring rapid activation of synthetic and catabolic processes. Inhibitor studies have demonstrated that germination and early tube growth are largely independent of transcription, but strictly dependent on translation (Hoekstra and Bruinsma, 1979; Mascarenhas, 1975). Native protein profiles of *in vitro* grown pollen tubes are very similar to those obtained from *in vitro* translation of mRNA isolated from mature pollen. Furthermore, stored transcripts may persist in the growing pollen tube for up to 20 hours after germination (Ursin *et al.*, 1989; Brown and Crouch, 1990; Weterings *et al.*, 1992). These results provide compelling evidence that many mRNAs are stored in mature pollen for use during germination and tube growth (Mascarenhas, 1993; Twell, 1994).

An important question regarding the storage of late pollen transcripts is whether they are also translated during pollen maturation. Numerous examples of late pollen-specific mRNAs, particularly those encoding pollen allergens, demonstrate their translation prior to dehydration (table 4.1). Thus, typically late pollen transcripts appear to be translated concomitantly with their appearance in the cytoplasm. Furthermore, it may be an advantage for certain mRNAs to retain high rates of translation in an increasingly hostile cellular environment during dehydration. In order to maintain translational efficiency under such conditions, which involves an increase in intracellular ion concentration, one strategy would be to maintain an unstructured 5′-untranslated region (5′UTR). Indeed, the 5′-UTRs of many cloned late pollen transcripts possess very low indices of secondary structure and support efficient translation during pollen maturation (Twell, 1994). Analysis of the role of the *LAT52* 5′UTR demonstrated its function as a pollen-specific translational enhancer and that specific mechanisms exist to mediate the activity of this enhancer during the final stages of pollen maturation and during pollen tube growth (Bate *et al.*, 1996). In contrast, sequences within the 5′UTR of the *LAT59* gene inhibited expression of reporter genes through a transcriptional mechanism (Curie and McCormick, 1997).

Regulation of pollen translation could occur at the level of ribosome recruitment. In this regard, pollen-expressed genes encoding translation initiation factors eIF4A and eIF4E have been isolated from tobacco (Owttrim *et al.*, 1991; table 4.1). One isoform of eIF4A was expressed specifically during pollen maturation (Brander and Kuhlemeier, 1995) and is phosphorylated during pollen germination (Op den Camp and Kuhlemeier, 1998). Since eIF4A possesses ATP-dependent RNA helicase activity in association with the cap binding complex eIF4F, increased levels of eIF4A could contribute to an increased translational efficiency during pollen maturation by unwinding of 5′UTRs.

The role of mRNA stability in regulating pollen-expressed genes has also been investigated in tobacco (Ylstra and McCormick, 1999). Analysis of the stabilities of 12 pollen-expressed mRNAs showed that many are indeed long-lived, but that at least some mRNAs undergo decay during pollen maturation. Thus, there is no overall cytoplasmic condition in pollen that stabilizes mRNAs, suggesting the operation of mRNA-specific turnover mechanisms.

Although efficient translation of many mRNAs occurs both during and after maturation, there is also convincing evidence that specific transcripts are translationally repressed during pollen maturation and subsequently translated during pollen tube growth (Storchová *et al.*, 1994). The tobacco pollen tube wall-specific glycoprotein p69 is the major newly synthesized protein in germinating pollen and pollen tubes (Capková *et al.*, 1988, 1994) and is encoded by the late pollen-specific gene *Ntp303* (Wittink *et al.*, 2000). Despite very high levels of p69 mRNA in mature pollen, only small amounts of p69 have been detected, using sensitive *in situ* immunolocalization techniques (Wittink *et al.*, 2000). Furthermore, abundant synthesis of p69 in pollen tubes was not influenced by inhibition of RNA synthesis (Capková *et al.*, 1988), clearly showing the phenomenon of translational repression of stored *Ntp303* transcripts. Recently, the subcellular compartmentation of *NTP303* mRNA was compared with that of the tobacco *LAT52* homologue *NTP52*, which is efficiently translated during pollen maturation (Honys *et al.*, 2000). During pollen maturation, *NTP303* mRNA and the *NTP52* mRNA were predominantly associated with the polysomal fraction. However, a novel subfraction, resistant to polysome destabilizing conditions, was identified that contained a significant portion of *Ntp303* mRNA but did not contain *Ntp52* mRNA. This differential mRNA distribution pattern highlights the potential role of ribosome-associated ribonucleoprotein complexes as a storage compartment facilitating rapid translation upon pollen activation and germination.

Further evidence for mRNA storage and translational repression during pollen maturation is provided by the *Arabidopsis* AtSUC1 protein, which encodes a plasma membrane H^+-sucrose symporter (Stadler *et al.*, 1999). *AtSUC1* mRNA was expressed in anther connective tissue, in funiculi and in mature pollen; however, despite high *AtSUC1* mRNA levels in mature pollen, no protein was detected. AtSUC1 was only detected in activated pollen and growing pollen tubes, suggesting translation of pre-existing AtSUC1 mRNA.

4.8.6   Gene expression in generative and sperm cells

In contrast to the body of information about genes expressed in the vegetative cell, relatively little is known about genes expressed within the gametic cells. However, recent progress promises exciting avenues for further research. This includes the use of gametic-specific genes as molecular cell fate markers and their application to understanding gametic cell functions.

Early work revealed incorporation of ^{32}P into generative cell RNA of *Tradescantia* and incorporation of (^3H)uridine into sperm or generative nuclei in *Seceale cereale* (Haskell and Rogers, 1985). More recently, Saito *et al.* (1998) used high-resolution *in situ* hybridization to demonstrate strong rRNA hybridization in the generative cell. Important developments have included refinement of procedures to isolate and culture generative and sperm cells that allowed RNA and protein synthesis to be monitored in *Plumbago*, maize and lily (Russell, 1991; Zhang *et al.*, 1993; Ueda and Tanaka, 1994; Xu *et al.*, 1998). Using these methods, core histone variants gH2A and gH3, present specifically in generative and sperm cell nuclei, were identified (Ueda and Tanaka, 1995a,b). Blomstedt *et al.* (1996) used such methods to demonstrate the presence of translatable mRNAs in GCs and to construct a cDNA library from isolated GCs in lily. One GC-expressed cDNA was isolated encoding a protein related to the human excision repair protein ERCC1, suggesting that DNA repair processes could play an important role in protecting the gametic genome against DNA damage (Xu *et al.*, 1998). A second gene, *LGC1*, was expressed exclusively in the male gametic cells (Xu *et al.*, 1999a). LGC1 was immunolocalized to the surface of male gametic cells, suggesting a possible role in sperm–egg interactions.

cDNA clones encoding the gametic-specific histones gH2A, gH2B and H3 were isolated using antibodies raised against purified generative cell-specific histones of lily (Ueda *et al.*, 2000) and by sequencing GC cDNA clones (Xu *et al.*, 1999b). Their deduced amino acid sequences show only 40–50% identity with somatic cell core histones, suggesting novel properties. Transcripts were first detected in bicellular pollen soon after PMI and *in situ* hybridization revealed specific expression in the cytoplasm of generative and sperm cells (Ueda *et al.*, 2000). The proteins accumulated in the generative nucleus in developing pollen and were most abundant in the sperm nuclei within the pollen tube. Gametic histones appear to be expressed throughout maturation, independently of DNA replication, suggesting a role in chromatin condensation or remodelling associated with repression of gene expression in male gametes (Ueda *et al.*, 2000). In *Arabidopsis*, where an essentially complete genome sequence is available, related histone variants appear to be absent, suggesting that in *Arabidopsis* chromatin remodelling is achieved through a different route (Twell, unpublished).

4.9 Male gametophytic gene functions

Genes that are specifically or preferentially expressed in the male gametophyte have been compiled and include over 150 unique mRNAs from 27 different species (table 4.1). Several large functional groups of genes emerge: pectin-degrading enzymes, cytoskeletal proteins, and regulatory proteins such as transcription factors and protein kinases. For 20 of these proteins there is evidence for functional activity (underlined in table 4.1) and in six examples (bold in table 4.1) an essential functional role has been demonstrated *in planta*. Most of these involve functions during the progamic phase. To date, the only example of a cloned gene shown to be essential for completion of pollen development is *Bcp1*. *Bcp1* encodes a small anther-specific protein from *Brassica campestris* and *Arabidopsis* that is expressed in the tapetum and in developing pollen. Antisense downregulation of *Bcp1* in *Arabidopsis* demonstrated an essential role in both sporophytic and gametophytic cells, leading to aborted pollen development and cell death (Xu *et al.*, 1995a). This could also suggest that BCP1 is an important structural component of the pollen wall contributed by both tapetum and microspore/pollen expression.

4.9.1 Pectin-degrading enzymes

Pollen-specific mRNAs encoding proteins with similarity to polygalacturonase, pectin methylesterase and pectate lyase have been isolated from a wide range of species (table 4.1). These mRNAs represent some of the most abundant pollen-specific mRNAs, which suggests an important role for pectin modification during pollen maturation and/or pollen tube growth. mRNAs persist throughout germination and pollen tube growth and all possess secretory signals, suggesting secretion at the pollen tube tip. Furthermore, there is evidence for secretion of polygalacturonase activity in mature pollen for a wide range of species (Pressey and Reger, 1989; Pressey, 1991).

Immunolocalization studies of polygalacturonic acid (unesterified pectin) and methyl-esterified pectin demonstrate that pectin metabolism is carefully regulated during tip extension (Li *et al.*, 1994). Periodic ring-like deposits of pectin were found in species possessing solid styles and a more uniform pectin sheath in species with hollow or no styles. Esterified pectin, which prevents Ca^{2+}-induced gelification of pectate, was localized predominantly at the apex in all species. Pectin esterification at the apex may therefore maintain a 'loose' wall structure enabling localized cell wall expansion. De-esterification of pectin following tip expansion is thought to provide a more rigid structure in older parts of the pollen tube. Furthermore, periodic pectin ring structures may provide additional reinforcement for pollen tubes growing through the transmitting tissues in solid styles (Li *et al.*, 1994).

Degradation of stylar pectin may also be important for pollen tube penetration through the transmitting tract and in regulating adhesion of the pollen tube. Other pollen-specific isoforms of hydrolases, such as β-galactosidase (Singh *et al.*, 1985) are expressed in pollen, and cutinases have been isolated from pollen of *Tropaeolum majus* (Shaykh *et al.*, 1977; Maiti *et al.*, 1979) and *Brassica napus* (Hiscock *et al.*, 1994) that may function in penetration of the stigmatic cuticle.

4.9.2   Cytoskeleton-related proteins

Pollen tubes elongate on the basis of actin-dependent targeted secretion at the tip, and functions such as cytoplasmic streaming, male germ unit transport and polar vesicle delivery to the pollen tube apex depend on actin. Several genes encoding isoforms of actin that are preferentially expressed in mature pollen have been identified (Thangavelu *et al.*, 1993; An *et al.*, 1996; Huang *et al.*, 1996a, 1997). These studies reveal that during pollen maturation there is a switch from vegetative to predominantly reproductive actin isoforms that may fulfil unique functions in vegetative and sperm cell development (Kandasamy *et al.*, 1999).

4.9.2.1   Actin-binding proteins
The regulation of pollen tube growth is also known to involve alterations in intracellular calcium levels, phosphoinositide signalling, and signalling pathways that operate through the actin cytoskeleton (Clarke *et al.*, 1998). In other eukaryotic cells, actin-binding proteins function as stimulus–response modulators, translating signals into alterations in cytoplasmic architecture. Pollen-specific regulators of F-actin assembly, including profilin and ADF (actin depolymerising factor)-like proteins, have been identified in a number of species (table 4.1). Profilin is a small (12–15 kDa) actin- and phospholipid-binding protein first

identified in plants as a birch pollen allergen (Valenta *et al.*, 1991, 1994). Two distinct profilin gene classes are present in higher plants; one is pollen-specific, another is constitutive (Staiger *et al.*, 1993; Yu *et al.*, 1998). Injection of maize profilin into plant cells promoted the disassembly of F-actin, suggesting that F-actin may be prevented from forming in the mature pollen grain by actin monomers binding to profilin. Profilin isoforms from maize were shown to have different effects on actin in living cells (Gibbon *et al.*, 1998), and their actin-sequestering activity was found to be dependent on the concentration of free calcium (Kovar *et al.*, 2000). Furthermore, native pollen profilin from *Papaver rhoeas* was shown to interact with soluble pollen components, resulting in dramatic alterations in the phosphorylation of several proteins (Clarke *et al.*, 1998). These data suggest that pollen profilin can regulate actin-based cytoskeletal protein assembly and protein kinase or phosphatase activity, indicating the involvement of profilin in signalling pathways that regulate pollen tube growth.

Pollen-specific ADF-like proteins, which have been isolated from *Lilium*, *B. napus* and maize, are also likely to play an important role in actin dynamics during pollen tube growth (Kim *et al.*, 1993; Lopez *et al.*, 1996). Pollen-specific ADF proteins are most similar to cofilin, which binds F-actin at neutral pH and reversibly controls actin depolymerization and polymerization in response to various stimuli. Pollen ADF was localized to electron-dense inclusions associated with storage bodies of mature pollen, suggesting a location from which ADF is released into the cytoplasm during pollen tube growth (Chung *et al.*, 1995). The pollen-specific maize ADF-like protein ZmABP3 bound monomeric actin (G-actin) and filamentous actin (F-actin) and decreased the viscosity of polymerized actin solutions, consistent with an ability to depolymerize filaments (Lopez *et al.*, 1996).

4.9.2.2 GTPases

Recent work has demonstrated an exciting pollen-specific regulation of actin assembly that operates via GTPases (Zheng and Yang, 2000). The Rho small GTP-binding proteins are versatile, conserved molecular switches in eukaryotic signal transduction. Plants contain a unique subfamily of Rho-GTPases called Rop (Rho-related GTPases from plants). Pea Rop1Ps is predominantly expressed in pollen and localized towards the apex and at the periphery of the generative cell (Lin *et al.*, 1996). This suggests that Rop1Ps may modulate both tip growth and actomyosin-mediated movement of the generative cell.

Direct evidence for the role of Rop1Ps in controlling tip growth was obtained by microinjection of anti-Rop1Ps antibodies, which diffused the Ca^{2+} gradient in the tip and blocked pollen tube growth (Lin and Yang, 1997). Overexpression of *Rop1At*, a pollen-specific *Arabidopsis* orthologue of Rop1Ps, led to its ectopic accumulation in the plasma membrane at the tip and depolarization of pollen tube growth (Li *et al.*, 1999). Conversely, antisense *Rop1At* inhibited tube growth, and pollen-specific expression of constitutively active *Rop1At* mutants induced isotropic growth of pollen tubes. Two other pollen-expressed *Arabidopsis* Rop genes, *Rop3At* and *Rop5At*, were also expressed in vegetative tissues (Li *et al.*, 1998).

Kost *et al.* (1999) have also characterized Rac-related Rho family proteins expressed in growing pollen tubes. Expression of a dominant-negative form of At-Rac2 reduced actin bundling and inhibited pollen tube growth, whereas expression of constitutively active Rac induced depolarized growth. Similarly, pollen tube Rac was found to accumulate at the tip plasma membrane and to physically associate with phosphatidylinositol monophosphate kinase (PtdIns P-K) activity. Furthermore, phosphatidylinositol 4,5-bisphosphate (PtdIns 4,5-P_2), the product of PtdIns P-Ks, showed a similar intracellular localization as Rac, suggesting that that Rac and PtdIns 4,5-P_2 act in a common pathway to control polar pollen tube growth. These studies provide strong evidence for a Rop GTPase-dependent tip growth pathway that couples the control of growth sites with the rate of tip growth, through the regulation of tip-localized extracellular Ca^{2+} influxes, and the formation of the tip-focused intracellular Ca^{2+} gradient in pollen tubes. For further discussion of the role of pollen GTPases and similarity of pollen tip growth with axon growth in animals, see Palanivelu and Preuss, (2000).

Another class of GTPases, the YPT/Rab family have been identified in plants (see Moore *et al.*, 1997). The predicted product of the *Arabidopsis At-Rab2* gene shares 79% identity with human Rab2 protein, which is required for vesicle traffic between the ER and the Golgi apparatus (Moore *et al.*, 1997). *At-Rab2* transcripts, and promoter-directed GUS activity, accumulated during pollen maturation and in rapidly growing organs of germinating seedlings. These results are consistent with a role for At-Rab2 in secretory activity in pollen tubes, which possess high rates of cell wall and membrane synthesis.

Heterotrimeric GTP-binding proteins (G proteins) are also important signal transducers in lower eukaryotes and in animal cells. G protein α subunits have been cloned from *Arabidopsis* (GPA1) and tomato (Weiss *et al.*, 1993). GPα1 was widely present in immature organs and during flower development, in dividing microspores, but not in mature pollen. However, GPα1 was present in growing pollen tubes, suggesting a dual role in signalling, both during development and pollen tube growth. Convincing evidence for their function was obtained in microinjection experiments, using G protein agonists and antagonists, which increased or decreased lily pollen tube growth rate respectively (Ma *et al.*, 1999). Furthermore, purified CaM, and the G protein agonist cholera toxin, both significantly activated GTPase activity in plasma membrane vesicles, which was completely inhibited by pertussis toxin and nonhydrolysable GTP analogues. This provides compelling evidence that heterotrimeric G proteins present in the plasma membrane of lily pollen are involved in the signal transduction of extracellular CaM in pollen tube growth.

4.9.2.3 Microtubules

Microtubules have a less defined role in pollen development and pollen tube growth than actin. For example, Mt depolymerization has been shown not to affect initial rates of pollen tube growth *in vitro*. However, Mts are implicated in transport of the male germ unit through interactions with myosin located on the surface of the vegetative nucleus and generative and sperm cells (Heslop-Harrison and Heslop-Harrison, 1989). Specific longitudinal arrays of bundled Mts are present in the cortical cytoplasm of the generative cell and are thought to maintain cell shape (section 4.7.2). Isoforms of both α- and β-tubulin have been identified that are specifically expressed during pollen maturation and could be present in such specific Mt arrays (Hussey *et al.*, 1990; Carpenter *et al.*, 1992; Rogers *et al.*, 1993). Isoforms of tubulin could also be specialized for interaction with microtubule-binding proteins involved in cross-linking or with Mt-dependent myosin motors such as kinesin that have been detected in pollen tubes (Tiezzi *et al.*, 1992; Cai *et al.*, 2000).

4.9.3 Transcriptional regulators

Given the central role of transcription factors in development, the identification of gametophytically expressed transcription factors remains an important goal. Despite detailed knowledge of several different pollen-specific *cis*-regulatory elements (section 4.8.4), their corresponding gametophytic transcription factors remain unknown. However, genes encoding several classes of transcription factors expressed in pollen have been isolated.

Pollen-specific genes encoding zinc-finger proteins of the LIM class have been isolated from sunflower and tobacco (Baltz *et al.*, 1992a,b; Sweetman *et al.*, 2000). Although recombinant sunflower PLIM binds DNA *in vitro*, antibodies co-localize PLIM with actin halos at sites of apertures *in vivo*, suggesting a cytoplasmic role during pollen germination (Baltz *et al.*, 1996, 1999). Anther-specific genes encoding a different class of zinc-finger proteins have been isolated from *Petunia*, members of which are expressed at different times during anther development. One member showed a similar late expression profile to *lat52*, suggesting that it may be gametophytically expressed (Kobayashi *et al.*, 1998).

Anther- and pollen-specific genes encoding MADS-box transcription factors, *DEFH125* and *ZmMADS2*, have been identified in snapdragon and maize respectively. (Zachgo *et al.*, 1997; Heuer *et al.*, 2000). Interestingly, DEFH125 is located to the vegetative cytoplasm, but after pollination it is also found in the nuclei of cells in the transmitting tissue. This could suggest export from the pollen tube or pollination-induced expression in transmitting tissue cells. Recently, an unrelated MADS-box gene, *AGL18*, was shown to be expressed in endosperm and developing male and female gametophytes in *Arabidopsis* (Alvarez-Buylla *et al.*, 2000). *AGL18* was initially expressed during the tetrad stage, suggesting a potential role in regulating microspore-expressed genes.

We have recently characterized anther-specific members of the myb family in tobacco (Yang *et al.*, 2001). *NtMybAS1* is expressed in tapetal cells of young anthers, but is also strongly expressed in maturing pollen. Recombinant NtMybAS1 was shown to bind two classes of myb-binding sites present in tobacco phenylalanine ammonia lyase (*PAL*) gene promoters. NtmybAS1 *trans*-activated tobacco *PAL* promoters in protoplasts and PAL expression when ectopically expressed in transgenic plants. Co-localization of *NtmybAS1* and *PAL* transcripts in the tapetum of young anthers suggests that NtMybAS1 may be an important regulator of PAL expression and phenylpropanoid metabolism in young anthers. Despite abundant *NtmybAS1* expression in tobacco pollen, *PAL* gene transcripts are not gametophytically expressed (S.A. Amirsadeghi and D. Twell, unpublished). Therefore, NtMybAS1 may also regulate gametophytically expressed genes harbouring myb-binding sites.

4.9.4   Protein kinases

Genes encoding several different classes of protein kinases that are preferentially expressed during pollen development have been isolated. Although much more work is needed to identify their relative positions within signal transduction cascades, several have been shown to encode functional protein kinases, and there is evidence that some are important for pollen development and/or pollen tube growth.

4.9.4.1   Receptor-like kinases
PRK1, a petunia pollen receptor-like kinase 1 gene, is preferentially expressed after PMI and persists in pollen tubes (Mu *et al.*, 1994a). The extracellular domain contains leucine-rich repeats and the protein autophosphorylates on serine and tyrosine, suggesting that PRK1 is a dual-specificity kinase. Antisense downregulation of petunia *PRK1*, led to microspore abortion at PMI (Lee *et al.*, 1996) and arrested embryo sac development (Lee *et al.*, 1997). This suggests a gametophytic role for PRK1 in regulating events associated with the progression from microspore to pollen development and in megagametogenesis. However, PRK1 is most abundant in mature pollen, and could also function during pollen tube growth.

Two structurally related receptor-like kinases genes from tomato, *LePRK1* and *LePRK2*, are similar to *PRK1* but are expressed later during pollen development (Muschietti *et al.*, 1998). The abundance of *LePRK2* increases upon pollen germination, but *LePRK1* remains constant. LePRK1/2 were localized to the plasma membrane/cell wall of growing pollen tubes. In phosphorylation assays with pollen membrane preparations, LePRK2, but not LePRK1, was phosphorylated, and the addition of style (but not leaf) extracts to these membrane preparations resulted in partial dephosphorylation of LePRK2. Therefore, LePRK1 and LePRK2 may play different roles, and LePRK2 in particular may mediate signal perception from the pistil.

Three receptor-like kinase cDNAs expressed in flowers were isolated from *Arabidopsis* (Takahashi *et al.*, 1998). *RKF1* mRNA is highly expressed in stamens, and the *RKF1* promoter directed high GUS expression in pollen grains. The putative extracellular domain of RKF1 contains 13 tandem repeats of leucine-rich sequences, and recombinant RKF1 was found to have kinase activity with serine/threonine specificity.

4.9.4.2 MAP kinases

Mitogen-activated protein kinases (MAPKs) are components of a kinase module that plays a central role in the transduction of diverse extracellular stimuli including osmotic stress. Various elements of the mitogen-activated protein kinase (MAP kinase) module have been isolated in plants, some of which are encoded by pollen-specific or pollen-expressed genes. Wilson *et al.* (1993, 1995) first isolated several active MAP kinase-type cDNAs from tobacco. *NTF4* transcripts and p45^{Ntf4} protein were shown to be synthesized after pollen mitosis I and throughout pollen maturation (Wilson *et al.*, 1997). However, the MAP kinase was present in an inactive form in the mature, dry pollen grain. Kinase activation was very rapid after rehydration, peaking at approximately 5 min. The kinetics of activation suggest that the MAP kinase encoded by *NTF4* plays a role in the activation of the pollen grain upon hydration rather than during pollen tube growth. The recent localization of MAPK proteins and mRNAs in vacuolate microspores prior to entry into mitosis suggests that pollen-expressed MAPKs could also play a role in transducing signals that maintain microspore proliferation (Prestamo *et al.*, 1999).

Putative pollen-expressed upstream regulators in the MAP signal transduction cascade have also been identified. Among 14 unique *Arabidopsis* cDNAs and genes encoding putative MAP kinase kinase kinases (MAP3Ks) related to the MEKK/STE11 and RAF protein kinases, AtMAP3Kγ was significantly expressed in pollen (Jouannic *et al.*, 1999).

Genes and cDNAs encoding plant protein kinases highly homologous to the animal GSK-3/shaggy subfamily have been isolated from several species using the petunia *PSK6* GSK-3/shaggy related cDNA as a probe (Tichtinsky *et al.*, 1998). All were shown to be predominantly expressed in developing pollen, contain the catalytic domain of GSK-3/shaggy protein kinases, and form an isolated PSK6 group. All possess an amino-terminal extension implicated in mitochondrial targeting. Furthermore, a related tobacco GSK3/shaggy kinase cDNA, *NtK-4*, was shown to be expressed in all sporophytic tobacco tissues tested, as well as in gametophytic and embryogenic pollen (Einzenberger *et al.*, 1995).

4.9.4.3 Ca^{2+}-regulated protein kinases

Given the established link between elevation of cytosolic Ca^{2+} at the pollen tube tip and its growth (reviewed in Franklin-Tong, 1999), Ca^{2+} binding proteins may expected to be synthesized in developing and growing pollen tubes. A calcium-dependent calmodulin-independent protein kinase (CDPK) has been identified in maize, with an N-terminal catalytic kinase domain and a C-terminal domain resembling calmodulins (Estruch *et al.*, 1994). CDPK mRNA expression was restricted to late stages of pollen development and abundant CDPK protein was present at pollen germination. *In vitro* germination and pollen tube growth were impaired upon addition of a calmodulin antagonist, CDPK inhibitors and antisense oligonucleotides directed against CDPK mRNA. These observations indicate that the pollen-specific maize CDPK protein is required for germination and pollen tube growth.

A different class of anther-expressed genes encoding chimeric $Ca^{2+}/$ calmodulin-dependent protein kinases (CCaMK) has been identified in lily and tobacco (Liu *et al.*, 1998; Poovaiah *et al.*, 1999). CCaMK mRNA and protein appear to be sporophytically expressed in pollen mother cells and in tapetal cells. CCaMK protein showed Ca^{2+}-dependent autophosphorylation and Ca^{2+}/calmodulin-dependent substrate phosphorylation (Liu *et al.*, 1998). Therefore, CCaMK has dual modes of regulation by Ca^{2+} and Ca^{2+}/calmodulin. CCaMK could play a role in sensing transient changes in free Ca^{2+} concentration in target cells, thereby controlling developmental events in the anther (Poovaiah *et al.*, 1999).

Pollen tube reorientation is also regulated by cytosolic free calcium ($[Ca^{2+}]_c$) (Trewavas and Malho, 1998). Using confocal ratio imaging of BODIPY FL bisindolylmaleimide, protein kinase activity was localized to the apical region of growing pollen tubes, whereas nongrowing cells showed a uniform distribution (Moutinho *et al.*, 1998). Modification of growth direction by diffusion of inhibitors/activators from a micropipette showed the spatial redistribution of kinase activity to predict the new growth orientation. Therefore, the tip-localized gradient of kinase activity promotes Ca^{2+}-mediated exocytosis and may act to regulate Ca^{2+} channel activity (Moutinho *et al.*, 1998).

Given that the action of Ca^{2+} is primarily mediated by Ca^{2+}-binding proteins such as calmodulin (CaM), Safadi *et al.* (2000) used an interaction screen to identify a CaM-binding proteins from maize pollen (MPCBP). Pollen-specific expression of MPCBP, its CaM-binding properties, and the presence of tetratricopeptide motifs suggest a role in Ca^{2+}-regulated events during pollen germination and tube growth.

4.9.5   Pollen allergens

A rapidly increasing number of pollen-specific mRNAs encoding important pollen allergens have been identified (Knox *et al.*, 1998). Pollen allergens form a diverse group, linked largely by their potential for inducing an allergic response. For example, some allergens encode putative secreted proteins such as pectin-degrading enzymes, others are related to disease resistance proteins, Ca^{2+}-binding proteins or unrelated structural proteins (table 4.1).

Certain pollen allergens appear to have essential gametophytic functions. The major olive pollen allergen OleI is closely related to the products of pollen-specific genes *lat52* (Twell *et al.*, 1989) and *Zm13* (Hanson *et al.*, 1989), from tomato and maize respectively, that also share limited similarities to the kunitz trypsin inhibitor family (McCormick, 1991). OleI was found to accumulate from the early microspore stage onwards and was immunolocalized to the ER cisternae and within the pollen wall and in the tapetum (De Dios Alche *et al.*, 1999). Antisense downregulation of *LAT52* resulted in a gametophytic lethal phenotype in tomato (Muschietti *et al.*, 1994). Pollen matured correctly but failed to grow effective pollen tubes *in vivo*. The defect also affected hydration properties *in vitro*, suggesting that LAT52 may play a structural role in the developing pollen tube wall.

In contrast, downregulation of Lol p 5, the major allergenic protein of ryegrass pollen, with an antisense construct targeted to *Lol p 5*, did not affect pollen development or fertility, suggesting redundant or conditional functions. This did, however, significantly reduce allergenicity of the transgenic ryegrass pollen, providing the potential to reduce pollen allergenicity in the field (Bhalla *et al.*, 1999).

The major allergens (group I) of grass pollen are structurally related to expansins. A pollen-specific homologue (*Zea mI*) of the *Lol p I* gene, encoding the major allergen of ryegrass pollen, is expressed at a low level prior to pollen mitosis I and at a high level in mature pollen (Broadwater *et al.*, 1993). In somatic cells expansins act to loosen the network of wall polysaccharides, permitting turgor-driven cell enlargement. Group I allergens were able to induce extension (creep) of plant cell walls (Cosgrove *et al.*, 1997). Furthermore, extracts of maize pollen possessed potent expansin-like activity that was selective for grass cell walls. Therefore, group I allergens are thought to facilitate invasion of the pollen tube into the maternal tissues by loosening the cell walls of the grass stigma and style (Cosgrove, 2000).

The *Brassica rapa* pollen allergen Bra r 1 encodes a Ca^{2+}-binding protein specifically expressed in anthers (Okada *et al.*, 2000). When expressed in tobacco, the highest accumulation of Bra r 1 protein was observed in mature pollen, with slightly more intense signals in the pollen tube tip. Another family of pollen allergens encode low molecular weight proteins with two EF-hand Ca^{2+}-binding motifs. This family includes BPC1 and APC1 (*B. napus* and *Arabidopsis* pollen calcium-binding protein 1), which both interact with calcium. BPC1 was found in the cytosol of mature pollen and in the pollen wall upon hydration. BPC1 was also concentrated at the surface of the elongating pollen tube, which suggest its function as a calcium-sensitive signal molecule (Rozwadowski *et al.*, 1999).

4.9.6   Pollen wall proteins

Genes encoding pollen tube wall-associated proteins that are likely to have a role during the progamic phase include the pollen-specific *Pex1* gene from maize (Rubinstein *et al.*, 1995a,b) and the pollen-specific *Ntp303* gene from tobacco (Capková *et al.*, 1988, 1994; Weterings *et al.*, 1992). *PEX1* is an extensin-like glycoprotein localized to the intine of mature pollen and to the callose wall of pollen tubes. *NTP303* is a 69 kDa glycoprotein located at the vegetative face of the pollen tube plasma membrane, in callose plugs and around vegetative cell membrane surrounding the GC during pollen maturation and the sperm cells during pollen tube growth (Wittink *et al.*, 2000). These proteins may function either as structural elements of the pollen tube plasma membrane and wall or in pollen–pistil interactions.

4.10   Advances in gametophytic genetics

The use of genetic analysis to reveal gametophytic factors controlling pollen development and function has a century-old history. In 1900, Correns first reported data suggesting gametophytic expression of pollen-lethals in *Oenothera*. Gametophytic segregation of *glutinous* and *waxy* phenotypes provided further evidence for the postmeiotic function of genes in pollen (Parnell, 1921; Brink and MacGillivray, 1924). Subsequently, a number of examples of distorted mendelian segregation ratios, resulting from reduced genetic transmission through pollen, were reported, including those affecting pollen development or tube growth (reviewed in Ottaviano and Mulcahy, 1989). Moreover, strong male gametophytic selection against most chromosomal deficiencies in maize and *Arabidopsis* further indicated that genes required for male gametophyte development are both numerous and dispersed throughout the genome (Stadler, 1933; Stadler and Roman, 1948; Kindiger *et al.*, 1991; Vizir *et al.*, 1994).

In their review of gametophytic genetics a decade ago, Ottaviano and Mulcahy, (1989) recognized the need to isolate gametophytic mutations to dissect male gametophyte development. However, they also concluded that 'due to the limited research and to difficulties in devising efficient methods for mutant selection and characterization, suitable material for these studies is not plentiful'. In recent years there has been renewed interest in the genetic dissection of pollen development and function, and new approaches have been devised to target functionally important gametophytic genes. Furthermore, the use of gene-tagging strategies and of *Arabidopsis* as a genetic model has had a major impact on the number of essential gametophytic genes identified.

4.10.1   Pollen morphogenesis screens

The first screens were developed to identify gametophytic genes involved in the control of asymmetric cell division and pollen cell fate. Given the stereotypic organization of mature *Arabidopsis* pollen grains, pollen was screened from physically or chemically mutagenized populations for mutants with deviations from the wild-type tricellular morphology (Chen and McCormick, 1996; Park *et al.*, 1998). These screens have yielded a wide range of novel phenotypes resulting from gametophytic mutations, many of which are still being characterized. Mutations affecting division asymmetry (*sidecar pollen*, *gemini pollen*), cytokinesis at PMI (*two-in-one pollen*), generative cell division (*duo*) and male germ unit organization (*germ unit malformed, male germ unit diplaced*) have been identified. The phenotypes of these mutations have been discussed earlier (sections 4.6.5, 4.7.3, 4.7.4). The map positions of these and other known gametophytic mutants in *Arabidopsis* are given in figure 4.7.

Progress in maize has included the isolation of two male-specific gametophytic mutants affecting pollen maturation in transposon mutagenized populations. In *gaMS-1* (*gametophytic male sterile-1*), microspores develop normally until PMI, but bicellular spores abnormally accumulate starch (Sari-Gorla *et al.*, 1996). Mutant pollen remains small and fails to complete generative cell division, suggesting that *GaMS-1* is required for vegetative cell development soon after PMI. In *gaMS-2*, defects also occur after PMI, but a range of pollen phenotypes are observed. These include collapsed pollen with no nuclei, binucleate pollen with undifferentiated nuclei and multinucleate pollen, suggesting defects

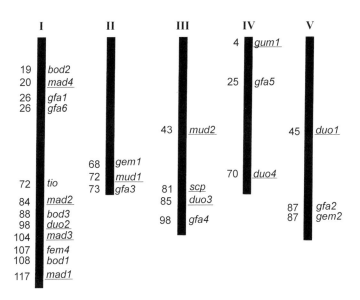

Figure 4.7 Map positions of gametophytic mutations affecting pollen development and function in *Arabidopsis*. Male-specific mutations are underlined and include *duo*; *gum* (*germ unit malformed*); *mad* (*male-defective*) and *mud* (*male germ unit displaced*). Other mutations affect both male and female transmission and include *bod* (*both defective*); *fem* (*female gametophyte*); *gfa* (*gametophytic factor*); *gem* (*gemini pollen*) and *tio* (*two-in-one*). Approximate map positions data are given in cM.

in regulating nuclear differentiation and the cell division cycle (Sari-Gorla *et al.*, 1997).

4.10.2   Segregation ratio distortion screens

Although pollen morphogenesis screens have proved successful in identifying gametophytic mutants, they are insufficient to identify mutations in gameto-phytic genes involved in postpollination events. In this regard, new inclusive strategies have been employed that select both developmental and progamic phase mutants by using marker segregation ratio distortion.

Most screens have employed a gene-tagging strategy that makes use of a dominant antibiotic or herbicide resistance marker carried by a DNA insertion (Bonhomme *et al.*, 1998; Christiansen *et al.*, 1998; Howden *et al.*, 1998). The rationale is that if a DNA insertion inactivates an essential gametophytic gene, then the ratio of resistant to sensitive progeny will deviate significantly below the expected 3:1 ratio. For example, this value would tend toward 1:1 for mutations in essential male- or female-specific genes. This approach is inclusive in that both male- and female-specific mutations can be recovered as well as those with important roles in both sexes. The frequencies of game-tophytic mutations, inferred from the number of lines that show segregation ratio distortion, have been found to be similar at about 1%, in both T-DNA and transposon insertion populations (E. Lalanne, U. Grossniklaus and D. Twell, unpublished). Therefore, systematic screening should lead to the identification of the majority of essential gametophytic genes. However, duplicated genes with redundant functions would be missed or would result in incompletely penetrant phenotypes.

Analysis of T-DNA segeregation in a population of >900 lines revealed 9% with strongly reduced inheritance of kanamycin resistance (Feldmann *et al.*, 1997). Genetic analysis of seven of these *gametophytic factor* (*gfa*) lines revealed reduced transmission through both male and female gametes, with five lines (*gfa3* to *gfa6*; *gf21*) showing a predominant effect in the male, and two lines (*gfa1*, *gfa2*) showing a moderate male effect, and no transmission through the female. Phenotypic analysis of *gfa* mutants and other mutants (*fem3*, *fem4*) affecting both male and female transmission revealed four mutants (*fem3*, *gfa3*, *gfa4*, *gfa5*) with developmental effects, including collapsed (*fem3*, *gfa4*, *gfa5*) and highly vacuolated (*gfa3*) pollen (Christensen *et al.*, 1998). *fem4* and *gfa2* showed normal pollen and therefore affect transmission through postpollination effects. Map positions of *gfa* and *fem* mutants are shown in figure 4.7.

Screening ~1000 promoter-trap T-DNA insertion lines for segregation dis-tortion produced eight mutants showing reduced transmission through male, female or both gametes (Howden *et al.*, 1998). These include the cellular morphogenesis mutant *limpet pollen*, which fails to complete generative cell migration, one male-specific progamic phase mutant, and four lines showing significant effects on male and female transmission. Similarly, Bonhomme *et al.* (1998), selected gametophytic mutants in T-DNA populations. They focused only on highly penetrant lines showing 1:1 segregation and scored 1.3% (207) of the 16 000 lines screened as good candidates. Genetic analysis of 38 of these T-DNA transmission defect (*Ttd*) lines confirmed eight as male-specific mutants, with 0–1% T-DNA transmission through the pollen. *Ttd6*, *Ttd17* and *Ttd38* all showed significant pollen abortion in mature anthers, suggesting a role during pollen development, whereas *Ttd7*, *Ttd8* and *Ttd26* showed viable pollen in mature anthers, suggesting a progamic phase defect. Among the progamic phase mutants, only *Ttd8* showed a defect in pollen tube elongation *in vitro* (Bonhomme *et al.*, 1998).

A different approach was devised by Grini *et al.* (1999), involving segregation distortion of nearby visible markers to screen for ems-induced gametophytic mutants. Using multiple markers on chromosome 1, seven lines were isolated that displayed a marked reduction in transmission through the male or through both male and female gametes. Three male-specific lines (*mad1*, *mad2*, *mad3*) showed defects during pollen development, one male-specific line (*mad4*) was specifically defective in pollen tube elongation, and three other lines showed variable defects in both male and female gametophyte development (*bod1*, *bod2*, *bod3*). The most common phenotype among the male-specific mutants was pollen, arrested at the bicellular stage. *mad3* appeared to act slightly later than *mad1* and *mad2*, after GC DNA replication had taken place. However, unlike *duo pollen* (Twell and Howden, 1998a,b), phenotypes were pleiotropic, suggesting delayed vegetative cell development indirectly affecting GC division. Interestingly, *mad1* also showed dividing walls at mid-bicellular stage reminiscent of phenotypes in *gem1* (Park *et al.*, 1998). Phenotypic analysis of *bod* mutants also revealed pleiotropic effects after PMI including collapse of the vegetative nucleus (*bod1*), abnormal vacuolation (*bod2*) and condensed GC chromosomes (*bod3*), suggesting the involvement of these genes in general cellular processes.

In summary, the 38 male effect mutants identified to date can be subdivided into two groups: morphological mutants, and progamic phase mutants that block male transmission but produce 100% morphologically normal pollen. Current effort, which is focused on the isolation of the genes affected by these mutations, is expected to reveal the molecular details of key processes involved in the differentiation and function of the male gametophyte, and will undoubtedly open up exciting avenues for further research.

4.11 Conclusions and perspective

The past decade has seen significant advances in our understanding of some of the key cellular processes that are essential for microgametogenesis and pollen fertility. The importance of sporophytic–gametophytic interactions and gametophytic control by the haploid genome is now well established, and general models have been proposed to account for microspore polarity, asymmetric division and cell fate determination. With regard to the molecular control of microgametogenesis, we currently have a broad overview of the diversity and modulation of haploid gene expression, but still only a tiny fraction of the haploid male transcriptome is known. In particular, key regulatory genes essential for the gametophytic transcription are yet to be identified. However, progress in the isolation of generative and sperm cells has led to the first reports of gametic cell-specific expression, such that the molecular analysis of the differentiation of the male gametes is now within reach.

The functional analysis of gametophytically expressed genes has progressed largely on a gene-by-gene basis, resulting in a rather anecdotal view of gametophytic gene functions. Most progress has been made in understanding the functions of pollen-expressed proteins that act during the progamic phase, such as signalling molecules that control actin microfilament organization, tip-directed secretion and pollen tube growth. More recently, genetic approaches for the selection of gametophytic mutants have been established, and a rapidly increasing number of mutants are beginning to define new gametophytic gene functions. Continued effort in this area is expected to lead to a comprehensive genetic map of the proteins that are required at different developmental stages and their genetic and cellular interactions.

As we enter the postgenomic era, new genome-wide molecular and genetic technologies will have a profound impact on the pace of discovery of pollen gene expression and functions. This detailed molecular information may then be used to discover patterns of molecular evolution that have shaped the male gametophytic generation. The increased integration of cellular and molecular approaches will also provide a new level of sophistication in the directed modification of the reproductive systems of crop plants. Benefits are therefore to be expected, which include improvements in existing technologies, such as hybrid seed and doubled haploid production, and novel applications yet to emerge.

Acknowledgements

I am particularly indebted to Tracy for many helpful discussions and for editing, and to current members of the laboratory for their effort and commitment to gametophytic genetics. Research in the author's laboratory is supported by BBSRC, The Royal Society and the Gatsby Charitable Foundation, and is gratefully acknowledged.

References

Aarts, M.G.M., Keijzer, C.J., Stiekema, W.J. and Pereira, A. (1995) Molecular characterization of the *CER1* gene of *Arabidopsis* involved in epicuticular wax biosynthesis and pollen fertility. *Plant Cell*, **7**, 2115-2127.

Aarts, M.G.M., Hodge, R., Kalantidis, K., *et al.* (1997) The *Arabidopsis* MALE STERILITY 2 protein shares similarity with reductases in elongation/condensation complexes. *Plant J.*, **12**, 615-623.

Albani, D., Robert, L.S., Donaldson, P.A., Altosaar, I., Arnison, P.G. and Fabijanski, S.F. (1990) Characterisation of a pollen-specific gene family from *Brassica napus* which is activated during early microspore development. *Plant Mol. Biol.*, **15**, 605-622.

Albani, D., Altosaar, I., Arnison, P.G. and Fabijanski, S.F. (1991) A gene showing sequence similarity to pectin esterase is specifically expressed in developing pollen of *Brassica napus*. Sequences in its 5' flanking region are conserved in other pollen-specific promoters. *Plant Mol. Biol.*, **16**, 501-513.

Albani, D., Sardana, R., Robert, L.S., Altosaar, I., Arnison, P.G. and Fabijanski, S.F. (1992) A *Brassica napus* gene family which shows sequence similarity to ascorbate oxidase is expressed in developing pollen—molecular characterization and analysis of promoter activity in transgenic plants. *Plant J.*, **2**, 331-342.

Allen, R.L. and Lonsdale, D.M. (1992) Sequence analysis of three members of the maize polygalacturonase gene family expressed during pollen development. *Plant Mol. Biol.*, **20**, 343-345.

Allen, R.L. and Lonsdale, D.M. (1993) Molecular characterization of one of the maize polygalacturonase gene family members which are expressed during late pollen development. *Plant J.*, **3**, 261-271.

Alvarez-Buylla, E.R., Liljegren, S.J., Pelaz, S., *et al.* (2000) MADS-box gene evolution beyond flowers: expression in pollen, endosperm, guard cells, roots and trichomes. *Plant J.*, **24**, 457-466.

An, Y-Q., Huang, S., McDowell, J.M., McKinney, E.C. and Meagher, R.B. (1996) Conserved expression of the *Arabidopsis* ACT1 and ACT3 actin subclass in organ primordia and mature pollen. *Plant Cell*, **8**, 15-30.

Asturias, J.A., Arilla, M.C., Gomez-Bayon, N., *et al.* (1998) Cloning and immunological characterization of the allergen Hel a 2 (profilin) from sunflower pollen. *Mol Immunol.*, **35**, 469-478.

Atanassov, I., Russinova, E., Antonov, L. and Atanassov, A. (1998) Expression of an anther-specific chalcone synthase-like gene is correlated with uninucleate microspore development in *Nicotiana sylvestris*. *Plant Mol. Biol.*, **38**, 1169-1178.

Baltz, R., Domon, C. and Steinmetz, A. (1992a) Characterization of a pollen-specific cDNA from sunflower encoding a zinc finger protein. *Plant J.*, **2**, 713-721.

Baltz, R., Evrard, J.L., Domon, C. and Steinmetz, A. (1992b) A LIM motif is present in a pollen-specific protein. *Plant Cell*, **4**, 1465-1466.

Baltz, R., Evrard, J-L., Bourdon, V. and Steinmetz, A. (1996) The pollen-specific LIM protein PLIM-1 from sunflower binds nucleic acids *in vitro*. *Sex. Plant Reprod.*, **9**, 264-268.

Baltz, R., Schmit, A.C., Kohnen, M., Hentges, F. and Steinmetz, A. (1999) Differential localization of the LIM domain protein PLIM-1 in microspores and mature pollen grains from sunflower. *Sex. Plant Reprod.*, **12**, 60-65.

Barakate, A., Martin, W., Quigley, F. and Mache, R. (1993) Characterization of a multigene family encoding an exopolygalacturonase in maize. *J. Mol. Biol.*, **229**, 797-801.

Bate, N. and Twell, D. (1998) Functional architecture of a late pollen promoter: pollen specific transcription is developmentally regulated by multiple stage-specific and co-dependent activator elements. *Plant Mol. Biol.*, **37**, 859-869.

Bate, N.J., Orr, J., Ni, W., *et al.* (1994) Quantitative relationship between phenylalanine ammonia-lyase levels and phenylpropanoid accumulation in transgenic tobacco identifies a rate-determining step in natural product synthesis. *Proc. Natl. Acad. Sci. USA*, **91**, 7608-7612.

Bate, N., Spurr, C., Foster, G.D. and Twell, D. (1996) Maturation-specific translational enhancement mediated by the 5'-UTR of a late pollen transcript. *Plant J.*, **10**, 101-111.

Bedinger, P. (1992) The remarkable biology of pollen. *Plant Cell*, **4**, 879-887.

Bedinger, P. and Edgerton, M.D. (1990) Developmental staging of maize microspore proteins. *Plant Physiol.*, **92**, 474-479.

Belostotsky, D.A. and Meagher, R.B. (1993) Differential organ-specific expression of three poly (A) binding protein genes from *Arabidopsis*. *Proc. Natl. Acad. Sci. USA*, **90**, 6686-6690.

Benito Moreno, R.M., Macke, F., Alwen, A. and Heberle-Bors, E. (1988) *In-situ* seed production after pollination with *in-vitro*-matured, isolated pollen. *Planta*, **176**, 145-148.

Bhalla, P.L., Swoboda, I. and Singh, M.B. (1999) Antisense-mediated silencing of a gene encoding a major ryegrass pollen allergen. *Proc. Natl. Acad. Sci. USA*, **96**, 11676-11680.

Bih, F.Y., Wu, S.S., Ratnayake, C., Walling, L.L., Nothnagel, E.A. and Huang, A.H. (1999) The predominant protein on the surface of maize pollen is an endoxylanase synthesized by a tapetum mRNA with a long 5' leader. *J. Biol. Chem.*, **274**, 22884-22894.

Blackmore, S. and Barnes, S.H. (1990) Pollen wall development in angiosperms, in *Microspores Evolutions and Ontogeny* (eds. S. Blackmore and R.B. Knox), Academic Press, San Diego, pp. 173-192.

Blackmore, S. and Crane, P.R. (1988) The systematic implications of pollen and spore ontogeny, in *Ontogeny and Sytematics* (ed. C.J. Humphries), Columbia University Press, New York, pp. 83-115.

Blomstedt, C.K., Knox, R.B. and Singh, M.B. (1996) Generative cells of *Lilium longiflorum* possess translatable mRNA and functional protein synthesis machinery. *Plant. Mol. Biol.*, **31**, 1083-1086.

Bonhomme, S., Horlow, C., Vezon, D., *et al.* (1998) T-DNA mediated disruption of essential gametophytic genes in *Arabidopsis* is unexpectedly rare and cannot be inferred from segregation distortion alone. *Mol. Gen. Genet.*, **260**, 444-452.

Brander, K.A. and Kuhlemeier, C. (1995) A pollen-specific DEAD-box protein related to translation initiation factor eIF-4A from tobacco. *Plant Mol. Biol.*, **27**, 637-649.

Breiteneder, H., Pettenburger, K., Bito, A., *et al.* (1989) The gene coding for the major birch pollen allergen BetvI, is highly homologous to a pea disease resistance response gene. *EMBO J.*, **8**, 1935-1938.

Brink, R.A. and MacGillivray, J.H. (1924) Segregation for the waxy character in maize pollen and differential development of the male gametophyte. *Am. J. Bot.*, **11**, 465-469.

Broadwater, A.H., Rubinstein, A.L., Chay, C.H., Klapper, D.G. and Bedinger, P.A. (1993) Zea mI, the maize homolog of the allergen-encoding Lol pI gene of rye grass. *Gene*, **15**, 227-230.

Brown, R.C. and Lemmon, B.E. (1991a) Pollen development in orchids. 3. A novel generative pole microtubule system predicts unequal pollen mitosis. *J. Cell Sci.*, **99**, 273-281.

Brown, R.C. and Lemmon, B.E. (1991b) The cytokinetic apparatus in meiosis: control of division plane in the absence of a preprophase band of microtubules, in *The Cytoskeletal Basis of Plant Growth and Form* (ed. C.W. Lloyd), Academic Press, New York, pp. 259-273.

Brown, R.C. and Lemmon, B.E. (1992) Pollen development in orchids. 4. Cytoskeleton and ultrastructure of the unequal pollen mitosis in *Phalaenopsis*. *Protoplasma*, **167**, 183-192.

Brown, R.C. and Lemmon, B.E. (1994) Pollen mitosis in the slipper orchid *Cypripedium fasiculatum*. *Sex. Plant Reprod.*, **7**, 87-94.

Brown, S.M. and Crouch, M.L. (1990) Characterization of a gene family abundantly expressed in *Oenothera organensis* pollen that shows sequence similarity to polygalacturonase. *Plant Cell*, **2**, 263-274.

Bucher, M., Brander, K.A., Sbicego, S., Mandel, T. and Kuhlemeier, C. (1995) Aerobic fermentation in tobacco pollen. *Plant Mol Biol.*, **28**, 739-750.

Burbulis, I.E., Iacobucci, M. and Shirley, B.W. (1996) A null mutation in the first enzyme of flavonoid biosynthesis does not affect male fertility in *Arabidopsis*. *Plant Cell*, **8**, 1013-1025.

Cai, G., Romagnoli, S., Moscatelli, A., *et al.* (2000) Identification and characterization of a novel microtubule-based motor associated with membranous organelles in tobacco pollen tubes. *Plant Cell.*, **12**, 1719-1736.

Capková, V., Hrabetová E. and Tupy, J. (1988) Protein synthesis in pollen tubes: preferential formation of new species independent of transcription. *Sex. Plant Reprod.*, **1**, 150-155.

Capková, V., Zbrozek, J. and Tupy, J. (1994) Protein synthesis in tobacco pollen tubes: preferential synthesis of cell-wall 69-kDa and 66-kDa glycoproteins. *Sex. Plant Reprod.*, **7**, 57-66.

Carpenter, J.L., Ploense, S.E., Snustad, D.P. and Silflow, C.D. (1992) Preferential expression of an alpha-tubulin gene of *Arabidopsis* in pollen. *Plant Cell*, **4**, 557-571.

Chaboud, A. and Perez, R. (1992) Generative cells and male gametes: isolation, physiology, and biochemistry. *Int. Rev. Cytol.*, **140**, 205-232.

Chang, C. and Meyerowitz, E.M. (1986) Molecular cloning and DNA sequence of the *Arabidopsis thaliana* alcohol dehydrogenase gene. *Proc. Natl. Acad. Sci. USA*, **83**, 1408-1412.

Charzynska, M., Murgia, M., Milanesi, C., Cresti, M. (1989) Origin of sperm cell association in the male germ unit of *Brassica* pollen. *Protoplasma*, **149**, 1-4.

Chaudhury, A.M. (1993) Nuclear genes controlling male fertility. *Plant Cell*, **5**, 1277-1283.

Chen, C.H., Oishi, K.K., Kloeckener-Gruissem, B. and Freeling, M. (1987) Organ-specific expression of maize Adh1 is altered after a Mu transposon insertion. *Genetics*, **116**, 469-477.

Chen, Y.-C. and McCormick, S. (1996) *sidecar pollen*, an *Arabidopsis thaliana* male gametophytic mutant with aberrant cell divisions during pollen development. *Development*, **122**, 3243-3253.

Christensen, H.E., Ramachandran, S., Tan, C.T., Surana, U., Dong, C.H. and Chua, N.-H. (1996) *Arabidopsis* profilins are functionally similar to yeast profilins: identification of a vascular bundle-specific profilin and a pollen-specific profilin. *Plant J.*, **10**, 269-279.

Christensen, C.A., Subramanian, S. and Drews, G.N. (1998) Identification of gametophytic mutations affecting female gametophyte development in *Arabidopsis*. *Dev. Biol.*, **202**, 136-151.

Chung, Y.-Y., Magnuson, N.S. and An, G. (1995) Subcellular localization of actin depolymerizing factor in mature and germinating pollen. *Mol. Cells*, **5**, 224-229.

Clarke, S.R., Staiger, C.J., Gibbon, B.C. and Franklin-Tong, V.E. (1998) A potential signaling role for profilin in pollen of *Papaver rhoeas*. *Plant Cell*, **10**, 967-979.

Correns, K. (1900) Uber den Einfluss, welchen die Zahl der zur Bestaubung verwendeten Pollenkorner auf die Machkommenschaft hat. *Ber. Bot. Ges.*, **18**, 422-435.

Cosgrove, D.J. (2000) Loosening of plant cell walls by expansins. *Nature*, **407**, 321-326.

Cosgrove, D.J., Bedinger, P. and Durachko, D.M. (1997) Group I allergens of grass pollen as cell wall-loosening agents. *Proc. Natl. Acad. Sci. USA*, **94**, 6559-6564.

Curie, C. and McCormick, S. (1997) A strong inhibitor of gene expression in the 5′ untranslated region of the pollen-specific *LAT59* gene of tomato. *Plant Cell*, **9**, 2025-2036.

Custers, J.B., Oldenhof, M.T., Schrauwen, J.A., Cordewener, J.H., Wullems, G.J. and van Lookeren Campagne, M.M. (1997) Analysis of microspore-specific promoters in transgenic tobacco. *Plant Mol. Biol.*, **35**, 689-699.

De Dios Alche, J., Castro, A.J., Olmedilla, A., *et al.* (1999) The major olive pollen allergen (Ole e I) shows both gametophytic and sporophytic expression during anther development, and its synthesis and storage takes place in the RER. *J. Cell Sci.*, **112**, 2501-2509.

De Vlaming, P., Cornu, A., Farcy, E., Gerats, A.G.M., Malzonier, D., Wiering, H. and Wijsman, H.J.W. (1984) *Petunia hybrida*: a short description of the action of 91 genes, their origin and their map location. *Plant Mol. Biol. Rep.*, **2**, 21-42.

De Vlaming, P. and Kho, K.F.F. (1976) 4.2,4.6-Tetrahydroxychalcone in pollen of *Petunia hybrida*. *Phytochemistry*, **15**, 348-349.

Dickinson, H.G. (1994) The regulation of alternation of generation in flowering plants. *Biol. Rev.*, **69**, 419-442.

Dickinson, H.G., Elleman, C.J. and Doughty, J. (2000) Pollen coatings—chimaeric genetics and new fucntions. *Sex Plant Reprod.*, **12**, 302-309.

Dodds, P.N., Bonig, I., Du, H., Rodin, J., Anderson, M.A., Newbigin, E. and Clarke, A.E. (1993) S-RNase gene of *Nicotiana alata* is expressed in developing pollen. *Plant Cell*, **5**, 1771-1782.

Doughty, J., Dixon, S., Hiscock, S.J., Willis, A.C., Parkin, I.A. and Dickinson, H.G. (1998) PCP-A1, a defensin-like *Brassica* pollen coat protein that binds the S locus glycoprotein, is the product of gametophytic gene expression. *Plant Cell.*, **10**, 1333-1347.

Dumas, C., Knox, R.B., McConchie, C.A. and Russell, S.D. (1984) Emerging physiological concepts in fertilisation. *What's News in Plant Physiology*, **15**, 17-20.

Dumas, C., Knox, R.B. and Gaude, T. (1985) The spatial association of the sperm cells and vegetative nucleus in the pollen grain of *Brassica*. *Protoplasma*, **124**, 168-174.

Dumas, C., Berger, F., Faure, J.-E. and Matthys-Rochon, E. (1998) Gametes, fertilization and early embryogenesis in flowering plants. *Adv. Bot. Res.*, **28**, 231-261.

Eady, C., Lindsey, K. and Twell, D. (1994) Differential activation and conserved vegetative-cell-specific activity of a late pollen promoter in species with bi- and tricellular pollen. *Plant J.*, **5**, 543-550.

Eady, C., Lindsey, K. and Twell, D. (1995) The significance of microspore division and division symmetry for vegetative cell-specific transcription and generative cell differentiation. *Plant Cell*, **7**, 65-74.

Einzenberger, E., Eller, N., Heberle-Bors, E. and Vicente, O. (1995) Isolation and expression during pollen development of a tobacco cDNA clone encoding a protein kinase homologous to shaggy/glycogen synthase kinase-3. *Biochim. Biophys. Acta*, **1260**, 315-319.

Erdtman, G. (1952) Pollen and spore morphology, in *Pollen Morphology and Plant Taxonomy— Angiosperms (An Introduction to Palymology I)* Almqvist and Wiksell, pp. 11-24.

Estruch, J.J., Kadwell, S., Merlin, E. and Crossland, L. (1994) Cloning and characterization of a maize pollen-specific calcium-dependent calmodulin-independent protein kinase. *Proc. Natl. Acad. Sci. USA*, **91**, 8837-8841.

Evans, D.E., Taylor, P.E., Singh, M.B. and Knox, R.B. (1992) The interrelationship between the accumulation of lipids, protein and the level of acyl carrier protein during the development of *Brassica napus* L. pollen. *Planta*, **186**, 343-354.

Eyal, Y., Curie, C. and McCormick, S. (1995) Pollen specificity elements reside in 30 bp of the proximal promoters of two pollen-expressed genes. *Plant Cell*, **7**, 373-384.

Feldmann, K.A., Coury, D.A. and Christianson, M.L. (1997) Exceptional segregation of a selectable marker (Kan[R]) in *Arabidopsis* identifies genes important for gametophytic growth and development. *Genetics*, **147**, 1411-1422.

Foster, G.D., Robinson, S.W., Blundell, R.P., *et al.* (1992) A *Brassica napus* messenger RNA encoding a protein homologous to phospholipid transfer proteins is expressed specifically in the tapetum and developing microspores. *Plant Sci.*, **84**, 187-192.

Franklin-Tong, V.E. (1999) Signalling and the modulation of pollen tube growth. *Plant Cell*, **11**, 727-738.

Friedman, W.E. (1999) Expression of the cell cycle in sperm of *Arabidopsis*: implications for understanding patterns of gametogenesis and fertilization in plants and other eukaryotes. *Development*, **126**, 1065-1075.

Fujita, T., Maggio, A., Garcia-Rios, M., Bressan, R.A. and Csonka, L.N. (1998) Transcriptional regulation of two evolutionarily divergent genes for Delta1-pyrroline-5-carboxylate is not important for the osmotic or pollen-specific regulation of proline synthesis in tomato. *Plant Physiol.*, **118**, 661-674.

Futamura, N., Mori, H., Kouchi, H. and Shinohara, K. (2000) Male flower-specific expression of genes for polygalacturonase, pectin methylesterase and β-1,3-glucanase in a dioecious willow (*Salix gilgiana* Seemen). *Plant Cell Physiol.*, **41**, 16-26.

Gerats, A.G., Vrijlandt, E., Wallroth, M. and Schram, A.W. (1985) The influence of the genes An1, An2, and An4 on the activity of the enzyme UDP-glucose:flavonoid 3-*O*-glucosyltransferase in flowers of *Petunia hybrida*. *Biochem Genet.*, **23**, 591-598.

Gervais, C., Simmonds, D.H. and Newcomb, W. (1994) Actin microfilament organization during pollen development of *Brassica napus* cv. Topas. *Protoplasma*, **183**, 67-76.

Gibbon, B.C., Zonia, L.E., Kovar, D.R., Hussey, P.J. and Staiger, C.J. (1998) Pollen profilin function depends on interaction with proline-rich motifs. *Plant Cell*, **10**, 981-993.

Gregerson, R., McLean, M., Beld, M., Gerats, A.G.M. and Strommer, J. (1991) Structure, expression, chromosomal location and product of the gene encoding ADH1 in *Petunia. Plant Mol. Biol.*, **17**, 37-48.

Griffith, I.J., Smith, P.M., Pollock, J., *et al.* (1991) Cloning and sequencing of *Lol p*I, the major allergenic protein of rye-grass pollen. *FEBS Lett.*, **279**, 210-215.

Grini, P.E., Schnittger, A., Schwarz, H., *et al.* (1999) Isolation of ethyl methanesulfonate-induced gametophytic mutants in *Arabidopsis thaliana* by a segregation distortion assay using the multimarker chromosome 1. *Genetics*, **151**, 849-863.

Guerrero, F.D., Crossland, L., Smutzer, G.S., Hamilton, D.A. and Mascarenhas, J.P. (1990) Promoter sequences from a maize pollen-specific gene direct tissue-specific transcription in tobacco. *Mol. Gen. Genet.*, **224**, 161-168.

Guyon, V.N., Astwood, J.D., Garner, E.C., Dunker, A.K. and Taylor, L.P. (2000) Isolation and characterization of cDNAs expressed in the early stages of flavonol-induced pollen germination in petunia. *Plant Physiol.*, **123**, 699-710.

Hamilton, D.A., Roy, M., Rueda, J., Sindhu, R.K., Sanford, J. and Mascarenhas, J.P. (1992) Dissection of a pollen-specific promoter from maize by transient transformation assays. *Plant Mol. Biol.*, **18**, 211-218.

Hamilton, D.A., Schwarz, Y.H. and Mascarenhas, J.P. (1998) A monocot pollen-specific promoter contains separable pollen-specific and quantitative elements. *Plant Mol Biol.*, **38**, 663-669.

Hanson, D.D., Hamilton, D.A., Travis, J.I., Bashe, D.M. and Mascarenhas, J.P. (1989) Characterization of a pollen-specific cDNA clone from *Zea mays* and its expression. *Plant Cell*, **1**, 173-179.

Haskell, D.W. and Rogers, O.M. (1985) RNA synthesis by vegetative and sperm nuclei of trinucleate pollen. *Cytologia*, **50**, 805-809.

Hause, G., Hause, B. and van Lammeren, A.A.M. (1991) Microtubular and actin filament configurations during microspore and pollen development in *Brassica napus* L. cv. Topas. *Can. J. Bot.*, **70**, 1369-1376.

Heizmann, P., Luu, D.-T. and Dumas, C. (2000) The clues to species specificity of pollination among *Brassicaceae. Sex. Plant Reprod.*, **13**, 157-161.

Heslop-Harrison, J. (1968) Synchronous pollen mitosis and the formation of the generative cell in massulate orchids. *J. Cell Sci.*, **3**, 457-466.

Heslop-Harrison, J. (1989) Actomyosin and movement in the angiosperm pollen tube. *Sex. Plant Reprod.*, **2**, 199-207.

Heslop-Harrison, J. (1971) Wall pattern formation in angiosperm microsporogenesis. *Symp. Soc. Exp. Biol.*, **25**, 277-300.

Heslop-Harrison, J. and Heslop-Harrison, Y. (1989) Myosin associated with the surfaces of organelles, vegetative nuclei and generative cells in angiosperm pollen grains and tubes. *J. Cell Sci.*, **94**, 319-325.

Heslop-Harrison, J., Heslop-Harrison, Y., Cresti, M., Tiezzi, A. and Moscatelli, A. (1988) Cytoskeletal elements, cell shaping and movement in the angiosperm pollen tube. *J. Cell Sci.*, **91**, 49-60.

Heuer, S., Lorz, H. and Dresselhaus, T. (2000) The MADS box gene *ZmMADS2* is specifically expressed in maize pollen and during maize pollen tube growth. *Sex. Plant Reprod.*, **13**, 21-27.

Hiscock, S.J., Dewey, F.M., Coleman, J.O.D. and Dickinson, H.G. (1994) Identification and localization of an active cutinase in the pollen of *Brassica napus* L. *Planta*, **193**, 377-384.

Hoekstra, F.A. and Bruinsma, J. (1979) Protein synthesis of binucleate and trinucleate pollen and its relationship to tube emergence and growth. *Planta*, **146**, 559-566.

Hoekstra, F.A., Crowe, L.M. and Crowe, J.H. (1989) Differential dessication sensitivity of corn and *Pennisetum* pollen linked to their sucrose contents. *Plant Cell Environ.*, **12**, 83-91.

Honys, D., Combe, J.P., Twell, D. and Capková, V. (2000) The translationally repressed pollen-specific *ntp303* mRNA is stored in non-polysomal mRNPs during pollen maturation. *Sex. Plant Reprod.*, **13**, 135-144.

Howden, R., Park, S.K., Moore, J.M., Orme, J., Grossniklaus, U. and Twell, D. (1998) Selection of T-DNA-tagged male and female gametophytic mutants by segregation distortion in *Arabidopsis*. *Genetics*, **149**, 621-631.

Hu, S. and Yu, H.S. (1988) Preliminary observations on the formation of the male germ unit in pollen tubes of *Cyphomandra betacea* Sendt. *Protoplasma*, **147**, 55-63.

Huang, J.C., Lin, S.M. and Wang, C.S. (2000) A pollen-specific and desiccation-associated transcript in *Lilium longiflorum* during development and stress. *Plant Cell Physiol.*, **41**, 477-485.

Huang, S., An, Y.Q., McDowell, J.M., McKinney, E.C. and Meagher, R.B. (1996a) The *Arabidopsis thaliana* ACT4/ACT12 actin gene subclass is strongly expressed throughout pollen development. *Plant J.*, **10**, 189-202.

Huang, S., McDowell, J.M., Weise, M.J. and Meagher, R.B. (1996b) The *Arabidopsis* profilin gene family. Evidence for an ancient split between constitutive and pollen-specific profilin genes. *Plant Physiol.*, **111**, 115-126.

Huang, S., An, Y.Q., McDowell, J.M., McKinney, E.C. and Meagher, R.B. (1997) The *Arabidopsis* ACT11 actin gene is strongly expressed in tissues of the emerging inflorescence, pollen, and developing ovules. *Plant Mol. Biol.*, **33**, 125-139.

Hülskamp, M., Kopezak, S.D., Horejsi, T.F., Kihl, B.K. and Pruitt, R.E. (1995) Identification of genes required for pollen-stigma recognition in *Arabidopsis thaliana*. *Plant J.*, **8**, 703-714.

Hülskamp, M., Parekh, N.S., Grini, P., *et al.* (1997) The *STUD* gene is required for male-specific cytokinesis after telophase II of meiosis in *Arabidopsis thaliana*. *Dev. Biol.*, **187**, 114-124.

Hussey, P.J., Haas, N., Hunsperger, J., Larkin, J., Snustad, D.P. and Silflow, C.D. (1990) The beta-tubulin gene family in *Zea mays*: two differentially expressed beta-tubulin genes. *Plant Mol. Biol.*, **15**, 957-972.

Izhar, S. and Frankel, R. (1971) Mechanisms of male sterility in *Petunia*: the relationship between pH, callase activity in the anthers and the breakdown of microsporogenesis. *Theor. Appl. Genet.*, **41**, 104-108.

Jackson, R.C., Skvarla, J.J. and Chissoe, W.F. (2000) A unique pollen wall mutation in the family *Compositae*: ultrastructure and genetics. *Am. J. Bot.*, **87**, 1571-1577.

John, M.E. and Peterson, M.W. (1994) Cotton (*Gossypium hirsutum* L.) pollen-specific polygalacturonase mRNA: tissue and temporal specificity of its promoter in transgenic tobacco. *Plant Mol. Biol.*, **26**, 1989-1993.

Jouannic, S., Hamal, A., Leprince, A.S., Tregear, J.W., Kreis, M. and Henry, Y. (1999) Characterisation of novel plant genes encoding MEKK/STE11 and RAF-related protein kinases. *Gene*, **229**, 171-181.

Kandasamy, M.K., McKinney, E.C. and Meagher, R.B. (1999) The late pollen-specific actins in angiosperms. *Plant J.*, **18**, 681-691.

Kaul, L.H. (1988) *Male Sterility in Higher Plants*. Springer-Verlag, Berlin.

Keller, W.A. and Armstrong, K.C. (1979) Stimulation of embryogenesis and haploid production in *Brassica campestris* anther cultures by elevated temperature treatments, *Theor. Appl. Genet.*, **55**, 65-67.

Kim, S.-R., Kim, Y. and An, G. (1993) Molecular cloning and characterization of anther-preferential cDNA encoding a putative actin depolymerizing factor. *Plant Mol. Biol.*, **21**, 39-45.

Kindiger, B., Beckett, J.B. and Coe, E.H. (1991) Differential effects of specific chromosomal deficiencies on the development of the maize pollen grain. *Genome*, **34**, 579-594.

Kishitani, S., Yomoda, A., Konno, N. and Tanaka, Y. (1993) Involvement of phenylalanine ammonia-lyase in the development of pollen in broccoli (*Brassica oleracea* L.). *Sex. Plant Reprod.*, **6**, 244-248.

Knox, R.B. (1971) Pollen wall proteins: localisation, enzymatic and antigenic activity during development in *Gladiolus* (*Iridaceae*). *J. Cell Sci.*, **9**, 209-217.

Knox, R.B. and Heslop-Harrison, J. (1970) Pollen wall proteins: localisation and enzymatic activity. *J. Cell Sci.*, **6**, 1-27.

Knox, R.B., Taylor, P., Ladiges, P., Nelson, G. and Suphioglu, C. (1998) Pollen alleregns: molecular and immunological analysis and implications for systematics, in *Reproductive Biology* (eds. S.J. Owens and P.J. Rudall). Royal Botanic Gardens, Kew, pp. 449-463.

Kobayashi, A., Sakamoto, A., Kubo, K., Rybka, Z., Kanno, Y. and Takatsuji, H. (1998) Seven zinc-finger transcription factors are expressed sequentially during the development of anthers of *Petunia*. *Plant J.*, **13**, 571-576.

Kost, B., Lemichez, E., Spielhofer, P., *et al.* (1999) Rac homologues and compartmentalized phosphatidylinositol 4,5-bisphosphate act in a common pathway to regulate polar pollen tube growth. *J. Cell. Biol.*, **145**, 317-330.

Kovar, D.R., Drobak, B.K. and Staiger, C.J. (2000) Maize profilin isoforms are functionally distinct. *Plant Cell*, **12**, 583-598.

Krauter-Canham, R., Bronner, R., Evrard, J.-L. Hahne, G., Friedt, W. and Steinmetz, A. (1997) A transmitting tissue- and pollen-expressed protein from sunflower with sequence similarity to the human RTP protein. *Plant Sci.*, **1129**, 191-202.

Kulikauskas, R. and McCormick, S. (1997) Identification of the tobacco and *Arabidopsis* homologues of the pollen-expressed LAT59 gene of tomato. *Plant Mol. Biol.*, **34**, 809-814.

Kyo, M., Miyatake, H., Mamezuka, K. and Amagata, K. (2000) Cloning of cDNA encoding NtEPc, a marker protein for the embryogenic dedifferentiation of immature tobacco pollen grains cultured *in vitro*. *Plant Cell Physiol.*, **41**, 129-137.

Lalanne, E., Mathieu, C., Roche, O., Vedel, F. and De Pape, R. (1997) Structure and specific expression of a *Nicotiana sylvestris* putative amino-acid transporter gene in mature and *in vitro* germinating pollen. *Plant Mol. Biol.*, **35**, 855-864.

Lalanne, E., Mathieu, C., Vedel, F. and De Paepe, R. (1998) Tissue-specific expression of genes encoding isoforms of the mitochondrial ATPase b-subunit in *Nicotiana sylvestris Plant Mol. Biol.*, **38**, 885-888.

Lansac, A.R., Sullivan, C.Y. and Johnson, B.E. (1996) Accumulation of free proline in sorghum (*Sorghum bicolor*) pollen. *Can J. Bot.*, **74**, 40-45.

Lee, H.-S., Chung, Y.-Y., Das, C., *et al.* (1997) Embryo sac development is affected in *Petunia inflata* plants transformed with an antisense gene encoding the extracellular domain of receptor kinase PRK1. *Sex. Plant Reprod.*, **10**, 341-350.

Lee, H-S., Karunanandaa, B., McCubbin, A., Gilroy, S. and Kao, T-H. (1996) PRK1, a receptor-like kinase of *Petunia inflata*, is essential for postmeiotic development of pollen. *Plant J.*, **9**, 613-624.

Li, H., Wu, G., Ware, D., Davis, K.R. and Yang, Z. (1998) *Arabidopsis* Rho-related GTPases: differential gene expression in pollen and polar localization in fission yeast. *Plant Physiol.*, **118**, 407-417.

Li, H., Lin, Y., Heath, R.M., Zhu, M.X. and Yang, Z. (1999) Control of pollen tube tip growth by a Rop GTPase-dependent pathway that leads to tip-localized calcium influx. *Plant Cell.*, **11**, 1731-1742.

Li, Y.Q., Chen, F., Linskins, H.F. and Cresti, M. (1994) Distribution of unesterified and esterified pectins in cell walls of pollen tubes of flowering plants. *Sex. Plant Reprod.*, **7**, 145-152.

Lin, Y. and Yang, Z. (1997) Inhibition of pollen tube elongation by microinjected anti-Rop1 Ps antibodies suggests a crucial role for Rho-type GTPases in the control of tip growth. *Plant Cell*, **9**, 1647-1659.

Lin, Y., Wang, Y., Zhu, J. and Yang, Z. (1996) Localization of a Rho GTPase implies a role in tip growth and movement of the generative cell in pollen tubes. *Plant Cell*, **8**, 293-303.

Liu, Z., Xia, M. and Poovaiah, B.W. (1998) Chimeric calcium/calmodulin-dependent protein kinase in tobacco: differential regulation by calmodulin isoforms. *Plant Mol. Biol.*, **38**, 889-897.

Lopez, I., Anthony, R.G., Maciver, S.K., *et al.* (1996) Pollen specific expression of maize genes encoding actin depolymerizing factor-like proteins. *Proc. Natl. Acad. Sci. USA*, **93**, 7415-7420.

Lord, E. (2000) Adhesion and cell movement during pollination: cherchez la femme. *Trends Plant Sci.*, **5**, 368-369.

Ma, L., Xu, X., Cui S. and Sun, D. (1999) The presence of a heterotrimeric G protein and its role in signal transduction of extracellular calmodulin in pollen germination and tube growth. *Plant Cell*, **11**, 1351-1364.

Maheshwari, P. (1950) *An Introduction to the Embryology of the Angiosperms.* McGraw-Hill, New York.

Maiti, I.B., Kolattukudy, P.E. and Shaykh, M. (1979) Purification and characterization of a novel cutinase from Nasturtium (*Tropaeolum majus*) pollen. *Arch. Biochem. Biophys.*, **196**, 412-423.

Malho, R. (1998) Pollen tube guidance—the long and winding road. *Sex. Plant Reprod.*, **11**, 242-244.

Mandaron, P., Niogret, M.F., Mache, R. and Monegar, F. (1990) *In vitro* protein synthesis in isolated microspores of *Zea mays* at several stages of development. *Theor. Appl. Genet.*, **80**, 134-138.

Mariani, C., de Beuckeleer, M., Truettner, J., Leemans, J. and Goldberg, R.B. (1990) Induction of male sterility in plants by a chimeric ribonuclease gene. *Nature*, **347**, 737-741.

Mascarenhas, J.P. (1975) The biochemistry of angiosperm pollen development. *Bot. Rev.*, **41**, 259-314.

Mascarenhas, J.P. (1990) Gene activity during pollen development. *Annu. Rev. Plant Physiol. Plant Mol. Biol.*, **41**, 317-338.

Mascarenhas, J.P. (1992) Pollen gene expression, in *International Review of Cytology: Sexual Reproduction in Flowering Plants* (eds. S.D. Russell and C. Dumas), Academic Press, San Diego, pp. 3-18.

Mascarenhas, J.P (1993) Molecular mechanisms of pollen tube growth and differentiation. *Plant Cell*, **5**, 1303-1314.

Matsuda, N., Tsuchiya, T., Kishitani, S., Tanaka, Y. and Toriyama, K. (1996) Partial male sterility in transgenic tobacco carrying antisense and sense PAL cDNA under the control of a tapetum-specific promoter. *Plant Cell Physiol.*, **37**, 215-222.

Mayfield, J.A. and Preuss, D. (2000) Rapid initiation of *Arabidopsis* pollination requires the oleosin-domain protein GRP17. *Nature Cell Biology*, **2**, 128-130.

McCormick, S. (1991) Molecular analysis of male gametogenesis in plants. *Trends Genet.*, **7**, 298-303.

Miller, K.D., Guyon, V., Evans, J.N., Shuttleworth, W.A. and Taylor, L.P. (1999) Purification, cloning, and heterologous expression of a catalytically efficient flavonol 3-*O*-galactosyltransferase expressed in the male gametophyte of *Petunia hybrida. J. Biol. Chem.*, **274**, 3401-3419.

Mittermann, I., Swoboda, I., Pierson, E., *et al.* (1995) Molecular cloning and characterization of profilin from tobacco (*Nicotiana tabacum*): increased profilin expression during pollen maturation. *Plant Mol. Biol.*, **27**, 137-146.

Mo, Y.Y., Nagel, C. and Taylor, L.P. (1992) Biochemical complementation of chalcone synthase mutants defines a role for flavonols in functional pollen. *Proc. Natl. Acad. Sci. USA*, **89**, 7213-7217.

Mogensen, H.L. (1992) The male germ unit: concept, composition and signification. *Int. Rev Cytol.*, **140**, 129-147.

Moore, I., Diefenthal, T., Zarsky, V., Schell, J. and Palme, K. (1997) A homolog of the mammalian GTPase Rab2 is present in *Arabidopsis* and is expressed predominantly in pollen grains and seedlings. *Proc. Natl. Acad. Sci. USA*, **94**, 762-767.

Moutinho, A., Trewavas, A.J. and Malho, R. (1998) Relocation of a Ca^{2+}-dependent protein kinase activity during pollen tube reorientation. *Plant Cell*, **10**, 1499-1510.

Mousavi, A., Hiratsuka, R., Takase, H., Hiratsuka, K. and Hotta, Y. (1999) A novel glycine-rich protein is associated with starch grain accumulation during anther development. *Plant Cell Physiol.*, **40**, 406-416.

Mu, J.H., Lee, H.S. and Kao, T.H. (1994a) Characterization of a pollen-expressed receptor-like kinase gene of *Petunia inflata* and the activity of its encoded kinase. *Plant Cell*, **6**, 709-721.

Mu, J.H., Stains, J.P. and Kao, T.H. (1994b) Characterization of a pollen-expressed gene enocoding a putative pectin esterase of *Petunia inflata. Plant Mol. Biol.*, **25**, 539-544.

Murgia, M., Charzynska, M., Rougier, M. and Cresti, M. (1991) Secretory tapetum of *Brassica oleracea* L.: polarity and ultrastructural features. *Sex. Plant Reprod.*, **4**, 28-35.

Muschietti, J., Dircks, L., Vancanneyt, G. and McCormick, S. (1994) LAT52 protein is essential for tomato pollen development: pollen expressing antisense *LAT52* RNA hydrates and germinates abnormally and cannot achieve fertilization. *Plant J.*, **6**, 321-338.

Muschietti, J., Eyal, Y. and McCormick, S. (1998) Pollen tube localization implies a role in pollen–pistil interactions for the tomato receptor-like protein kinases LePRK1and LePRK2. *Plant Cell*, **10**, 319-330.

Nagata, N., Saito, C., Sakai, A., Koroiwa, H. and Kuroiwa, T. (1999) The selective increase or decrease of organellar DNA in generative cells just after pollen mitosis one controls cytoplasmic inheritance. *Planta*, **209**, 53-65.

Napoli, C.A., Fahy, D., Wang, H.Y. and Taylor, L.P. (1999) *white anther*: a petunia mutant that abolishes pollen flavonol accumulation, induces male sterility, and is complemented by a chalcone synthase transgene. *Plant Physiol.*, **120**, 615-622.

Nasrallah, J.E. (2000) Cell–cell signalling in the self-incompatibility response. *Curr. Opin. Plant Biol.*, **3**, 368-373.

Niogret, M.F., Dubald, M., Mandaron, P. and Mache, R. (1991) Characterization of pollen polygalacturonase encoded by several cDNA clones in maize. *Plant Mol. Biol.*, **17**, 1155-1164.

Oakeley, E.J., Podesta, A. and Jost, J.P. (1997) Developmental changes in DNA methylation of the two tobacco pollen nuclei during maturation. *Proc. Natl. Acad. Sci. USA*, **94**, 11721-11725.

Okada T, Sasaki Y, Ohta R, Onozuka N. and Toriyama K. (2000) Expression of Bra r 1 gene in transgenic tobacco and Bra r 1 promoter activity in pollen of various plant species. *Plant Cell Physiol.*, **41**, 757-766.

Oldenhof, M.T., de Groot, P.F., Visser, J.H., Schrauwen, J.A. and Wullems, G.J. (1996) Isolation and characterization of a microspore-specific gene from tobacco. *Plant Mol. Biol.*, **31**, 213-225.

Olmedilla, A., Schrauwen, J.A.M. and Wullems, G.J. (1991) Visualization of starch-synthase expression by *in situ* hybridization during pollen development. *Planta*, **184**, 182-186.

Olsen, E., Zhand, L., Hill, R.D., Kisil, F.T., Sehon, A.H. and Mohapatra, S.S. (1991) Identification and characterization of the *Poa p* IX group of basic allergens of Kentucky Bluegrass pollen. *J. Immunol.*, **147**, 205-211.

Op den Camp, R.G. and Kuhlemeier, C. (1997) Aldehyde dehydrogenase in tobacco pollen. *Plant Mol. Biol.*, **35**, 355-365.

Op den Camp, R.G. and Kuhlemeier, C. (1998) Phosphorylation of tobacco eukaryotic translation initiation factor 4A upon pollen tube germination. *Nucleic Acids Res.*, **26**, 2058-2062.

Ottaviano, E. and Mulcahy, D.L. (1989) Genetics of angiosperm pollen. *Adv. Genet.*, **26**, 1-64.

Owen, H.A. and Makaroff, C.A. (1995) Ultrastructure of microsporogenesis and microgametogenesis in *Arabidopsis thaliana* (L.) Heynh. ecotype Wassilewskija (*Brassicaceae*). *Protoplasma*, **185**, 7-21.

Owttrim, G.W., Hofman, S. and Kuhlemeier, C. (1991) Divergent genes for translation initiation factor eIF-4A are coordinately expressed in tobacco. *Nucleic Acids Res.*, **19**, 5491-5496.

Pacini, E. (1990) Tapetum and microspore function, in *Microspores: Evolution and Ontogeny* (eds. S. Blackmore and R.B. Knox), Academic Press, London, pp. 213-237.

Pacini, E. (1997) Tapetum character states: analytical keys for tapetum types and activities. *Can. J. Bot.*, **75**, 1448-1459.

Palevitz, B.A. and Liu, B. (1992) Microfilaments (F-actin) in generative cells and sperm: an evaluation. *Sex. Plant Reprod.*, **5**, 89-100.

Palanivelu, R. and Preuss, D. (2000) Pollen tube targeting and axon guidance: parallels in tip growth mechanisms. *Trends Cell Biol.*, **10**, 517-524.

Palevitz, B.A. and Tiezzi, A. (1992) Organization, composition, and function of the generative cell and sperm cell cytoskeleton. *Int. Rev. Cytol.*, **140**, 149-185.

Palanivelu, R., Belostotsky, D.A. and Meagher, R.B. (2000) Conserved expression of *Arabidopsis thaliana* poly (A) binding protein 2 (PAB2) in distinct vegetative and reproductive tissues. *Plant J.*, **22**, 199-210.

Park, S.K. and Twell, D. (2001) Novel patterns of ectopic cell plate growth and lipid body distribution in the *Arabidopsis gemini pollen1* mutant. *Plant Physiol.*, **126**, 899-909.

Park, S.K., Howden, R. and Twell, D. (1998) The *Arabidopsis thaliana* gametophytic mutation *gemini pollen 1* disrupts microspore polarity, division asymmetry and pollen cell fate. *Development*, **125**, 3789-3799.

Parnell, F.R. (1921) Note on the detection of segregation by examination of the pollen of rice. *J. Genet.*, **11**, 209-212.

Paxson-Sowders, D.M., Owen, H.A. and Makaroff, C.A. (1997) A comparative ultrastructural analysis of exine pattern in wild type *Arabidopsis* and a mutant defective in pattern formation. *Protoplasma*, **198**, 53-65.

Pe, M.E., Frova, C., Colombo, L., *et al.* (1994) Molecular cloning of genes expressed in pollen of *Sorghum bicolor*. *Maydica*, **39**, 107-113.

Perez, M., Ishioka, G.Y., Walker, L.E. and Chesnut, R.W. (1990) cDNA cloning and immunological characterization of the rye grass allergen *Lol p* I. *J. Biol. Chem.*, **265**, 16210-16215.

Poovaiah, B.W., Xia, M., Liu, Z., *et al.* (1999) Developmental regulation of the gene for chimeric calcium/calmodulin-dependent protein kinase in anthers. *Planta*, **209**, 161-171.

Pressey, R. (1991) Polygalacturonase in tree pollens. *Phytochemistry*, **30**, 1753-1755.

Pressey, R. and Reger, B.J. (1989) Polygalacturonase in pollen from corn and other grasses. *Plant Sci.*, **59**, 57-62.

Prestamo, G., Testillano, P.S., Vicente, O., *et al.* (1999) Ultrastructural distribution of a MAP kinase and transcripts in quiescent and cycling plant cells andpollen grains. *J. Cell Sci.*, **112**, 1065-1076.

Preuss, D., Lemieux, B., Yen, G. and Davis, R.W. (1993) A conditional sterile mutation eliminates surface components from *Arabidopsis* pollen and disrupts cell signalling during fertilization. *Genes Dev.*, **7**, 974-985.

Preuss, D., Rhee, S.Y. and Davis, R.W. (1994) Tetrad analysis possible in *Arabidopsis* with mutation of the *QUARTET* (*QRT*) genes. *Science*, **264**, 1458-1460.

Pruitt, R.E. (1999) Complex sexual signals for the male gametophyte. *Curr. Opin. Plant Biol.*, **2**, 419-422.

Qiu, X. and Erickson, L. (1996) A pollen-specific polygalacturonase-like cDNA from alfalfa. *Sex. Plant Reprod.*, **9**, 123-124.

Rafnar, T. Griffith, I.J., Kuo, M., Bond, J.F., Rogers, B.L. and Klapper, D.G. (1991) Cloning of *Amb a* I (antigen E), the major allergen family of short ragweed pollen. *J. Biol. Chem.*, **266**, 1229-1236.

Raghavan, V. (1989) mRNAs and a cloned histone gene are differentially expressed during anther and pollen development in rice (*Oryza sativa* L.). *J. Cell Sci.*, **92**, 217-229.

Reddy, J.T., Dudareva, N., Evrard, J.L., Krauter, R.R., Steinmetz, A. and Pillay, D.T.N. (1995) A pollen-specific gene from sunflower encodes a member of the leucine-rich-repeat protein superfamily. *Plant Sci.*, **111**, 81-93.

Reijnen, W.H., van Herpen, M.M.A., de Groot, P.F.M., *et al.* (1991) Cellular localization of a pollen-specific mRNA by *in situ* hybridization and confocal laser scanning microscopy. *Sex. Plant Reprod.*, **4**, 254-257.

Rhee, S. and Somerville, C.R. (1998) Tetrad pollen formation in quartet mutants of *Arabidopsis thaliana* is associated with persistence of pectic polysaccharides of the pollen mother cell wall. *Plant J.*, **15**, 79-88.

Robert, L.S., Allard, S., Gerster, J.L., Cass, L. and Simmonds, J. (1993) Isolation and characterization of a polygalacturonase gene highly expressed in *Brassica napus* pollen. *Plant Mol. Biol.*, **23**, 1273-1278.

Roberts, M.R., Robson, F., Foster, G.D., Draper, J. and Scott, R.J. (1991) A *Brassica napus* mRNA expressed specifically in developing microspores. *Plant Mol. Biol.*, **17**, 295-299.

Roberts, M.R., Foster, G.D., Blundell, R.P., *et al.* (1993) Gametophytic and sporophytic expression of an anther-specific *Arabidopsis thaliana* gene. *Plant J.*, **3**, 111-120.

Rogers, B.L., Morgenstern, J.P., Griffith, I.J., *et al.* (1991) Complete sequence of the allergen Amb a II. *J. Immunol.*, **147**, 2547-2552.

Rogers, H.J., Harvey, A. and Lonsdale, D.M. (1992) Isolation and characterization of a tobacco gene with homology to pectate lyase which is specifically expressed during microsporogenesis. *Plant Mol. Biol.*, **20**, 493-502.

Rogers, H.J., Greenland, A.J. and Hussey, P.J. (1993) Four members of the maize beta-tubulin gene family are expressed in the male gametophyte. *Plant J.*, **4**, 875-882.

Rogers, H.J., Bate, N., Combe, J., *et al.* (2001) Functional analysis of *cis*-regulatory elements within the promoter of the tobacco late pollen gene *g10*. *Plant Mol. Biol.*, **45**, 577-585.

Roman, H. (1948) Directed fertilization in maize. *Proc. Natl. Acad. Sci. USA*, **34**, 36-42.

Ross, J.H.E. and Murphy, D.J. (1996) Characterisation of anther-expressed genes encoding a major class of extracellular oleosin-like proteins in the pollen coat of *Brassicaceae*. *Plant J.*, **9**, 625-637.

Rougier, M., Jnoud, N., Said, C., Russell, S. and Dumas, C. (1991) Male gametophyte development and formation of the male germ unit in *Populus deltoides* following compatible pollination. *Protoplasma*, **162**, 140-150.

Rousselin, P., Lepingle, A., Faure, J.D., Bitoun, R. and Caboche, M. (1990) Ethanol-resistant mutants of *Nicotiana plumbaginifolia* are deficient in the expression of pollen and seed alcohol dehydrogenase activity. *Mol. Gen. Genet.*, **222**, 409-415.

Rozwadowski, K., Zhao, R., Jackman, L., *et al.* (1999) Characterization and immunolocalization of a cytosolic calcium-binding protein from *Brassica napus* and *Arabidopsis* pollen. *Plant Physiol.*, **120**, 787-798.

Rubinelli, P., Hu, Y. and Ma, H. (1998) Identification, sequence analysis and expression studies of novel anther-specific genes of *Arabidopsis thaliana*. *Plant Mol. Biol.*, **37**, 607-619.

Rubinstein, A.L., Broadwater, A.H., Lowrey, K.B. and Bedinger, P.A. (1995a) Pex1, a pollen-specific gene with an extensin-like domain. *Proc. Natl. Acad. Sci. USA*, **92**, 3086-3090.

Rubinstein, A.L., Marquez, J., Suarez-Cervera, M. and Bedinger, P.A. (1995b) Extensin-like glycoproteins in the maize pollen tube wall. *Plant Cell*, **7**, 2211-2225.

Ruiter, R.K., van Eldik, G.J., van Herpen, M.M.A., Schrawen, J.A.M. and Wullems, G.J. (1997) Characterization of oleosins in the pollen coat of *Brassica oleracea*. *Plant Cell*, **9**, 1621-1631.

Russell, S.D. (1985) Preferential fertilisation in *Plumbago*: ultrastructural evidence for gamete level recognition in an angiosperm. *Proc. Natl. Acad. Sci. USA*, **82**, 6129-6132.

Russell, S.D. (1991) Isolation and characterisation of sperm cells in flowering plants. *Annu. Rev. Plant Physiol. Plant Mol. Biol.*, **42**, 189-204.

Russell, S.D. (1996) Attraction and transport of male gametes for fertilization. *Sex. Plant Reprod.*, **9**, 337-342.

Safadi, F., Reddy, V.S. and Reddy, A.S. (2000) A pollen-specific novel calmodulin-binding protein with tetratricopeptide repeats. *J. Biol. Chem.*, **275**, 35457-35470.

Saito, C., Fujie, M., Sakai, A., Kuroiwa, H. and Kuroiwa, T. (1998) Detection and quantification of rRNA by high-resolution *in situ* hybridization in pollen grains. *J. Plant Res.*, **111**, 45-52.

Sanders, P.M., Bui, A.Q., Weterings, K., *et al.* (1999) Anther developmental defects in *Arabidopsis thaliana* male sterile mutants. *Sex. Plant Reprod.*, **11**, 297-322.

Sanger, J.M. and Jackson, W.T. (1971) Fine structure study of pollen development in Haemunthus katerinae Baker II. Microtubules and elongation of the generative cells. *J. Cell Sci.*, **8**, 303-315.

Sari-Gorla, M. and Pe, M.E. (1999) Genetic control of pollen development and function, in *Fertilization in Higher Plants: Molecular and Cytological Aspects* (eds. M. Cresti, G. Cai and A. Moscatelli), Springer-Verlag, Berlin, pp. 217-233.

Sari-Gorla, M., Ferrario, S., Villa, M. and Pe, M.E. (1996) gaMS-1, a gametophytic expressed male sterile mutant of maize. *Sex. Plant Reprod.*, **9**, 216-220.

Sari-Gorla, M., Gatti, E., Villa, M. and Pe, M.E. (1997) A multinucleate male sterile mutant of maize with gametophytic expression. *Sex. Plant Reprod.*, **10**, 22-26.

Sauter, M., von Wiegen, P., Lorz, H. and Kranz, E. (1998) Cell cycle regulatory genes from maize are differentially controlled during fertilization and first embryonic cell division. *Sex. Plant Reprod.*, **11**, 41-48.

Schiefthaler, U., Balasubramanian, S., Sieber, P., Chevalier, D., Wisman, E. and Schneitz, K. (1999) Molecular analysis of NOZZLE, a gene involved in pattern formation and early sporogenesis during sex organ development in *Arabidopsis thaliana*. *Proc. Natl. Acad. Sci. USA*, **96**, 11664-11669.

Schopfer, C.R., Nasrallah, M.E. and Nasrallah, J.B. (1999) The male determinant of self incompatibility in *Brassica*. *Science*, **286**, 1697-1700.

Schrauwen, J.A.M., de Groot, P.F.M., van Herpen, M.M.A., *et al.* (1990) Stage-related expression of mRNAs during pollen development in lily and tobacco. *Planta*, **182**, 298-304.

Schwacke, R., Grallath, S., Breitkreuz, K.E., *et al.* (1999) LeProT1, a transporter for proline, glycine betaine, and γ-amino butyric acid in tomato pollen. *Plant Cell*, **11**, 377-391.

Scott, R.J. (1994) Pollen exine-the sporopollenin enigma and the physics of pattern, in *Molecular and Cellular Aspects of Plant Reproduction* (eds. R.J. Scott and M.A. Stead), Cambridge University Press, Cambridge, pp. 49-81.

Scott, R., Dagless, E., Hodge, R., Paul, W., Soufleri, I. and Draper, J. (1991a) Patterns of gene expression in developing anthers of *Brassica napus*. *Plant Mol. Biol.*, **17**, 195-207.

Scott, R., Hodge, R., Paul, W. and Draper, J. (1991b) The molecular biology of anther differentiation. *Plant Sci.*, **80**, 167-191.

Shaykh, M., Kolattukudy, P.E. and Davis, R. (1977) Production of a novel extracellular cutinase by the pollen and chemical composition and ultrastructure of the stigma cuticle of Nasturtium (*Tropaeolum majus*). *Plant Physiol.*, **60**, 907-915.

Silvanovich, A., Astwood, J., Zhang, L., *et al.* (1991) Nucleotide sequence analysis of three cDNAs coding for *Poa p*IX isoallergens of Kentucky bluegrass pollen. *J. Biol. Chem.*, **266**, 1204-1210.

Singh, M.B., O'Neill, P. and Knox, R.B. (1985) Initiation of postmeiotic β-galactosidase synthesis during microsporogenesis in oilseed rape. *Plant Physiol.*, **77**, 231-237.

Singh, M.B., Hough, T., Theerakulpisut, P., *et al.* (1991) Isolation of a cDNA encoding a newly identified major allergenic protein of ryegrass pollen: intracellular targeting to the amyloplast. *Proc. Natl. Acad. Sci. USA*, **88**, 1384-1388.

Smykal, P., Janotova, I. and Pechan, P. (2000a) A novel *Brassica napus* L. pollen-specific gene belongs to a nucleic-acid-binding protein family. *Sex. Plant Reprod.*, **13**, 127-134.

Smykal, P., Masin, J., Hrdy, I., Konopasek, I. and Zarsky, V. (2000b) Chaperone activity of tobacco HSP18, a small heat-shock protein, is inhibited by ATP. *Plant J.*, **23**, 703-713.

Spielman, M.L., Preuss, D., Li, F.L., Browne, W.E., Scott, R.J. and Dickinson, H.G. (1997) *TETRASPORE* is required for male meiotic cytokinesis in *Arabidopsis thaliana*. *Development*, **124**, 2645-2657.

Stadler, L.J. (1933) On the genetic nature of induced mutations in plants. II. A haplo-viable deficiency in maize. *Res. Bull. Mo. Agric. Exp. Stn.*, **204**, 29.

Stadler, L.J. and Roman, H. (1948) The effect of X-rays upon mutation of a gene A in maize. *Genetics*, **33**, 273-303.

Stadler, R., Truernit, E., Gahrtz, M. and Sauer, N. (1999) The AtSUC1 sucrose carrier may represent the osmotic driving force for anther dehiscence and pollen tube growth in *Arabidopsis*. *Plant J.*, **19**, 269-278.

Staiger, C.J., Goodbody, K.C., Hussey, P.J., Valenta, R., Drobak, B.K. and Lloyd, C.W. (1993) The profilin multigene family of maize: differential expression of three isoforms. *Plant J.*, **4**, 631-641.

Steiglitz, H. (1977) Role of β-1-3-glucanase in postmeiotic microspore release. *Dev. Biol.*, **57**, 87-97.

Stephenson, A.G., Doughty, J., Dixon, S., Elleman, C., Hiscock, S. and Dickinson, H.G. (1997) The male determinant of self incompatibility in *Brassica oleracea* is located in the pollen coating. *Plant J.*, **12**, 1351-1359.

Stinson, J.R., Eisenberg, A.J., Willing, R.P., Pe, M.P., Hanson, D.D. and Mascarenhas, J.P. (1987) Genes expressed in the male gametophyte of flowering plants and their isolation. *Plant Physiol.*, **83**, 442-447.

Storchová, H., Capková, V. and Tupy, J. (1994) A *Nicotiana tabacum* mRNA encoding a 69-kDa glycoprotein occurring abundantly in pollen tubes is transcribed but not translated during pollen development in the anthers. *Planta*, **192**, 441-445.

Sweetman, J., Spurr, C., Eliasson, A., Gass, N., Steinmetz, A. and Twell, D. (2000) Isolation and characterisation of two pollen-specific LIM domain protein cDNAs from *Nicotiana tabacum*. *Sex. Plant Reprod.*, **12**, 339-345.

Swoboda, I., Dang, T.C.H., Heberle-Bors, E. and Vicente, O. (1995) Expression of Bet v I, the major birch pollen allergen, during anther development: an *in situ* hybridization study. *Protoplasma*, **187**, 103-110.

Takahashi, T., Mu, J.H., Gasch, A. and Chua, N.H. (1998) Identification by PCR of receptor-like protein kinases from *Arabidopsis* flowers. *Plant Mol. Biol.*, **37**, 587-596.

Takayama, S., Shiba, H., Iwano, M., *et al.* (2000a) Isolation and characterization of pollen coat proteins of *Brassica campestris* that interact with S locus-related glycoprotein 1 involved in pollen–stigma adhesion. *Proc. Natl. Acad. Sci. USA*, **97**, 3765-3770.

Takayama, S., Shiba, H., Iwano, M., *et al.* (2000b) The pollen determinant of self incompatibility in *Brassica campestris*. *Proc. Natl. Acad. Sci. USA*, **97**, 1920-1925.

Tanaka, I. (1997) Differentiation of generative and vegetative cells in angiosperm pollen. *Sex. Plant Reprod.*, **10**, 1-7.

Tanaka, I. and Ito, M. (1980) Induction of typical cell division in isolated microspores of *Lilium longiflorum* and *Tulipa gesneriana*. *Plant Sci. Lett.*, **17**, 279-285.

Tanaka, I. and Ito, M. (1981) Control of division patterns in explanted microspores of *Tulipa gesneriana*. *Protoplasma*, **108**, 329-340.

Tanaka, I., Ono, K. and Fukuda, T. (1998) The developmental fate of angiosperm pollen is associated with a preferential decrease in the level of histone H1 in the vegetative nucleus. *Planta*, **206**, 561-569.

Taylor, L.P. and Hepler, P.K. (1997) Pollen germination and tube growth. *Annu. Rev. Plant Physiol. Plant Mol. Biol.*, **48**, 461-491.

Taylor, L.P. and Jorgensen, R. (1992) Conditional male fertility in chalcone synthase deficient *Petunia*. *J. Hered.*, **83**, 11-17.

Taylor, P.E., Glover, J.A., Lavithis, M., *et al.* (1998) Genetic control of male fertility in *Arabidopsis thaliana*: structural analyses of postmeiotic developmental mutants. *Planta*, **205**, 492-505.

Terada, R., Nakajima, M., Isshiki, M., Okagaki, R.J., Wessler, S.R. and Shimamoto, K. (2000) Antisense waxy genes with highly active promoters effectively suppress waxy gene expression in transgenic rice. *Plant Cell Physiol.*, **41**, 881-888.

Tebbutt, S.J., Rogers, H.J. and Lonsdale, D.M. (1994) Chracterization of a tobacco gene encoding a pollen-specific polygalacturonase. *Plant Mol. Biol.*, **25**, 283-297.

Terasaka, O. and Niitsu, T. (1987) Unequal cell division and chromatin differentiation in pollen grain cells I. Centrifugal, cold and caffeine treatments. *Bot. Mag. Tokyo*, **100**, 205-216.

Terasaka, O. and Niitsu, T. (1990) Unequal cell division and chromatin differentiation in pollen grain cells II. Microtubule dynamics associated with the unequal cell division. *Bot. Mag. Tokyo*, **103**, 133-142.

Terasaka, O. and Niitsu, T. (1995) The mitotic apparatus during unequal microspore division observed by a confocal laser scanning microscope. *Protoplasma*, **189**, 187-193.

Terasaka, O. and Tanaka, R. (1974) Cytological studies on the nuclear differentiation in microspore division of some angiosperms. *Bot. Mag. Tokyo*, **87**, 209-217.

Thangavelu, M., Belostotsky, D., Bevan, M.W., Flavell, R.B., Rogers, H.J. and Lonsdale, D.M. (1993) Partial characterization of the *Nicotiana tabacum* actin gene family: evidence for pollen-specific expression of one of the gene family members. *Mol. Gen. Genet.*, **240**, 290-295.

Theerakulpisut, P., Xu, H., Singh, M.B., Pettitt, J.M. and Knox, R.B. (1991) Isolation and developmental expression of *Bcp1*, an anther-specific cDNA clone in *Brassica campestris*. *Plant Cell*, **3**, 1073-1084.

Tichtinsky, G., Tavares, R., Takvorian, A., Schwebel-Dugue, N., Twell, D. and Kreis, M. (1998) An evolutionary conserved group of plant GSK-3/shaggy-like protein kinase genes preferentially expressed in developing pollen. *Biochim Biophys Acta*, **1442**, 261-273.

Tiezzi, A., Moscatelli, A., Cai, G., Bartalesi, A. and Cresti, M. (1992) An immunoreactive homolog of mammalian kinesin in *Nicotiana tabacum* pollen tubes. *Cell Motil. Cytoskeleton.*, **21**, 132-137.

Ting, J.T., Wu, S.S., Ratnayake, C. and Huang, A.H. (1998) Constituents of the tapetosomes and elaioplasts in *Brassica campestris* tapetum and their degradation and retention during microsporogenesis. *Plant J.*, **16**, 541-551.

Touraev, A., Lezin, F., Heberle-Bors, E. and Vicente, O. (1995) Maintenance of gametophytic development after symmetrical division in tobacco microspore culture. *Sex. Plant Reprod.*, **8**, 70-76.

Touraev, A., Pfosser, M., Vicente, O. and Heberle-Bors, E. (1996) Stress as the major signal controlling the developmental fate of tobacco microspores: towards a unified model of induction of microspore/pollen embryogenesis. *Planta*, **200**, 144-152.

Touraev, A., Vicente, O. and Heberle-Bors, E. (1997) Initiation of microspore embryogenesis by stress. *Trends Plant Sci.*, **2**, 297-302.

Treacy, B.K., Hattori, J., Prud'homme, I., *et al.* (1997) Bnm1, a *Brassica* pollen-specific gene. *Plant Mol. Biol.*, **34**, 603-611.

Trewavas, A.J. and Malho, R.(1998) Ca^{2+} signalling in plant cells: the big network! *Curr. Opin. Plant Biol.*, **1**, 428-433.

Truernit, E., Stadler, R., Baier, K. and Sauer, N. (1999) A male gametophyte-specific monosaccharide transporter in *Arabidopsis*. *Plant J.*, **17**, 191-201.

Tupy, J., Rihova, L. and Zarsky, V. (1991) Production of fertile tobacco pollen from microspores in suspension culture and its storage for *in situ* pollination. *Sex. Plant Reprod.*, **4**, 284-287.

Turcich, M.P., Hamilton, D.A. and Mascarenhas, J.P. (1993) Isolation and characterization of pollen-specific maize genes with sequence homology to ragweed allergens and pectate lyases. *Plant Mol. Biol.*, **23**, 1061-1065.

Turcich, M.P., Hamilton, D.A., Yu, X. and Mascarenhas, J.P. (1994) Characterisation of a pollen-specific gene from *Tradescantia paludosa* with an unusual cysteine grouping. *Sex. Plant Reprod.*, **7**, 201-202.

Twell, D. (1995) Diphtheria toxin-mediated cell ablation in developing pollen: vegetative cell ablation blocks generative cell migration. *Protoplasma*, **187**, 144-154.

Twell, D. (1992) Use of a nuclear-targeted β-glucuronidase fusion protein to demonstrate vegetative cell-specific gene expression in developing pollen. *Plant J.*, **2**, 887-892.

Twell, D. (1994) The diversity and regulation of gene expression in the pathway of male gametophyte development, in *Molecular and Cellular Aspects of Plant Reproduction* (eds. R.J. Scott and A.D. Stead), Cambridge University Press, Cambridge, pp. 83-135.

Twell, D. (1999) Mechanisms of microspore polarity and differential cell fate determination in developing pollen, in *Fertilization in Higher Plants. Molecular and Cytological Aspects* (eds. M. Cresti, G. Cai and A. Moscatelli), Springer-Verlag, Berlin, pp. 201-215.

Twell, D. and Howden, R. (1998a) Mechanisms of asymmetric division and cell fate determination in developing pollen, in *Androgenesis and Haploid Plants, 1967–1994. In memory of Jean-Pierre Bourgin* (eds. Y. Chupeau, M. Caboche and Y. Henry), INRA-Springer-Verlag, Berlin, pp. 69-103.

Twell, D. and Howden, R. (1998b) Cell polarity, asymmetric division and cell fate determination in developing pollen, in *Reproductive Biology* (eds. S.J. Owens and P.J. Rudall), Royal Botanic Gardens, Kew, pp. 197-218.

Twell, D., Park, S.K. and Lalanne, E. (1998) Asymmetric division and cell fate determination in developing pollen. *Trends Plant Sci.*, **3**, 305-310.

Twell, D., Wing, R.A., Yamaguchi, J. and McCormick, S. (1989) Isolation and expression of an anther-specific gene from tomato. *Mol. Gen. Genet.*, **217**, 240-245.

Twell, D., Yamaguchi, J. and McCormick, S. (1990) Pollen-specific expression in transgenic plants: coordinate regulation of two different promoters during microsporogenesis. *Development*, **109**, 705-713.

Twell, D., Yamaguchi, J., Wing, R.A., Ushiba, J. and McCormick, S. (1991) Promoter analysis of three genes that are coordinately expressed during pollen development reveals pollen-specific enhancer sequences and shared regulatory elements. *Genes Dev.*, **5**, 496-507.

Twell, D., Patel, S., Sorensen, A., *et al.* (1993) Activation and developmental regulation of an *Arabidopsis* anther-specific promoter in microspores and pollen of *Nicotiana tabacum*. *Sex. Plant Reprod.*, **6**, 217-224.

Ueda, K. and Tanaka, I. (1994) The basic proteins of male gametic nuclei isolated from pollen grains of *Lilium longiflorum*. *Planta*, **192**, 446-452.

Ueda, K. and Tanaka, I. (1995a) The appearance of male gamete-specific histones gH2B and gH3 during pollen development in *Lilium longiflorum*. *Dev. Biol.*, **169**, 210-217.

Ueda, K. and Tanaka, I. (1995b) Male gametic nucleus-specific H2B and H3 histones, designated gH2B and gH3, in *Lilium longiflorum*. *Planta*, **197**, 289-295.

Ueda, K., Kinoshita, Y., Xu, Z.-J., *et al.* (2000) Unusual core histones specifically expressed in male gametic cells of *Lilium longiflorum*. *Chromosoma*, **108**, 491-500.

Ursin, V.M., Yamaguchi, J. and McCormick, S. (1989) Gametophytic and sporophytic expression of anther-specific genes in developing tomato anthers. *Plant Cell*, **1**, 727-736.

Valenta, R., Duchene, M., Pettenburger, K., *et al.* (1991) Identification of profilin as a novel pollen allergen; IgE autoreactivity in sensitized individuals. *Science*, **253**, 557-560.

Valenta, R., Ball, T., Vrtala, S., Duchene, M., Kraft, D. and Scheiner, O. (1994) cDNA cloning and expression of timothy grass (*Phleum pratense*) pollen profilin in *Escherichia coli*: comparison with birch pollen profilin. *Biochem. Biophys. Res. Commun.*, **199**, 106-118.

Vallverdu, A., Asturias, J.A., Arilla, M.C., *et al.* (1998) Characterization of recombinant *Mercurialis annua* major allergen Mer a 1 (profilin). *J. Allergy Clin Immunol.*, **101**, 363-370.

Van der Meer, I.M., Stam, M.E., van Tunen, A.J., Mol, J.N. and Stuitje, A.R. (1992) Antisense inhibition of flavonoid biosynthesis in petunia anthers results in male sterility. *Plant Cell*, **4**, 253-262.

Van Eldik, G.J., Reijnen, W.H., Ruiter, R.K., van Herpen, M.M.A., Schrawen, J.A.M. and Wullems, G.J. (1997) Regulation of flavonol biosynthesis during anther and pistil development, and during pollen tube growth in *Solanum tuberosum*. *Plant J.*, **11**, 105-113.

Van Lammeren, A.A.M., Bednara, J. and Willemse, M.T.M. (1989) Organization of the actin cytoskeleton in *Gasteria verrucosa* (Mill.) H. Duval visualized with rhodamine phalloidin, *Planta*, **178**, 531-539.

Van Tunen, A.J., Mur, L.A., Recourt, K., Gerats, A.G.M. and Mol, J.N.M. (1991) Regulation and manipulation of flavonoid gene expression in anthers of *Petunia*: the molecular basis of the *Po* mutation. *Plant Cell*, **3**, 39-48.

Vauterin, M., Frankard, V. and Jacobs, M. (1999) The *Arabidopsis thaliana* dhdps gene encoding dihydrodipicolinate synthase, key enzyme of lysine biosynthesis, is expressed in a cell-specific manner. *Plant Mol. Biol.*, **39**, 695-708.

Vergné, P. and Dumas, C. (1988) Isolation of viable wheat male gametophytes of different stages of development and variation in their protein patterns. *Plant Physiol.*, **75**, 865-868.

Villalba, M., Batanero, E., Monsalve, R.I., Gonzalez de la Pena, M.A., Lahoz, C. and Rodrigues, R. (1994) Cloning and expression of Ole e I, the major allergen from olive tree pollen. *J. Biol. Chem.*, **269**, 15217-15222.

Villemur, R., Haas, N.A., Joyce, C.M., Snustad, P.D. and Silflow, C.D. (1994) Characterization of four new β-tubulin genes and their expression during male flower development in maize (*Zea mays* L.). *Plant Mol. Biol.*, **24**, 295-315.

Vizir, I.Y., Anderson, M.L., Wilson, Z.A. and Mulligan, B.J. (1994) Isolation of deficiencies in the *Arabidopsis* genome by gamma-irradiation of pollen. *Genetics*, **137**, 1111-1119.

Vogt, T., Wollenweber, E. and Taylor, L.P. (1995) The structural requirement of flavonols that induce pollen germination of conditionally male fertile petunia. *Phytochemistry*, **38**, 459-462.

Wakeley, P.R., Rogers, H.J., Rozycka, M., Greenland, A.J. and Hussey, P.J. (1998) A maize pectin methylesterase-like gene, ZmC5, specifically expressed in pollen. *Plant Mol. Biol.*, **37**, 187-192.

Weiss, C.A., Huang, H. and Ma, H. (1993) Immunolocalization of the G protein alpha subunit encoded by the GPA1 gene in *Arabidopsis*. *Plant Cell*, **5**, 1513-1528.

Weterings, K., Reijnen, W., van Aarssen, R., *et al.* (1992) Characterization of a pollen-specific cDNA clone from *Nicotiana tabacum* expressed during microgametogenesis and germination. *Plant Mol. Biol.*, **18**, 1101-1111.

Weterings, K., Schrauwen, J., Wullems, G. and Twell, D. (1995) Functional dissection of the promoter of the pollen-specific gene NTP303 reveals a novel pollen-specific, and conserved *cis*-regulatory element. *Plant J.*, **8**, 55-63.

Wilhelmi, L.K. and Preuss, D. (1999) The mating game: pollination and fertilization in flowering plants. *Curr. Opin. Plant Biol.*, **2**, 18-22.

Wilson, C., Eller, N., Gartner, A., Vicente, O. and Heberle-Bors, E. (1993) Isolation and characterization of a tobacco cDNA clone encoding a putative MAP kinase. *Plant Mol. Biol.*, **23**, 543-551.

Wilson, C., Anglmayer, R., Vicente, O. and Heberle-Bors, E. (1995) Molecular cloning, functional expression in *Escherichia coli*, and characterization of multiple mitogen-activated-protein kinases from tobacco. *Eur. J. Biochem.*, **233**, 249-257.

Wilson, C., Voronin, V., Touraev, A., Vicente, O. and Heberle-Bors, E. (1997) A developmentally regulated MAP kinase activated by hydration in tobacco pollen. *Plant Cell*, **9**, 2093-2100.

Wing, R.A., Yamaguchi, J., Larabell, S.K., Ursin, V.M. and McCormick, S. (1989) Molecular and genetic characterization of two pollen-expressed genes that have sequence similarity to pectate lyases of the plant pathogen *Erwinia*. *Plant Mol. Biol.*, **14**, 17-28.

Wisman, E., Koornneef, M., Chase, T., Lifshytz, E., Ramanna, M.S. and Zabel, P. (1991) Genetic and molecular characterization of an Adh-1 null mutant in tomato. *Mol. Gen. Genet.*, **226**, 120-128.

Wittink, F.R.A., Knuiman, B., Derksen, J., *et al.* (2000) The pollen-specific gene *Ntp303* encodes a 69 kD glycoprotein associated with the vegetative membranes and the cell wall. *Sex. Plant Reprod.*, **12**, 276-284.

Worrall, D., Hird, D.L., Hodge, R., Paul, W., Draper, J. and Scott, R. (1992) Premature dissolution of the microsporocyte callose wall causes male sterility in transgenic tobacco. *Plant Cell*, **4**, 759-771.

Xu, H., Davies, S.P., Kwan, B.Y., O'Brien, A.P., Singh, M. and Knox, R.B. (1993) Haploid and diploid expression of a *Brassica campestris* anther-specific gene promoter in *Arabidopsis* and tobacco. *Mol. Gen. Genet.*, **239**, 58-65.

Xu, H., Knox, R.B., Taylor, P.E. and Singh, M.B. (1995a) *Bcp1*, a gene required for male fertility in *Arabidopsis*. *Proc. Natl. Acad. Sci. USA*, **92**, 2106-2110.

Xu, H., Theerakulpisut, P., Goulding, N., Suphioglu, C., Singh, M.B. and Bhalla, P.L. (1995b) Cloning, expression and immunological characterization of Ory s 1, the major allergen of rice pollen. *Gene*, **164**, 255-259.

Xu, H., Swoboda, I., Bhalla, P.L., *et al.* (1998) Plant homologue of human excision repair gene ERCC1 points to conservation of DNA repair mechanisms. *Plant J.*, **13**, 823-829.

Xu, H., Swoboda, I., Bhalla, P.L. and Singh, M.B. (1999a) Male gametic cell-specific gene expression in flowering plants. *Proc. Natl. Acad. Sci. USA*, **96**, 2554-2558.

Xu, H., Swoboda, I., Bhalla, P.L. and Singh, M.B. (1999b) Male gametic cell-specific expression of H2A an H3 histone genes. *Plant Mol. Biol.*, **39**, 607-614.

Yang, S., Sweetman, J., Amirsadeshi, S., *et al.* (2001) Novel anther-specific myb genes from tobacco as putative regulators of phenylalanine ammonia lyase expression. *Plant Physiol.*, **126**, 1738-1753.

Yang, W.-C. and Sundaresan, V. (2000) Genetics of gametophyte biogenesis in *Arabidopsis*. *Curr. Opin. Plant Biol.*, **3**, 53-57.

Yang, W.-C., Ye, D., Xu, J. and Sundaresan, V. (1999) The *SPOROCYTELESS* gene of *Arabidopsis* is required for initiation of sporogenesis and encodes a novel nuclear protein. *Genes Dev.*, **13**, 2108-2117.

Ylstra, B. and McCormick, S. (1999) Analysis of mRNA stabilities during pollen development and in BY2 cells. *Plant J.*, **20**, 101-108.

Ylstra, B., Garrido, D., Busscher, J. and van Tunen, A.J. (1998) Hexose transport in growing petunia pollen tubes and characterization of a pollen-specific, putative monosaccharide transporter. *Plant Physiol.*, **118**, 297-304.

Ylstra, B., Touraev, A., Benito-Moreno, R.M., *et al.* (1992) Flavanols stimulate development, germination and tube growth of tobacco pollen. *Plant Physiol.*, **100**, 902-907.

Yu, L.X., Nasrallah, J., Valenta, R. and Parthasarathy, M.V. (1998) Molecular cloning and mRNA localization of tomato pollen profilin. *Plant Mol. Biol.*, **36**, 699-707.

Zabaleta, E., Heiser, V., Grohmann, L. and Brennicke, A. (1998) Promoters of nuclear-encoded respiratory chain complex I genes from *Arabidopsis thaliana* contain a region essential for anther/pollen-specific expression. *Plant J.*, **15**, 49-59.

Zachgo, S., Saedler, H. and Schwarz-Sommer, Z. (1997) Pollen-specific expression of DEFH125, a MADS-box transcription factor in *Antirrhinum* with unusual features. *Plant J.*, **11**, 1043-1050.

Zaki, M.A.M. and Dickinson, H.G. (1990) Structural changes during the first divisions of embryos resulting from anther and microspore culture in *Brassica napus*. *Protoplasma*, **156**, 149-162.

Zaki, M.A.M. and Dickinson, H.G. (1991) Microspore-derived embryos in *Brassica*: the significance of division asymmetry in pollen mitosis I to embryogenic development. *Sex. Plant Reprod.*, **4**, 48-55.

Zarsky, V., Garrido, D., Rihova, L., Tupy, J., Vicente, O. and Heberle-Bors, E. (1992) Derepression of the cell cycle by starvation is involved in the induction of tobacco pollen embryogenesis. *Sex. Plant Reprod.*, **5**, 189-194.

Zhang, H. and Croes, A.F. (1983) Protection of pollen germination from adverse temperatures: a possible role for proline. *Plant Cell Environ.*, **6**, 471-476.

Zhang, G., Gifford, D.J. and Cass, D.D. (1993) RNA and protein synthesis in sperm cells isolated from *Zea mays* L. pollen. *Sex. Plant Reprod.*, **6**, 239-243.

Zheng, Z.L. and Yang, Z. (2000) The Rop GTPase switch turns on polar growth in pollen. *Trends Plant Sci.*, **5**, 298-303.

Zinkl, G.M. and Preuss, D. (2000) Dissecting *Arabidopsis* pollen–stigma interactions reveals novel mechanisms that confer mating specificity. *Ann. Bot.*, **85** (suppl. A), 15-21.

Zinkl, G.M., Zwiebel, B.I., Grier, D.G. and Preuss, D. (1999) Pollen–stigma adhesion in *Arabidopsis*: a species-specific interaction mediated by lipophilic molecules in the pollen exine. *Development*, **126**, 5431-5440.

Zonia, L., Tupy, J. and Staiger, C. J. (1999) Unique actin and microtubule arrays coordinate the differentiation of microspores to mature pollen in *Nicotiana tabacum*. *J. Exp. Bot.*, **50**, 581-594.

Zou, J.T., Zhan, X.Y., Wu, H.M., Wang, H. and Cheung, A.Y. (1994) Characterization of a rice pollen-specific gene and its expression. *Am. J. Bot.*, **81**, 552-561.

5 Plant embryogenesis—the cellular design of a plant

Luis Perez-Grau

5.1 Introduction

In higher plants, embryogenesis is the first stage in the development of the sporophyte from a single-cell zygote. Embryogenesis is a unique phase of plant development in which a single cell, the fertilized egg cell, develops into a multicellular embryo that contains the primary plant tissues and the shoot and root meristem that form the seedling during germination (Raghavan, 1986; Steeves and Sussex, 1989).

A comprehensive description of embryogenesis in a wide variety of higher plants can be approximated if embryo development is divided into a sequence of developmental stages. These stages are characterized by the predominance of specific developmental processes that are largely conserved in the embryogenesis of many plants. The stages correspond approximately to proembryo/ suspensor stage, globular stage, heart stage, maturation and dormancy (Essau, 1977; Kaplan and Cooke, 1997; Yadegari and Goldberg, 1997). Embryogenesis begins with the cellular proliferation of the fertilized egg cell to form a multicellular undifferentiated proembryo, or embryo proper, and the suspensor, a terminally differentiated multicellular structure that nurtures the early embryo and connects it to the inner integuments of the seed. This initial stage of embryogenesis is characterized by rapid cell division and the segregation of developmental programs in the cellular descendants of the zygote to form the embryo proper and the suspensor (Raghavan, 1986; Schwartz *et al.*, 1997). Cell division and differentiation continues in the embryo proper that develops into a globular embryo in which the differentiation of the primary cell types of the embryo is initiated. Concurrent with cellular differentiation, there are regional specification processes that organize two main different domains in the globular embryo. The embryo domains or regions contribute specific parts of the mature embryo (Mayer *et al.*, 1991; Jürgens and Mayer, 1994). A typical globular dicot embryo has two domains: the apical domain that forms the shoot meristem and the cotyledons; and the basal domain that forms the hypocotyl, the primary root, and the root meristem. The individual elements of the mature embryo are formed largely independently in the different domains. After late globular stage, embryo development generally experiences a morphological transition from a globular to a heart-shaped polarized embryo that is caused by the differential growth of the apical domain to form the cotyledon primordia and the shoot meristem. The heart stage embryo is conserved in many dicot plants. At heart stage, embryos have already differentiated most of the tissues and cell types of the mature embryo as well as the shoot and root meristem. Most of the processes of cell differentiation and morphogenesis are complete in mature heart stage embryos.

A maturation stage, during which the embryo increases dramatically in size, frequently follows the heart stage. The growth of the embryo during maturation is mainly due to cellular expansion and the accumulation of storage products, primarily in the cotyledons. The maturation phase appears to be an evolutionarily recent addition to the developmental sequence of higher plant embryogenesis, and its actual duration is highly variable among different plant species as well as within a given plant depending on growth conditions. Plant embryogenesis ends with the differentiation of a state of quiescence that arrests the further development of the embryo for an extended period until development is continued during germination (Morton *et al.*, 1995; Harada, 1997; Wobus and Weber, 1999). The position of the quiescent period within the developmental sequence is variable in different plants and some plants lack this phase altogether (Harada, 1997).

There is a large variability in the duration of embryogenesis as well as the relative contributions of each of the stages described above to the complete process of embryo formation in different plant species. There is also variability in the relative contributions of each embryo domain to the formation of the mature embryo. The relative development of each embryo domain largely dictates the final morphology of the mature embryo that is characteristic for a plant (Raghavan, 1986; Kaplan and Cooke, 1997).

Monocot plants have a pattern of embryogenesis that has many elements in common with dicot embryogenesis. The development of the monocot embryo is also based on cell differentiation and regional specification processes that form the different primary tissues of the plant, the shoot and root meristems and the embryonic storage organs. Monocot plants also arrest embryonic development prior to germination (Essau, 1977; Raghavan, 1986; Steeves and Sussex, 1989).

This review will discuss mainly dicot embryogenesis and fundamentally will consider research in the dicot model plant *Arabidopsis* unless indicated otherwise. We will discuss the developmental aspects of embryogenesis, focusing this review on recent research on the molecular and cellular processes by which the single-celled zygote differentiates the multicellular organization that directs the development of the embryo into a plant.

Many molecular and cellular processes that direct plant embryogenesis are largely conserved among distant plant species, suggesting that the processes that regulate *Arabidopsis* embryogenesis are probably extensive to other plants (Goldberg *et al.*, 1989, 1994; Perez-Grau and Goldberg, 1989).

5.2   Embryo development in *Arabidopsis* is representative
of many dicot plants

Arabidopsis has been a model system for the study of plant biology and plant
development for a number of years. The short lifecycle and small genome
size of *Arabidopsis* are ideal for genetic studies and for the generation and
identification of mutations that affect specific aspects of embryo development
(Meinke, 1985, 1995). The ongoing studies of *Arabidopsis* development using a
variety of approaches, including mutant analysis, already provide an expanding
but still fragmented catalogue of genes and molecules that regulate different
aspects of embryogenesis (Jürgens, 1995; Harada, 1997; Laux and Jürgens,
1997; Yadegari and Goldberg, 1997). The sequence of the *Arabidopsis* genome
will further facilitate the complete identification of the main genes that control
the beginning of plant development during embryogenesis.

Embryo development in *Arabidopsis* is a condensed version of dicot
embryogenesis. The stages of higher-plant embryogenesis described above are
clearly recognizable in *Arabidopsis*, although their duration is abbreviated and
the processes of differentiation and morphogenesis are accomplished with very
few cells. The cellular complexity of the *Arabidopsis* embryo is relatively small.
The mature embryo has about 15 000 cells, comprising fewer than 20 different
cell types that are organized in six primary tissues (protoderm, cortex, endodermis, pericycle, xylem and phloem), characteristic of many plant embryos
(Essau, 1977; Jürgens and Mayer, 1994).

The development of the *Arabidopsis* embryo is rapid, taking place in about
10 days from fertilization. The mature embryo is small in size but has all the
morphological elements and primary tissues and cell types of a typical dicot
embryo. The *Arabidopsis* embryo consists of two organs, the cotyledons and
the embryonic axis. The axis contains the shoot meristem, the hypocotyl and the
embryonic root with the primary root meristem. The cotyledons and hypocotyl
contain most of the 15 000 cells of the mature embryo. The embryonic shoot
meristem contains about 110 cells and the primary root meristem contains probably less than half that number, although it has not been directly estimated (Irish
and Sussex, 1992). While inconspicuous in the mature embryo, the shoot and
root meristems provide the sources for most of the cells for the postembryonic
development of the plant.

Arabidopsis has an invariant pattern of embryogenesis very similar to *Capsela
bursa-pastoris* (Soueges, 1914; Mansfield and Briarty, 1991; Jürgens and Mayer,
1994). The initial multiplication of the *Arabidopsis* zygote follows a precise and
predictable sequence of cell divisions that reproducibly places specific embryo
cells at precise spatial locations in the early embryo. In *Arabidopsis* it is possible
to construct approximate cell descent maps that trace back individual cells of the
developing embryo to their progenitor cells at earlier stages. These maps are also
predictive of specific contributions of early embryo cells to the mature embryo

and seedling. This property of embryogenesis in *Arabidopsis* is very useful for the detection of deviations in the program of cell divisions of the zygote and also facilitates the analysis of cell type differentiation during embryo development.

The *Arabidopsis* embryo at the 16-cell stage is already organized in two domains that develop largely independently of each other and that contribute different structures to the mature embryo. The apical domain establishes the shoot meristem and the cotyledons. The basal domain forms the hypocotyl, the root meristem and the primary root. The early embryo is organized in three concentric primary tissue layers, the protoderm, the ground meristem and the procambium. Within these primary tissues additional cell types and tissue layers differentiate during embryogenesis. Cell differentiation in the ground meristem forms the cortex and the endodermis. The procambium tissue differentiates the pericycle layer and the vascular cells of the xylem and phloem. In *Arabidopsis* embryos the absence of a particular cell type or cell layer can be traced back to abnormal cell division and/or cell differentiation events (Scheres *et al.*, 1995; Yadegari and Goldberg, 1997). The morphogenesis of the *Arabidopsis* embryo is mostly complete by the end of the torpedo stage. The embryonic primary tissues and the cellular organization of the shoot and root meristem are established during early embryogenesis and become functional by late torpedo stage. During the maturation phase the embryo increases in size mainly by cell expansion to fill up the seed. The maturation phase ends with the differentiation of a state of quiescence that permits seed desiccation and dormancy until plant development is continued during germination (West and Harada, 1993; Koornneef and Karsen, 1994; Harada, 1997; Raz *et al.*, 2001).

5.3   Early embryogenesis integrates two developmental programs

Early after fertilization, the zygote proliferates and differentiates the suspensor and the embryo proper. The suspensor develops rapidly to form a terminally differentiated organ that supports and nurtures the early embryo. Among angiosperms there is a wide variability in the morphology and the extent of the development of the suspensor that contrasts with the similarities in the development of the embryo. The rapid differentiation of the suspensor and embryo proper after fertilization suggests that cytoplasmic determinants in the egg cell and/or morphogenetic gradients in the egg cell or the female gametophyte might regulate the initiation of the two developmental programs in early embryogenesis (Raghavan, 1986; Wardlaw, 1955).

In *Arabidopsis* the segregation of suspensor and embryo developmental programs occurs during the first cell division of the zygote. The *Arabidopsis* zygote is a highly polarized cell with an asymmetric distribution of cytoplasm and subcellular organelles. The first division of the zygote is transverse and asymmetric, and the precise position of the dividing cell wall produces a small and dense apical cell that develops into the embryo proper and a large vacuolated basal cell that forms the suspensor. Cell fate determinants or other molecules that specify suspensor and embryo developmental fates could be differentially segregated during the asymmetric division of the zygote (Mansfield *et al.*, 1991; Vernon and Meinke, 1994; Schwartz *et al.*, 1997). Little is known about the molecular mechanisms that regulate the assignment of these two initial cellular fates in plant embryos. In other systems, such as in the brown alga *Fucus*, the unequal distribution of cytoplasm determinants has been involved in the process of segregation of rhizoid cell fate and vegetative cell fate during the first cell division of the zygote (Goodner and Quatrano, 1993; Quatrano and Shaw, 1997). Similar processes might take place in *Arabidopsis* and other plants, but very little is known about the molecules involved in the process (Brownlee and Berger, 1995).

A number of mutations that affect early embryogenesis interfere with the developmental programs of the suspensor and the embryo. These early mutants can be sorted into two classes: a class with defects in the processes that support the cellular multiplication of the zygote, and a second class that manifest alterations in the segregation of the developmental programs of the embryo proper and the suspensor. Collectively the study of early embryogenesis mutations should provide information about the processes that regulate early zygote development during the first hours after fertilization. These processes include the transition from gametogenesis to embryogenesis, the segregation of suspensor and embryo proper developmental fates, and the regulation of suspensor development and early embryo development (Yadegari *et al.*, 1994; Schwarz *et al.*, 1997; Zhang and Somerville, 1997). The initiation of the differentiation programs of the suspensor and embryo proper in the zygote sister cells is reminiscent of the segregation of cell fates that occurs during cell type specification later in the embryo.

5.3.1 Control of cell division after fertilization requires zygotic gene expression

The first class of early embryo mutants is exemplified by the *titan* and *pilz* mutations. The *titan* and *pilz* mutants of *Arabidopsis* are disrupted in the progression through mitosis and lack cell cycle control (Liu and Meinke, 1998; Mayer *et al.*, 1999). The *titan* mutations manifest early after fertilization, suggesting that gamete cell division is not affected. In particular, in the *titan-5* mutant, mitotic cell division control is completely lacking after fertilization and the proliferation of the zygote results in highly abnormal embryonic structures. The processes of nuclear and cytoplasm division become uncoupled in the *titan-5* zygote (Liu and Meinke, 1998; McElver *et al.*, 2000). The *titan-5* mutant shows aberrant development of the zygote and the endosperm after fertilization. In *titan-5* seeds the cellular proliferation of the zygote generates a giant single cell with supernumerary nuclei. *titan* single-celled early embryos are multinuclear, suggesting that the fertilized egg cell has a cell autonomous capability to undergo several rounds of unchecked nuclear duplication that are followed by different extents of kariokynesis in absence of cytoplasm partitioning. Some *titan* single-celled giant embryos contain 50–100 nuclei, reflecting an autonomous capacity of the zygote for four or five rounds of nuclear replication that in normal embryogenesis will produce a globular embryo. It also suggests that in early embryos, nuclear division and cytokinesis can be uncoupled. This anomalous pattern of *titan-5* embryo cellular development is somewhat similar to the early development of endosperm before cellularization.

The *TITAN-5* gene encodes the ADP ribosylation protein factor 1 (ARF1) that is probably involved in vesicle transport required for cell plate formation. In *Arabidopsis* there are at least six additional genes that encode related ADP ribosylation factor proteins that could participate in cytoplasmic vesicular traffic and cell plate formation during mitosis; however, early embryos have a specific requirement for the *TITAN-5* gene (McElver *et al.*, 2000).

The *titan* mutants collectively suggest that there are different requirements of cell division control in gamete and embryo cell cycles. The pattern of inheritance of the titan mutations strongly suggests that both the maternal and paternal alleles are functional in the fertilized egg cell. This observation, which has also been made for a number of early embryo mutations, argues for the expression of the maternal and paternal genome in early embryogenesis.

5.3.2 *Early embryo and suspensor development are coordinately regulated*

A number of *Arabidopsis* mutants in early embryogenesis affect the development of the embryo and the suspensor in contrasting ways. In these mutants, embryo development arrests but the suspensor develops abnormal structures. The arrest in embryo development causes the loss of identity of the suspensor cells and the anomalous development of the suspensor following an embryogenesis pathway. These mutations suggest that differentiation programs of the embryro and the suspensor are coordinated, probably by active signaling processes that continue during early embryogenesis (Yadegari *et al.*, 1994; Schwarz *et al.*, 1997; Zhang and Somerville, 1997). Two classes of mutations can be distinguished by the stage at which embryo development arrests: a first class in which embryogenesis arrests very early after fertilization, as in *twin-2* type mutants where development of the embryo stops after one or two mitotic divisions of the apical cell; and a second class in which embryogenesis is initiated correctly and a multicellular early embryo is formed with normal morphology, as in the *raspberry* and *suspensor* mutants. In this second class, embryogenesis arrests at the globular to heart-stage transition when the embryo consists of about 100 cells and the regional specification processes that differentiate the apical and basal domains do not take place (Schwartz *et al.*, 1994, 1997; Yadegari *et al.*, 1994; Zhang and Somerville, 1997).

5.3.3 Early after fertilization the suspensor differentiation program can be remodeled into embryogenesis

In the early arrested *twin-2* embryos, cells of the fully differentiated suspensor switch their developmental program to embryogenesis and develop into mature embryos. The reprogramming of suspensor development occurs as if fate-determining molecules were missing in the suspensor cells or in the arrested embryo. In *twin-2*, the suspensor cells recapitulate the whole spectrum of developmental programs normally restricted to the embryo proper and form viable embryos that germinate into normal plants. The genetic defect that causes the *twin-2* mutation has recently been characterized. It affects the expression of the valyl-tRNA synthase gene that appears to be an essential component of the translation machinery in early embryos (Zhang and Somerville, 1997). The molecular defect of the *twin-2* mutant suggests that sustained levels of *de novo* protein synthesis are needed very early in embryogenesis. The arrest of apical cell development could be due to the insufficient synthesis of an essential regulator of embryo development in the egg cell, the zygote or the apical cell. Alternatively, a complete arrest or reduced rate of protein synthesis could bring *twin-2* apical cell development to a halt. A paradox of this mutation is that while the arrest of apical cell development indicates a requirement of the *TWIN-2* gene for early embryo development, the normal development of suspensor derived *twin-2* embryos argues against it. This paradox becomes particularly interesting because the *twin-2* mutation affects the regulation of the expression of the single-copy gene that encodes valyl-tRNA synthase in *Arabidopsis*. An interesting resolution is provided if the single copy valyl-tRNA synthase gene is integrated into multiple programs of gene expression that are independently regulated in the apical cell and the suspensor-derived early embryos. Suspensor embryogenesis then would not be an exact recapitulation of apical cell embryogenesis. An extension of this rationale would imply that multiple programs of gene expression can initiate and regulate early embryo development as probably occurs during other processes of ectopic and somatic embryogenesis (Perez-Grau and Goldberg, 1989; Lotan *et al.*, 1999).

The isolation of the *TWIN-2* gene does not provide a direct explanation about the molecules involved in cell fate segregation; however, it highlights early inductive interactions between apical and basal cells of the zygote. It also demonstrates that suspensor development is not a cell autonomous process of the basal cell but requires active and continued communication between the apical cell and the suspensor cells. A model of suspensor development that is consistent with the *twin-2* mutant was proposed earlier (Schwarz *et al.*, 1994, 1997).

It is also interesting to consider the possible effects of a hypothetical morphogenetic gradient specific to the proembryo, including suspensor and embryo proper. If the morphogen molecule is not utilized and diluted in the continued development of the embryo as may occur in arrested embryogenesis, the persistence or reshaping of the morphogenetic gradient could induce development of suspensor cells into embryogenic pathways. In this context it is interesting to note that the developmental potential of the suspensor cells is progressively restricted as embryogenesis advances as if a signal molecule were being depleted during early embryo development.

5.3.4 Cellular differentiation in the suspensor is regulated by the embryo

The developmental plasticity of the early suspensor is progressively restricted as embryogenesis advances. The development of the suspensor is also redirected in embryo mutants that arrest embryogenesis after the differentiation of the proembryo. However, in late embryo arrest mutants, the cells of the suspensor acquire embryo characteristics but do not reinitiate or recapitulate embryogenesis, in contrast to early arrest mutants such as *twin-2* (Meinke, 1995).

The *raspberry* and *suspensor* class of mutations arrest embryo development after the initial development of the zygote has formed an early embryo. In these mutants the suspensor and embryo differentiate without apparent deviation from the wild-type pattern up to the globular or early heart stage when embryo development arrests. In arrested embryos the suspensor loses its identity and redirects its development to an embryo program. In the *sus-1* mutant the basal region of the late globular embryo undergoes extraordinary cell divisions that form an abnormal elongated embryo before arresting development. Following the arrest of embryo development, the suspensor initiates abnormal cell divisions and forms a massive suspensor that can be as large as the arrested embryo. In the *sus-2* mutants, embryo arrest takes place at a slightly later stage than in *sus-1* (Yadegari *et al.*, 1994; Schwartz *et al.*, 1994).

Anomalies in embryonic cell type differentiation programs are apparent in the *sus-2* and *raspberry* mutants, particularly in the epidermal tissue layer. Although the specification of the epidermal tissue and initial morphological differentiation of the epidermal cell layer take place in the *sus-2* and *raspberry* mutants, some element of the differentiation or the maintenance of the differentiated state of the protoderm is defective. The defect is manifested morphologically as an abnormal swelling of the protoderm cells that precedes the arrest of the globular embryo and the abnormal differentiation of the suspensor cells (Yadegari *et al.*, 1994). Interestingly, the arrested embryos do not degenerate but maintain their cellular vitality and appear to continue embryo-specific differentiation programs that are uncoupled from the associated morphogenesis processes that normally accompany them in embryo development. Similarly, the arrested suspensor does not enter the apoptosis program of degeneration that is characteristic of its terminal differentiation (Schwartz *et al.*, 1994). By contrast, the enlarged suspensors of *raspberry* mutants acquire embryonic characteristics and initiate embryo-specific cell differentiation processes. The embryonic-like processes in the suspensor take place in close synchrony with the continuing cellular differentiation in the arrested *raspberry* embryo. These embryo mutants, although morphologically arrested in early development, manifest cell differentiation programs normally associated with late embryogenesis, as clearly evidenced by the expression of storage protein genes and other maturation differentiation markers. The expression of maturation genes in arrested embryos and suspensors could be due to cellular responses to late embryogenesis signaling molecules or to the timely progression of cell autonomous gene expression programs initiated in early embryogenesis (Perez-Grau and Goldberg, 1989; Quatrano *et al.*, 1993). These mutations also suggest that multiple processes might be involved in the continued regulatory interactions between embryo and suspensor that are initiated during the first division of the zygote but appear to be maintained long thereafter into advanced embryogenesis (Meinke, 1995; Schwartz *et al.*, 1997). Programs of gene expression specific of embryogenesis can also be observed in a variety of abnormal developmental situations, as in the roots of the *pickle* mutant (Ogas *et al.*, 1997).

Although the analysis of the *raspberry* and *suspensor-2* mutations has provided very interesting insights into the regulatory interactions and communication processes between suspensor and embryo, the cloning and identification of the mutant genes does not provide much information about the molecular processes that regulate the interactions between embryo and suspensor. The *RASPBERRY* gene encodes a putative chloroplast S1 ribosomal protein and the *SUSPENSOR-2* gene encodes a protein with homology to the yeast PRP8 spliceosome assembly factor required for processing of primary transcripts (Brown and Beggs, 1992; Schwartz *et al.*, 1997). Both genes appear to encode cellular housekeeping functions despite the fascinating embryo development phenotypes that are revealed by these mutations.

It is interesting to note that embryo development can take place up to advanced globular stage in mutants with seriously impaired cellular metabolism. Early embryogenesis might therefore to some extent be considered a cell autonomous process of the zygote, endowed by the megagametophyte with basic cellular components and nondiffusible metabolites to support early embryogenesis.

5.3.5   *Subcellular organelles may be required for the progression of embryo development*

A mutation that affects chloroplast differentiation displays a mutant phenotype that is very reminiscent of the *raspberry* and *suspensor* mutations described above. The mutation of a chloroplast glycyl-tRNA synthase gene affects the differentiation of the chloroplast and leads to an arrest in embryo development that has some similarities to the *raspberry* and *suspensor* mutants. This observation suggests that there is a coordination of chloroplast development and embryogenesis that may also indicate a need for chloroplast functions to support embryo cell metabolism and development (Uwer *et al.*, 1998).

5.4 Early embryo development requires control of cell plate formation and cell wall position

In many plants the development of the early embryo follows a precise sequence of cell divisions that appear to be coordinated spatially and temporally across the embryo. The cell division patterns of the zygote specify the relative position of individual cells in early embryos. Cell position in turn largely determines the cellular contribution to the mature embryo, given that regional specification processes begin very early after fertilization and plant embryo cells do not rearrange their positions following developmental cues (West and Harada, 1993; Raghavan, 1986; Laux and Jürgens, 1997; Yadegari and Goldberg, 1997).

In *Arabidopsis* the pattern of cell division in the early embryo is highly predictable. The apical daughter cell of the zygote undergoes three rounds of isometric cell divisions to form an octant stage embryo. Octant embryo cells are of similar shape and volume, suggesting the equivalent cleavage of the apical cell. The inner cell walls across the octant embryo are aligned in three continuous near-perpendicular planes. The cell wall alignment in octant embryos can result from synchronous cell division or be directed by postcytokinesis processes that position the nascent cell walls at specific locations in adjacent cells. The next round of cell division is periclinal and asymmetrically divides each octant embryo cell into an outer and an inner cell with different fates. The periclinal cell walls also appear aligned along neighboring cells to define a continuous outer cell layer that differentiates as protoderm and an inner cell core that will form the internal tissues of the embryo (Mansfield and Briarty, 1991; Jürgens and Mayer, 1994).

A number of mutations that disrupt the cell division sequence of the zygote and the architecture of the early embryo have different manifestations; initially they cause defects in embryo cell differentiation that later lead to deranged morphogenesis of the embryo. Some of the abnormalities in the embryo are highly suggestive of specific developmental defects and can be interpreted as embryonic pattern formation defects. The recent cloning and identification of some of the mutant genes show that these mutations collectively affect different molecules involved in the control of cell division patterns and cell plate formation in the zygote and early embryos (Assaad *et al.*, 1996; Lukowitz *et al.*, 1996; McElver *et al.*, 2000).

Early cell division mutants can be classified into two major classes. A cytokinesis class manifests defects in cell plate formation frequently causing incomplete separation of zygote daughter cells. A second class has no apparent defects in cytokinesis but exhibits altered patterns of cell division; in these mutants the cell walls are positioned in random abnormal patterns. Both classes of mutations have distinct defects early in specific cell differentiation processes, and later in the morphogenesis of the embryo and the seedling.

5.4.1   Cell plate formation is abnormal in cytokinesis mutants

In some cell division mutants, the zygote initiates embryogenesis but embryo development becomes abnormal after a few cell division cycles. Two such mutations that have been studied extensively are the *knolle* and the *keule* mutants. *knolle* and *keule* have incomplete cytokinesis and during early embryogenesis produce aberrant embryos with characteristic defects in cell differentiation and embryo morphogenesis. The *knolle* mutants have abnormal cell division patterns that already deviate from normal embryo development in the initial cell divisions of the zygote. *knolle* embryo cells have incomplete cell walls that are positioned erratically in random patterns. *knolle* embryos also have a high frequency of multinucleated cells that probably result from a lack of coordination of nuclear replication and cell division. The *KNOLLE* gene encodes a syntaxin protein required for the partitioning of the cytoplasm during cytokinesis but not for nuclear division. Syntaxin proteins are involved in vesicular traffic and cell plate formation (Lukowitz *et al.*, 1996; Lauber *et al.*, 1997). This process is also defective in the *titan-5* mutants discussed above (McElver *et al.*, 2000).

keule is a mutant with cell division defects very similar to *knolle* (Assaad *et al.*, 1996). At the subcellular level both *knolle* and *keule* mutations manifest an abnormal accumulation of individual vesicles in the division plane of embryo cells at telophase. In wild-type embryo cells the vesicles fuse to form the phragmoplast at the cell division plane, but in the mutants vesicle fusion is abnormal and cell plate formation is not consistently completed. The *KEULE* gene has not yet been cloned, but double mutant studies strongly suggest that the gene products of both *keule* and *knolle* act in the same process that promotes vesicle fusion during cell plate formation. The double mutant *knolle/keule* phenotype is more severe than in either single mutant. The proliferation of the zygote is arrested early and leads to the generation of giant embryo cells with supernumerary nuclei, often polyploid, that somewhat resemble some of the *titan* mutants discussed above. The double mutant embryos develop into club-shaped multinuclear structures without internal cell walls containing about 30 nuclei. In *keule/knolle* double mutants the zygote nuclei divide up to five rounds and generate an enlarged multinuclear single cell embryo (Waizenegger *et al.*, 2000). The autonomous capacity of the zygote to undergo a few unchecked cycles of nuclear replication suggest that some aspects of early embryo development up to the globular embryo stage might be self-supported by the zygote.

Other cell division processes, such as cytoskeleton organization and localization of cell division molecules to the nascent cell plate, are normal in *knolle/keule* embryos (Waizenegger *et al.*, 2000). The cell cycle-dependent organization of microtubule arrays is normal, and a dynamin-like protein (ADL1) used as a cytokinesis marker accumulates between pairs of *knolle/keule* daughter nuclei. The cell cycle-dependent organization of these markers suggests that the defect is specific to vesicle fusion and cell wall formation and not to other cytoplasmic processes of cytokinesis. The KNOLLE protein itself accumulates in the cell plate region of wild-type and *keule* embryos similarly to the cytokinesis marker protein ADL1, suggesting the independent contributions to cytokinetic vesicle fusion by *knolle* and *keulle* (Waizenegger *et al.*, 2000).

5.4.2   Cell type differentiation is defective in cytokinesis mutants

knolle and *keule* were initially described as radial pattern mutations. Cell type specification is defective in these mutants and causes abnormal tissue differentiation and morphogenetic defects in embryos and seedlings that are suggestive of radial patterning defects (Mayer *et al.*, 1991; Jürgens and Mayer, 1994; Lukowitz *et al.*, 1996).

The cell type differentiation programs are abnormal in *knolle* embryos; in particular, early differentiation of the protoderm layer does not take place. Other cell types, like root cells, differentiate in germinating cultured *knolle* embryos, suggesting that different cell types might have a differential control or specific requirements for cell division and differentiation. In addition, general cytokinesis defects are likely to be more pronounced in the extensive transversal cell walls of early protoderm cells. *knolle* embryos also manifest a deregulation of the spatial and temporal patterns of gene expression in the embryo. The AtLTP1 gene is a marker of epidermal cell differentiation in wild-type embryos. Its expression is restricted to the epidermal cell layer late in embryogenesis. In *knolle* embryos, the AtLTP1 is expressed in both outer and inner cells of early embryos. The spatial deregulation of AtLP1 expression in *knolle* embryos suggests that the AtLTP gene could be regulated by factors that can be exchanged between cells with incomplete walls. Interestingly, the lack of differentiation of the protoderm layer contrasts with the expression of the AtLTP gene, suggesting that epidermis-specific gene expression programs are initiated in *knolle* embryos (Lukowitz *et al.*, 1996; Vroemen *et al.*, 1996). This situation is reminiscent of the progression of embryonic patterns of gene expression in morphologically arrested embryos (Yadegari *et al.*, 1994).

5.4.3   Cytokinesis control molecules are cell cycle regulated

The pattern of expression of the *KNOLLE* gene in wild-type embryos is cell cycle regulated. *knolle* mRNA and protein accumulation in embryo cells is restricted to a brief period of the cell cycle just before cytokinesis. *In situ* hybridization shows that *knolle* mRNA accumulates in a patch pattern in octant stage embryos, which indicates cell cycle regulation of *knolle* expression and also strongly suggests that early embryo cells do not progress through the cell cycle synchronously (Lukowitz *et al.*, 1996). Cytological studies of cytoskeletal organization in wild-type and *keule* and *knolle* mutant embryo cells also provide interesting observations about the coordination of cell cycle and cell division in early embryos. In wild-type embryos, there are marked asynchronies in the progression of the cell cycle that can already be observed at the two-cell embryo stage when one cell is beginning the preprophase while its sister cell is already at metaphase, as seen in figure 2a in Waizenegger *et al.*, 2000. The asynchrony in cytoskeletal organization of embryo cells is also observed in octant embryos (Waizenegger *et al.*, 2000) and during the segregation of the periderm layer as shown in figure 5.1.

The nuclear divisions appear synchronous in *knolle* and *keule* embryos, in contrast to the differences in timing of the cell cycle in wild-type embryo cells.

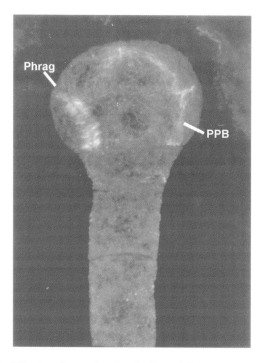

Figure 5.1 Early *Arabidopsis* embryo undergoing the first periclinal series of cell divisions that form the protoderm cell layer. Microtubules (Phrag, phragmoplast; PPB, preprophase band) stained by indirect immunofluorescense. Note the marked cell cycle asynchrony between neighbor cells. (This picture was kindly provided by Dr Hong Nguyen from Professor R. Brown's laboratory).

These observations suggest that the imperfect cell division processes in the *keule* and *knolle* mutants prevent the appearance of asynchronies in the cell cycle, and that complete cytokinesis is required for specific cell cycling differences that normally occur in wild-type embryos (Waizenegger *et al.*, 2000).

The marked cell cycle asynchrony in individual early embryo cells contrasts with the apparent regularity of the alignment of cell walls in embryo cells and with the progressive symmetric morphogenesis of the embryo up to the globular stage. It also argues for the existence of cellular processes actively controlling the position and alignment of cell walls in neighbor cells during early embryogenesis. The developmental significance, if any, of the differences in cell cycling in early embryos is not clear, but it might provide an early cell autonomous mechanism towards the initiation of cell-specific gene expression programs and also establish differential cell proliferation rates in early embryos.

5.4.4 Cell plate and cell wall positioning mutants—cell plate positioning is deregulated in the zygote of the gnom mutant

In the second class of early cell division mutants, cell plate formation and cell division is completed normally and the mutant zygote proliferates and forms a multicellular embryo. However, the early embryos develop morphological abnormalities as the patterns of cell division become erratic. In this second class of mutants, the positions of the cell plates that direct cell wall formation after cell division are abnormal. These mutations suggest that cell plate formation and positioning are somewhat independent processes. Cell plate position mutations have profound effects on embryo morphogenesis, presumably by deranging the spatial distribution of embryo cells within preformed sets of patterning axes or by altering cellular coordinates in morphogenetic fields that might guide pattern formation and morphogenesis during early embryogenesis. Abnormal cell division patterns could also lead to defects in short-range cell–cell interactions and result in abnormal regional specification processes and defective morphogenesis of the embryo.

gnom is a characteristic mutant of this class that has been extensively studied. *gnom* was isolated as a pattern formation mutant with defects in the establishment of the apical–basal pattern of development in the embryo and during postembryonic development (Mayer *et al.*, 1993). *gnom* seedlings lack a root system, and are also defective in the coordinated alignment of cells in the vascular tissue and have variable reduction of apical seedling structures that can be traced back to defects in early embryogenesis. In *gnom* the placement and orientation of the cell wall during the first division of the zygote is already defective. Early *gnom* embryos have abnormal morphogenesis because of the failure to coordinate the patterns of cell division across neighbor embryo cells. *gnom* embryos at octant stage frequently have extranumerary cell tiers similar to *bodenlos*, a pattern formation mutant that specifically affects root formation during embryogenesis and that will be discussed in a later section of this review (Hamann *et al.*, 1999). *gnom* embryos do not form genuine globular stage embryos. The differentiation of the suspensor is also anomalous and the hypophysis cell is not formed in *gnom* embryos (Mayer *et al.*, 1993). At the subcellular level the *gnom* mutants have severe defects in cell wall ultrastructure and composition. *gnom* cell walls have an anomalous distribution of pectin that probably causes defects in cell wall position and also impairs the adhesion between neighbor cells. Some cellular interactions in the *gnom* mutant are probably abnormal as well (Shevell *et al.*, 1994, 2000).

The *GNOM* gene encodes a large 165 kDa multifunctional protein that is associated with cellular membranes. The GNOM protein is a functional ADP ribosylation factor with a domain that catalyzes guanine nucleotide exchange on small GTP-binding proteins, strongly suggesting its participation in vesicle traffic control and secretion during various cellular processes. However, the GNOM protein is multifunctional and probably also interacts with other partners, making possible GNOM participation in a number of functions in the cell (Busch *et al.*, 1996; Grebe *et al.*, 2000). A specific partner of GNOM that has been identified is cyclophilin 5. Cyclophylin 5 (Cyp5) catalyzes protein folding at prolylpeptidyl bonds and has a potential regulatory role in signal transduction pathways. Cyp5 might be required for proper folding of GNOM in wild-type embryos or might be a regulator of activity by modulation of GNOM dimerization (Grebe *et al.*, 2000).

The pattern of expression of *GNOM* precludes a function that is restricted to embryogenesis. The *GNOM* gene is expressed in all cells of the early embryo. The GNOM and Cyp5 proteins are detected in all plant organs. The Cyp5 protein is also distributed uniformly in globular embryos, but at the early heart stage Cyp5 is detected at lower levels in the epidermal cell layer. In late heart stage embryos the pattern of expression of *GNOM* becomes restricted mainly to provascular cells in embryos (Grebe *et al.*, 2000).

The experimental manipulation of auxin levels in isolated early *Brassica* embryos phenocopies some characteristic of gnom mutant phenotypes, suggesting that auxin might be involved in pattern formation in conjunction with gnom (Liu *et al.*, 1993; Hadfi *et al.*, 1998; Steinmann *et al.*, 1999).

Another mutant with altered positioning of cell walls is *fass*, also known as *ton*. This mutant does not form the organized microtubule networks associated with cytokinesis in plant cells, and cellular division occurs in highly disorganized patterns. Embryo morphogenesis is highly abnormal in *fass*. However, like *gnom*, the *fass* mutation does not prevent some cellular differentiation and tissue formation processes from taking place in highly abnormal embryos (Torres-Ruiz and Jürgens, 1994; Trass *et al.*, 1995). These observations suggest that the precise cell division patterns of *Arabidopsis* embryos are not the only elements regulating embryo morphogenesis. Probably short-range cell communication processes and cell autonomous processes that still take place in morphologically abnormal embryos like *fass* and *gnom* regulate local aspects of embryo development that result in partial histogenesis and partial morphogenesis.

5.5 Regional specification processes define embryo domains with different developmental roles

A number of *Arabidopsis* embryo development mutants collectively define separate spatial domains in the early embryo that have different contributions to the morphogenesis of the mature embryo and the seedling (Mayer *et al.*, 1991; Meinke, 1995; Yadegari and Goldberg, 1997). Some of these developmental embryo mutations have been interpreted as pattern formation mutants defining three pattern elements in the embryo, the apical, central and basal elements (Mayer *et al.*, 1991; Mayer and Jürgens, 1998). In this review the central and basal pattern elements are discussed together as part of a larger basal domain since a number of recent observations suggest the coordinate development of both elements in early embryogenesis (Hardtke and Berleth, 1998; Hamann *et al.*, 1999). The organization of early plant embryo cells in developmental domains suggests that embryogenesis integrates multiple developmental processes that exhibit some independence of each other. The developmental programs taking place in individual regions appear to be regulated locally by short-range cellular interactions that are not propagated throughout the entire embryo. While regional specification processes might initially be regulated independently of each other, the progress of embryogenesis depends on their integration and coordination to form a mature embryo. The integral development of the embryo is probably mediated by signaling gradients of plant hormones that are established very early in embryogenesis.

5.5.1 Developmental embryo domains are established in the initial divisions of the zygote

The regional differentiation of the *Arabidopsis* embryo is initiated during the first few cell divisions of the zygote. The cells of the octant stage embryo are already organized into an upper tier and a lower tier with different developmental roles. The descendants of the upper tier form the apical domain of the embryo that is essentially a cellular clone of a single early daughter cell of the zygote. The descendants of the lower cell tier, together with the uppermost cell of the suspensor, form the basal domain. The uppermost cell of the suspensor is integrated back into embryogenesis at the 16-cell embryo stage to form the hypophysis that functions as an organizing center for the root meristem and the larger basal domain. At the 16-cell stage the embryo is already organized into an apical and a basal domain that will have different contributions to the morphogenesis of the mature embryo and the seedling. The cells in the apical domain are isometric and organized in a stratified fashion that precedes the cellular organization of the shoot apical meristem. The cells in the basal domain are elongated in an apical basal axis and form concentric cell layers that are characteristic of the hypocotyl and the primary root (Barton and Poethig, 1993; Jürgens and Mayer, 1994; Scheres *et al.*, 1994).

It is not yet known whether the initial specification of different developmental fates in the embryo cell tiers reflects a prepattern inherited from the female gametophyte, or whether it is initiated in the zygote by processes specific to embryogenesis. A number of mutations that affect the pattern of cell division of the zygote also fail to establish regional differentiation domains in the embryo (Mayer *et al.*, 1991; Hardtke and Berleth, 1998; Hamann *et al.*, 1999). These mutations also affect the two-tiered organization of the octant embryo and have abnormal early embryogenesis. The *gnom* mutation alters the formation of the phragmoplast and the positioning of the cell plate that leads to abnormal patterns of cell division of the zygote. The irregular cytokinesis of the *gnom* zygote often produces a larger than normal apical cell that develops into embryos with extra cell tiers at the octant stage. The extra cell tiers of *gnom* apparently cause the abnormal differentiation of the apical domain and an arrest of morphogenesis. The basal domain of the embryo is also abnormal since the *gnom* mutation also affects the pattern of cell divisions of the suspensor. The basal daughter cell of the *gnom* zygote is often smaller than normal and develops abnormally into a suspensor of reduced size and cell number. In extreme *gnom* alleles the hypophysis cell is not formed and the basal domain of the embryo does not differentiate, similarly to other mutants with defects in the basal domain of the embryo (Hamann *et al.*, 1999). *gnom* embryos have abnormal morphogenesis and manifest a range of cone-shaped phenotypes with severe reductions in embryo cell numbers and small embryo sizes (Mayer *et al.*, 1993; Shevel *et al.*, 1994).

gurke is a mutation that causes general defects in the late differentiation of the entire apical domain but not its establishment in early embryos. *gurke* embryos develop normally until globular stage, but then fail to make the transition into heart stage and fail to initiate the differentiation of the cotyledons or the shoot meristem. *gurke* embryos have a severely reduced or absent apical region at maturity but differentiate the elements of the basal domain, the hypocotyl and the primary root (Mayer *et al.*, 1991). The identification of the *GURKE* gene function should provide information on the processes that establish the apical domain. *gurke* mutants are allelic with emb22. The emb22 alleles of *gurke* have more extensive defects in embryogenesis that are not restricted to the apical domain (Evans and Barton, 1997).

A mutation that produces extranumerary structures at the apical domain has recently been isolated. The *hydra* mutant has defects in the control of cell shape in early embryos and has an abnormal morphogenesis of the entire embryo. The apical domain of *hydra* differentiates supernumerary cotyledon-like structures of reduced size (Topping *et al.*, 1997).

5.5.2 The apical domain of the embryo forms the shoot meristem and the cotyledons

The apical domain of the embryo forms two different structures, the cotyledons and the shoot apical meristem. The cotyledons are terminally differentiated storage organs that support the germination of the seed. The shoot apical meristem is the primary growth organizing center for the aerial organs of the plant during postembryonic development. Traditionally, the cotyledons have been considered to be the first differentiated products of the embryonic shoot meristem; however, recent observations suggest that in *Arabidopsis* the differentiation of the cotyledons precedes the establishment of a functional shoot apical meristem. The differentiation of the cotyledons and the shoot meristem is probably regulated by a sequential developmental process that is initiated early after fertilization. This process appears first to specify the cotyledons and then continues to specify the shoot meristem in the remaining cells of the apical embryo region (Laux and Mayer, 1998; Long and Barton, 1998; Woodrich *et al.*, 2000). The development of the cotyledons is initiated at late globular stage and continues into late embryogenesis, while the shoot apical meristem is not completely established until the end of the heart embryo stage. In *Arabidopsis* the cotyledons contain about 6000 cells, or approximately half of the cells of the embryo. In mature embryos the shoot apical meristem contains only about 110 cells (Irish and Sussex, 1992; Jürgens and Mayer, 1994).

5.5.2.1 Spatially regulated gene expression programs in early embryo cells specify the shoot apical meristem

The vegetative apical development of the plant is organized at the shoot meristem. The shoot meristem is a dynamic self perpetuating multicellular system that is established by multiple short-range cell interactions and is maintained by complex cellular communication processes within the apical region of the shoot. In recent years it has been demonstrated that many of the cellular interactions that differentiate and maintain the shoot apical meristem functional throughout vegetative development are initiated in the early embryo (Endrizzi *et al.*, 1996; Laux and Mayer, 1998; Long and Barton, 1998; Mayer *et al.*, 2000; Schoof *et al.*, 2000).

The cell-specific programs of gene expression that specify the shoot apical meristem in *Arabidopsis* begin in the quadrant stage embryo. The establishment of the shoot apical meristem is regulated by a cascade of gene expression in which early genes delimit and regulate the cellular expression patterns of late genes (Laux and Mayer, 1998; Long and Barton, 1998; Mayer *et al.*, 1998; Bowman and Eshed, 2000). Early shoot meristem genes follow specific spatial and temporal patterns of expression that progressively define subdomains of late shoot meristem gene expression in the apical region of the embryo. The expression of the late shoot meristem genes establishes a central zone of the apical domain that stops cell proliferation and differentiates the shoot apical meristem and a peripheral zone in which continued cell division forms the cotyledons. Further differentiation within the central zone defines a group of stem cells that maintain the meristem and a differentiation zone that regulates lateral organ formation. The system of cellular interactions and molecular processes that sets up the shoot meristem and maintains its organization has recently been described in detail (Evans and Barton, 1997; Aida *et al.*, 1998; Laux and Mayer, 1998; Long and Barton, 1998; Bowman and Eshed, 2000).

Mutations in a number of shoot meristem genes completely abolish the formation of a functional shoot meristem without apparently affecting other regional specification processes during embryogenesis. Two examples are the *wuschel* and the *shoot meristemless* mutants. *WUSCHEL* is an early shoot apical meristem organizing gene. The expression program of *WUSCHEL* begins at the 16-cell embryo or earlier. *WUSCHEL* is expressed in all the internal cells of the apical domain of the 16-cell embryo. The program of expression of the *WUSCHEL* gene is inherited asymmetrically during cell division. The cells expressing *WUSCHEL* divide to form daughter cells towards the periphery that cease to express the *WUSCHEL* gene, while the progenitor cells remaining in the original position maintain *WUSCHEL* expression. The expression of the *WUSCHEL* gene is maintained in the four cells occupying the center of the apical domain of the embryo during most of embryo development. The embryonic pattern of expression of *WUSCHEL* is maintained essentially unchanged during vegetative development and it is only modified when the shoot apical meristem enters the flower differentiation program (Mayer *et al.*, 2000). *wuschel* mutants have defects in the maintenance of the shoot meristem during postembryonic development but do not affect embryo morphogenesis. The differentiation of the embryo apical domain in *wuschel* is normal.

The *SHOOT MERISTEMLESS* gene is expressed late in the gene expression program that establishes the shoot meristem. *SHOOT MERISTEMLESS* expression marks the spatial location of the central region of the meristem in the embryo. Mutations in the *SHOOT MERISTEMLESS* gene also fail to organize a functional shoot apical meristem and similarly to *wuschel* do not have any manifestations in the morphogenesis of the embryo that develops to maturity. *shoot meristemless* and *wuschel mutants* develop morphologically normal cotyledons and mature embryos. These mutations indicate a complete independence of shoot meristem specification and embryo development (Barton and Poethig, 1993, 1994; Long *et al.*, 1996). They also suggest that key molecules required for the complex interactions regulating the gene expression programs and the cellular communication and differentiation processes that establish the shoot apical meristem are not necessary for other aspects of embryo development.

5.5.2.2   Mutations in some early shoot meristem specification genes cause defects in embryogenesis

The apparent independence between the processes of shoot apical meristem specification and embryo development suggested by the absence of an embryo phenotype in the *shoot meristemless* and the *zwille* mutants is not supported by mutations in other shoot meristem specification genes. The *pinhead* mutant, also known as *zwille*, has anomalous postembryonic development of the shoot apical meristem, but *pinhead* seems to play a role in early embryo development. Strong *pinhead* alleles completely ablate the shoot apical meristem without affecting the morphogenesis of the cotyledons or the mature embryo, similarly to *shoot meristemless* and *wuschel* (McDonell and Barton, 1995; Laux *et al.*, 1996; Moussian *et al.*, 1998). The *PINHEAD* gene can be considered an early regulator of shoot meristem formation that regulates shoot meristemless expression. The *PINHEAD* gene encodes a translation initiation factor that is already expressed in the apical and basal domain of quadrant stage embryos preceding expression of the *ZWILLE* gene. *pinhead* has defects in early embryogenesis. Early *pinhead* embryos have abnormalities in the basal domain and extranumerary cells in the suspensor. However, mature *pinhead* embryos are morphologically normal, suggesting that the early defects are corrected during embryogenesis, probably by functional complementation by a redundant gene (Lynn *et al.*, 1999).

The *ARGONAUTE 1* gene has been proposed to functionally complement *pinhead* function during early embryogenesis in *pinhead* mutants. The *ARGONAUTE 1* gene encodes a protein very similar to *pinhead*, and *ARGONAUTE* is expressed in early embryos in a pattern that overlaps with *pinhead*. Unlike *pinhead*, mutations in the *ARGONAUTE1* gene have persistent embryonic defects in the late development of the apical region. During maturation the expansion of the cotyledons takes place at an angle in *argonaute* embryos, the cotyledons fail to come together and have abnormal bending in mature embryos (Lynn *et al.*, 1999). *argonaute1* mutants also have postembryonic phenotypes of altered polarity in leaves (Bohmert *et al.*, 1997).

Embryogenesis is defective in double mutants *pinhead/argonaute*; double mutant embryos arrest embryo development at the globular stage at about 100 cells. This intriguing phenotype brings together shoot apical meristem specification and embryo morphogenesis, suggesting that the early regulatory processes of SAM specification, particularly those contributed by *pinhead* and *argonaute*, are also necessary for embryo development (Lynn *et al.*, 1999). The analysis of the *pinhead* and *argonaute1* mutations also suggests that functional redundancy probably complicates the genetic identification of molecular and cellular processes shared during meristem specification and early embryogenesis (Lynn *et al.*, 1999).

PINHEAD is also expressed at all stages of development in the provascular tissue and vascular cells, suggesting that *pinhead* function may also be required for histogenesis (Lynn *et al.*, 1999).

5.5.2.3 Some late shoot meristem specification genes regulate cotyledon initiation

Mutations in the genes that establish the peripheral subdomain of the shoot apical meristem affect the development of the apical region of the embryo. *cup shaped cotyledon* (*cuc*) mutants have defects in the shoot apical meristem and the cotyledons. In *Arabidopsis* there are two *cuc* genes, *cuc1* and *cuc2*, that are required for shoot meristem formation. Both *CUC* genes encode NAC domain proteins with unknown cellular function. These genes are homologous to the *Petunia NO APICAL MERISTEM* genes. The *CUC* genes interact with *shoot meristemless* to form the shoot apical meristem and delineate the region of the apical embryo domain that will form the cotyledons (Aida *et al.*, 1999). In *cuc1/cuc2* double mutants the cotyledons do not separate but remain fused and the shoot apical meristem is not properly formed. The *cuc1/cuc2* double mutant phenotype manifests at the transition between globular and heart stage embryos when the cells in the presumptive shoot apical meristem region of the apical domain elongate vertically and in some cases produce one extra cell layer at the meristem. The *CUC* genes are expressed in the apical domain shoot meristem region preceding, but partially overlapping with, *SHOOT MERISTEMLESS* expression. The expression of the *CUC2* gene marks the boundaries of the apical domain that contribute cells to the cotyledons and to the shoot apical meristem. While *CUC1* and *CUC2* genes are both required for the expression of *SHOOT MERISTEMLESS*, *SHOOT MERISTEMLESS* is in turn required to regulate the proper spatial expression of *CUC2* to separate the cotyledons, suggesting a regulatory loop (Aida *et al.*, 1999).

A mutant with a variable phenotype that specifically affects the development of the entire apical domain has been described as *topless*. Some *topless* embryos develop hypocotyl and root apical meristem without cotyledons or shoot apical meristem (Evans and Barton, 1997).

5.5.3 The basal domain—histogenesis and formation of the primary root in the embryo

The basal domain has a key role in the differentiation of the primary embryonic tissues and the establishment of the apical basal pattern of embryo morphogenesis. The basal domain of the early embryo has three differentiated tissue layers, the epidermis, the ground meristem and the procambium. All the tissues and cell types of the embryo derive from these three primary tissues by progressive cell differentiation processes. The basal domain integrates the hypocotyl, the primary root and the root meristem. The basal domain, together with the shoot apical meristem, forms the embryo axis that directs the postembryonic development of the plant. In *Arabidopsis* the embryonic axis contains about 8000 cells or about half of the cells of the mature embryo.

5.5.3.1 Steroid molecules may regulate initial embryo histogenesis

Tissue differentiation in *Arabidopsis* appears to be a sequential process. Three primary tissues are established during early embryogenesis: the epidermis, the ground meristem and the vascular procambium. As embryogenesis advances, secondary tissues differentiate from the ground meristem and the procambium layers. The procambium differentiates one additional tissue layer that forms the pericycle at heart stage. The ground meristem layer also undergoes cell division and becomes two-celled in early heart stage embryos. After heart stage most of the cells in the embryonic axis originate from the root initials that serve as stem cells for the primary root. The root initials also provide most of the cells for the initial growth of the seedling during germination (Scheres *et al.*, 1994).

Recent observations suggest that steroid molecules regulate the differentiation of the primary tissues of the embryo. Two mutations that affect the cellular organization of the embryo have been recently characterized at the gene level; one is *fackel* and the second is a mutation in a sterol methyltransferase gene (Jürgens and Mayer, 1994; Diener *et al.*, 2000; Jang *et al.*, 2000; Schrick *et al.*, 2000). Both mutations affect the sterol metabolism pathway and suggest that sterol molecules might be involved as signal molecules regulating the cellular differentiation processes that establish the primary tissue layers of the early embryo.

fackel mutants have defects in cell division and cell elongation. The mutation is apparent in globular embryos when the cells in the central region fail to elongate and divide. *fackel* embryos also have aberrant morphogenesis that results in reduced cotyledons and variable positioning of the presumptive shoot meristem, as indicated by the expression of the *SHOOT MERISTEMLESS* gene. The mature embryos have shortened or altogether absent hypocotyls but have a root meristem (Jürgens and Mayer, 1994). Two groups have recently cloned the *FACKEL* gene. *FACKEL* encodes a C_{14} reductase enzyme of the sterol biosynthesis pathway. *fackel* mutants accumulate intermediates in sterol biosynthesis that are consistent with a block in the sterol biosynthesis pathway at the level of the C_{14} reductase (Jang *et al.*, 2000; Schrick *et al.*, 2000).

A mutation in a different step of the sterol biosynthesis pathway, at the level of the sterol methyltransferase 1 gene, also causes defects in cell division and cell expansion during embryogenesis and displays mutant phenotypes very similar to *fackel* (Diener *et al.*, 2000). The process by which defective sterol biosynthesis causes abnormal cell differentiation in *fackel* and methyltransferase mutants during embryogenesis is not entirely clear. Sterols are significant components of cellular membranes; a structural role for these molecules suggests that a deficiency in membrane steroids might interfere with cell expansion during cell differentiation in embryogenesis. However, other sterol metabolism mutants do not manifest anomalies in embryo development. These two mutations suggest that specific sterol-derived molecules might have a signaling function in embryogenesis at the level of histodifferentiation and radial patterning of the embryo. The possible involvement of sterol molecules in plant embryogenesis is very exciting, particularly given the widespread role of steroid molecules in gene regulation and inductive regulation of development in other systems (Clouse, 2000).

5.5.3.2   The root meristem is established during early embryogenesis

The root meristem is also established during early embryogenesis similarly to the shoot apical meristem. The establishment of both meristems forms the embryonic axis in early embryos. The root meristem contributes a significant number of cells to the embryo axis in contrast to the shoot meristem (Scheres *et al.*, 1994)

The organization of the root meristem contrasts with that of the shoot meristem. The root meristem is organized around a founder cell, the hypophysis, that forms the central cells of the quiescent center. The quiescent center induces the differentiation of a layer of root cell initials that act as root stem cells and become the main cell sources for the development of the primary root (Van den Berg *et al.*, 1997). The root initials provide strong lineage specifications that limit the future cellular fates of root cells and maintain a template of the cellular organization of the root at the level of the initials (Steeves and Sussex, 1989; Scheres *et al.*, 1994).

A number of mutants that have abnormal root development during germination can be traced back to defects in the differentiation of the basal domain of the embryo. These mutations can be classified into two groups. A group with severely reduced or completely absent roots that has defects in the establishment of the root meristem, and a second group with abnormal root morphology and growth that is associated with defects in the cellular organization of the root (Scheres *et al.*, 1995).

5.5.3.3   The hypophysis cell is required to establish the root meristem

A number of mutations that show defects in the basal embryo domain also have abnormal differentiation of the hypophysis cell. An additional aspect of these mutations is that they suggest that auxin is involved in the establishment and differentiation of the basal domain and the establishment of the apical basal embryo axis.

The *bodenlos* and *hobbit* mutations have defects in the establishment of the root meristem during embryogenesis and do not form the primary root of the seedling during germination. Both mutations interfere with the differentiation of the hypophysis cell and have abnormalities in embryo morphogenesis, particularly in the organization of the basal embryo domain (Scheres *et al.*, 1995; Willemsen *et al.*, 1998; Hamann *et al.*, 1999). The *bodenlos* seedlings have variable phenotypes with two extreme morphologies. In the weak phenotype, the primary root and root meristem are absent, but the hypocotyl and cotyledons are well differentiated. In the strong phenotype, the hypocotyl, primary root and root meristem are all absent. In addition, the apical embryo domain is also defective, and the cotyledons are very reduced and have abnormal vascular tissues. The first departure from normal development in *bodenlos* occurs very early in embryogenesis when the apical daughter cell of the zygote divides horizontally instead of vertically. In octant and heart stage *bodenlos* embryos two types of cellular defects can be distinguished: double octant embryos that develop distinct elongated provascular cells at heart stage, and octant embryos with additional cell files that result in small-sized heart stage embryos in which the provascular cells do not differentiate. The two different cellular organizations of the octant *bodenlos* embryos can be correlated with the weak and strong seedling phenotypes (Hamann *et al.*, 1999). The hypophysis cell does not differentiate in any *bodenlos* embryo. It is not known whether the uppermost cell of the suspensor that normally differentiates the hypophysis follows a different differentiation path in *bodenlos*. It could form an additional tier of cells in the octant embryo, or the extra cells of the octant embryo could interfere with the cellular differentiation of the hypophysis that in *bodenlos* remains a suspensor cell.

The bodenlos mutant is very similar to *monopteros* (Hardtke and Berleth, 1998). *monopteros* fails to differentiate the entire basal embryo domain and has a severely reduced embryonic axis. In *monopteros* the cotyledons develop abnormally and are either completely absent or only one cotyledon is formed. *monopteros* also has defects in root formation probably because of defective differentiation of the hypophysis. The *MONOPTEROS* gene encodes an auxin-regulated transcription factor (Ulmasov *et al.*, 1997). *monopteros* also mediates the differentiation of the vascular system during postembryonic development (Przemeck *et al.*, 1996). *bodenlos* seedlings, contrary to *monopteros*, are able to form secondary roots and grow into viable mature plants. Secondary roots form from the pericycle in *bodenlos* seedlings as in wild-type plants, apparently by a process that does not require *BODENLOS* gene function.

The double mutant *bodenlos/monopteros* does not show a clear epistatic effect of either gene in early embryogenesis, but displays a novel phenotype after heart stage. After heart stage the *bodenlos/monopteros* embryos have abnormal morphogenesis: they do not develop cotyledon primordia and remain wedge-shaped until maturity. The mature embryo has a severely reduced apical domain structure, and the hypocotyl and the root are completely absent, suggesting a concerted action of both gene functions to coordinate the integral development of the entire embryo. The synergistic effect of *bodenlos* and *monopteros* suggests that auxin might be involved in the mode of action of *bodenlos*. The double mutant phenotype also has similarities to *gnom* (Mayer *et al.*, 1993; Hamann *et al.*, 1999).

auxin resistant 6 is a recently described mutant with an embryo phenotype very similar to *bodenlos*. *auxin resistant 6* is insensitive to the hormone auxin, further supporting the role of auxin in the early differentiation of the basal domain and the hypophysis cell (Hobbie *et al.*, 2000). The similarities between the *monopteros*, *bodenlos* and *auxin resistant 6* mutants strongly suggest that auxin is involved in the specification of the hypophysis cell and the establishment of the apical basal pattern of differentiation of the early embryo.

The *hobbit* mutants, similar to *bodenlos*, do not differentiate the root meristem during embryogenesis. The *hobbit* mutation also interferes with the differentiation of the hypophysis cell by a process that appears different from *bodenlos*. The *hobbit* mutation can be traced back to the anomalous division of the uppermost cell of the suspensor and the failure to differentiate the hypophysis cell. The upper cell of the suspensor divides vertically in *hobbit* embryos (in wild type the division is horizontal) and appears to form an abnormal additional tier of cells at octant stage. *hobbit* embryos lack the hypophysis cell and also lack the cells of the root cap and columella that are derived directly from the hypophysis, strongly suggesting a lineage relationship between these root cells. In *hobbit* embryos the root meristem initials do not differentiate. The *hobbit* hypocotyl develops to variable extents and has a normal cellular pattern in the absence of the root initials, although it has a reduced number of cells (Scheres *et al.*, 1995; Willemsen *et al.*, 1998).

The *hobbit*, *bodenlos* and *auxin resistant 6* mutations could be considered cell wall positioning mutations similar to *gnom*. In these mutants the early partitioning of the zygote occurs abnormally. It is interesting to speculate that auxin might be involved in the regulation of cell plate positioning in the zygote by some yet unknown mechanism.

The *hobbit* and *bodenlos* mutations suggest that at least two events are required for the specification of the hypophysis cell. *hobbit* might encode a cell autonomous determinant of the hypophysis cell that initiates the differentiation of the hypophysis, and *bodenlos* might be facilitate a later process of hypophysis differentiation. These mutations also show that the hypophysis cell is the precursor cell of the quiescent center and that it also functions as a source of inductive signals to differentiate the root initials and the root meristem region. The *monopteros* mutation is epistatic to *hobbit* in the basal region of the embryo, suggesting that auxin signaling might also mediate the function of the *HOBBIT* gene similar to *bodenlos*.

5.5.3.4   *The cellular organization of the root is established*
in the embryo

In *Arabidopsis* the root is organized as a set of concentric cylinders corresponding to the epidermis, ground meristem or cortex, and vascular procambium primary tissues. Each primary tissue integrates different cell types. The ground tissue has an outer layer of cortex cells and an inner endoderm cell layer. The vascular procambium differentiates the pericycle, a single cell layer that contacts the cortex endodermis and envelops the provascular tissue in the center of the root (Dolan *et al.*, 1993). The primary origin of the cells of the root are the root initial cells that acquire stem cell properties through their apical contact with the central cells of the quiescent center. The root initials provide not only a cell source for the development of the root but also a spatial template for the radial cellular organization of the primary root that remains invariant during *Arabidopsis* development (Scheres *et al.*, 1994; Van den Berg *et al.*, 1997).

The invariant cellular organization of the *Arabidopsis* root is altered in a group of seedling root mutants. The *scarecrow*, *short-root* and *pinocchio* mutants have defects in the cellular differentiation of the cortex tissue. The *gollum* and *woodenleg* mutants have cellular defects during the differentiation of the vascular tissue. The mutant defects are restricted to single tissue layers, suggesting that cell differentiation may be controlled by independent mechanisms in each primary tissue layer (Benfey *et al.*, 1993; Scheres *et al.*, 1994).

5.5.3.5 *Asymmetric cell divisions during embryogenesis establish the radial cellular organization of the root and shoot*

In the *short-root* and *scarecrow* mutants the alterations in the cellular pattern of the root can be traced back to cellular differentiation defects in the ground meristem tissue of the embryo. The cellular defects are apparent at heart stage of embryogenesis when a periclinal round of cell division that doubles the ground meristem layer does not take place in the mutants. In *shortroot* embryos the endodermis cell layer is not present, whereas in *scarecrow* the ground meristem lacks the cortex cells but differentiates the endodermis cell layer. The mutant phenotypes suggest that *scarecrow* participates in the control of cell division in the cortex tissue, while *short-root* is required for the specification of endodermal cell fate. The cellular defects of *scarecrow* and *short-root* are also observed in secondary roots, suggesting that the embryonic processes of cell type specification are not corrected during vegetative root development or during secondary root formation (Scheres *et al.*, 1995).

The *SHORT-ROOT* and *SCARECROW* genes have been cloned. Interestingly, both genes are very similar and encode putative transcription factors of the GRAS family (Di Laurencio *et al.*, 1996; Pysh *et al.*, 1999; Helariutta *et al.*, 2000; Wysocka-Diller *et al.*, 2000). Both genes are expressed in early embryos in nonoverlapping patterns. *SCARECROW* expression is first detected in late globular stage embryos in the hypophysis cell; in heart stage embryos *SCARECROW* is also expressed in the cortex cell initials and the endodermis cell layer (Wysocka-Diller *et al.*, 2000). *SHORT-ROOT* gene expression also begins at globular embryo stage in the provascular cells of the procambium. The patterns of expression of both genes do not overlap at any stage of development (Helariutta *et al.*, 2000; Wysocka-Diller *et al.*, 2000). *SHORT-ROOT* is required for endodermis cell specification and for cell division in the cortex, although *SHORT-ROOT* itself is not expressed in the cortex cells. *short-root* is epistatic to *scarecrow* and positively regulates *SCARECROW* expression in a non-cell-autonomous manner that is reminiscent of some shoot meristem gene interactions. A simple model for the coordinated differentiation of cortex and procambium mediated by the *SHORT-ROOT* and *SCARECROW* genes has been proposed (Helariutta *et al.*, 2000). *SHORT-ROOT* is expressed in the pericycle and vascular procambium and signals the adjacent cortex cells to divide. *short-root* also specifies the cortex cells in direct contact with the pericycle to differentiate as endodermis. *short-root* acts by regulating *SCARECROW* expression in a non-cell-autonomous manner since *short-root* is not expressed in cortex cells where its regulatory actions mediated by *scarecrow* take place. *scarecrow* is required to set up the asymmetric cell division pattern of the cortex cells of the ground meristem layer. *scarecrow* and *short-root* mutants also have phenotypes in the shoot, suggesting that both genes could also be involved in cellular differentiation of the ground tissue system throughout the plant (Helariutta *et al.*, 2000).

5.5.3.6   The vascular system differentiates progressively during embryogenesis

The differentiation of the vascular system is initiated in a coordinated fashion across the apical and basal domain of the embryo. The vascular system in *Arabidopsis* consists of two cell lineages originating from the procambium tissue, the xylem and the phloem (Essau, 1977; Steeves and Sussex, 1989). These lineages are specified sequentially during embryogenesis, with the specification of the xylem cells preceding that of the phloem cells (Scheres *et al.*, 1995).

Two mutants that have defective differentiation of the vascular tissue are *gollum* and *wooden leg*. In these mutants the vascular differentiation program is initiated but not completed. The mutations manifest as a reduced complement of cell types and cell numbers in the vascular procambium. In *gollum* disorganized cell division in the procambium tissue forms an incomplete pericycle layer and an aberrant arrangement of the xylem and phloem cells. In *wooden leg* the vascular tissue has a deficit of cells and the phloem elements do not differentiate. These mutations have defects that are specific to different cell layers of the root but, interestingly, some of the defects also extend into the hypocotyl and the shoot (Scheres *et al.*, 1995).

The differentiation of phloem elements in the embryo requires the expression of *wooden leg*, a novel two-component signal transducer receptor (Mahonen *et al.*, 2000). *wooden leg* mutants are deficient in vascular cells and specifically lack the phloem cells. The number of vascular initials in *wooden leg* is reduced, suggesting that *wooden leg* activity is required for a set of formative cell division in the procambial embryonic layer. The expression of the *WOODEN LEG* gene is asymmetrically inherited in procambial cells; *WOODEN LEG* expression is only maintained in the progenitor cells undergoing transversal cell division but is lost in the daughter cells. The persistence of wooden leg expression after the asymmetric cell division is reminiscent of the expression pattern of *SCARE-CROW* gene that controls periclinal cell divisions in ground tissue development. The *WOODEN LEG* gene is expressed at globular stage in the vascular precursor cells of the procambium. *WOODEN LEG* expression continues through embryogenesis in provascular elements and during primary root elongation. Interestingly, the expression pattern of *WOODEN LEG* is not affected in the mutant, suggesting that *wooden leg* is not auto-regulated. The regulation of phloem cell differentiation by a signal transduction pathway reveals a new mechanism by which signal molecules organize the patterns of cell division in embryogenesis and vascular histogenesis. The expression pattern of *WOODEN LEG* suggest that molecular sensing processes between phloem and xylem may direct the spatial patterning of the vascular elements (Mahonen *et al.*, 2000). *WOODEN LEG* encodes a plant-specific two component signal transducer that has recently been involved in cytokinin sensing (Inoue *et al.*, 2001). The finding that cytokinin sensing regulates procambial cell differentiation suggests that a common mechanism establishes the vascular system across embryo domains and in the root and shoot.

The expression of *WOODEN LEG* can be traced back to octant stage embryos, much earlier than any morphological differentiation of vascular elements can be detected. The early expression of *WOODEN LEG* and other vascular differentiation genes, like *MONOPTEROS*, suggests that vascular cell specification and differentiation could be regulated by a multicomponent system of cellular communication that is progressively set up during embryogenesis and spans the root and the shoot.

5.6   Integrative development of the embryo

The observations discussed in the previous sections show that the differentiation of the shoot apical meristem, cotyledons, hypocotyl and root meristem is locally regulated by regional specification processes in early plant embryos. Many of the embryo mutant phenotypes show that development of the mature embryo requires integration of the different regional differentiation processes into a seamless developmental program, and that when integration fails to take place either embryo development arrests and/or aberrant morphogenesis occurs. The molecular analysis of the embryo mutant shows that the regional differentiation processes are probably regulated by a combination of cell-specific gene expression programs and short-range cellular interactions that are beginning to be described at the molecular level.

How are the regional differentiation processes integrated in the development of the entire embryo? Recent observations strongly suggest that the plant hormone auxin is involved in coordinating the integrative development of the embryo. The role of auxin as a plant hormone involved in long-range signaling and the regulation of many aspects of vegetative plant development is well known (Thimann, 1977; Davis, 1995; Guilfoyle *et al.*, 1998; Souter and Lindsey, 2000; Berleth and Sachs, 2001). The requirement for polar transport of auxin during vascular differentiation and for the maintenance of polarized apical growth during postembryonic development is particularly well known (Lomax *et al.*, 1995). Recent observations suggest that the role of auxin as an integrative signaling molecule of plant development starts during embryo development. During embryogenesis, auxin seems to act at multiple levels similar to its postembryonic roles. In early embryos, auxin is probably involved in establishing the apical–basal pattern of the embryonic axis, perhaps as early as the first divisions of the zygote (Steinman *et al.*, 1999). Auxin is also involved in the differentiation of the apical embryo domain and the establishment of the root meristem in the basal domain. During embryogenesis, auxin acts at multiple levels from cellular differentiation to regulation of growth, probably by regulating cell division rates. Auxin also functions as a regulatory molecule for the coordination of gene expression programs in embryo cells, particularly during provascular tissue differentiation processes (Hardtke and Berleth, 1998; Guilfoyle *et al.*, 1998). The observations that auxin concentration gradients appear to be established in early embryos, and that the manipulation of auxin transport *in vitro* affects embryo morphogenesis and phenocopies embryo mutant phenotypes, suggest that auxin acts in a concentration-dependent manner similarly to a classical morphogen. Auxin's proposed action as a plant embryo morphogen brings together the multiple involvement of auxin in cellular signaling processes, regulation of gene expression and coordination of cell differentiation (Liu *et al.*, 1993; Hadfi *et al.*, 1998).

5.6.1   Auxin signaling during embryogenesis

In recent years the isolation and characterization of a number of mutants with altered auxin homeostasis has facilitated the identification of different auxin transporters and auxin-responsive genes. These are being used as tools to understand the cellular and molecular aspects of auxin function in plant development (Bennett *et al.*, 1998; Palmer and Galweiler, 1999; Ulmasov *et al.*, 1999; Hardtke and Berleth, 1998).

Although the quantitative measurement of auxin in individual plant cells is not yet possible, two indirect approaches can be taken to predict the distribution of auxin in plant tissues. In one approach the distribution of auxin transporters in plant tissues and cells is considered indicative of possible routes of auxin fluxes in plant tissues and cells. The second approach is based on a recently developed ingenious method that displays the distribution of auxin in plant tissues. This method relies on the auxin-activated expression of a synthetic reporter gene. The *DR5* gene construct combines auxin-responsive transcriptional elements fused to the reporter gene β-glucuronidase (GUS). In transgenic plants containing the *DR5* reporter gene construct the distribution of glucuronidase enzyme activity correlates with the levels of auxin, permitting the visualization of auxin distribution in plant tissues. In transgenic plants containing the *DR5* construct the level of expression of the reporter gene permits the qualitative visualization of auxin distribution in plant embryos and roots. This method also has the potential to allow relative measurements of active auxin levels in plant cells (Sabatini *et al.*, 1999).

Both approaches have been applied to *Arabidopsis* embryos, and they demonstrate that concentration gradients of auxin are formed in the developing embryo. First, the analysis of the distribution of PIN1, a polar auxin transporter, is suggestive of the existence of directional auxin fluxes in embryos (Steinmann *et al.*, 1999). Second, the qualitative visualization of auxin distribution facilitated by the *DR5* synthetic auxin-inducible gene expression system shows that auxin is concentrated in the basal region of the embryo. This observation agrees with the direction of the auxin flux suggested by the distribution of the PIN1 polar auxin efflux carrier (Galweiler *et al.*, 1998; Sabatini *et al.*, 1999).

Auxin also appears to be implicated in the abnormal embryo morphogenesis manifested by many of the embryo development mutants. This aspect of the regulation of embryo development has been elegantly illustrated *in vitro* by the experimental manipulation of auxin metabolism and transport in isolated embryos. These experiments clearly show that the manipulation of auxin levels and auxin transport precisely phenocopies several embryo development mutant phenotypes that are associated with a deregulation of embryo morphogenesis (Liu *et al.*, 1993; Hadfi *et al.*, 1998). These experiments suggest that auxin mediates multiple processes during early embryo morphogenesis.

5.6.2   The distribution of PIN1 auxin transporter is regulated during early embryogenesis

The directional transport of auxin in plant cells is actively mediated by a set of auxin transporters (Lomax *et al.*, 1995). The *PIN1* component of the auxin efflux carrier has been cloned (Galweiler *et al.*, 1998). Using an antibody against PIN1 it was found that PIN1 localizes specifically at the basal end of vascular cells, suggesting that the distribution of the PIN1 protein is predictive of the routes and direction of auxin transport in plant tissues (Galweiler *et al.*, 1998). Both the pattern of expression and the subcellular localization of PIN1 are regulated during early embryogenesis. PIN1 is initially present in all cells of early embryos, but as embryo development advances PIN1 accumulation is progressively restricted to specific embryo cells. Simultaneously the subcellular distribution of PIN1 is also restricted to the basal end of the cell. In early globular embryos PIN1 is accumulated at the inner cell boundaries of all the cells of the 16-cell stage embryo. The distribution of PIN1 is actively reorganized during the next round of cell division to form a polarized pattern in globular embryos. In mid-globular stage embryos the PIN1 protein accumulates preferentially in the four innermost cells of the basal domain of the embryo. The PIN1 protein is specifically localized at the basal boundaries of the provascular cells in direct contact with the hypophyseal cell. However, PIN1 is not detected in the hypophyseal cell itself. The redistribution of PIN1 in late globular embryos is accompanied by a new subcellular localization of the auxin efflux carrier, which might suggest the redistribution of auxin in late globular embryos. During the development of the heart stage embryo, the PIN1 protein is progressively restricted to vascular precursor cells both in the embryo axis and in the cotyledons. Within the provascular cells PIN1 is asymmetrically accumulated at the basal cell boundaries. Interestingly, the subcellular distribution of PIN1 appears to be different in the cells of the epidermis, where the opposite pattern of accumulation at the apical cell boundaries is observed. The distribution of PIN1 suggests that auxin transport pathways spanning the entire embryo are established in early embryos. At a later stage in embryogenesis PIN1 accumulation is progressively restricted to the provascular tissue, suggesting that auxin polar transport becomes restricted to the provascular cells in late embryogenesis. The active reorganization of the PIN1 auxin transport channels suggests that auxin may have different roles in early embryo stages and late embryogenesis. The distribution of PIN1 also suggests that a basipetal system of auxin transport by provascular cells is already differentiated during embryogenesis (Steinmann *et al.*, 1999).

The distribution of auxin efflux carrier protein in wild-type embryos is suggestive of organized auxin transport; perhaps auxin is circulated in opposite directions in the epidermis and prevascular tissues in early embryogenesis. The gradual establishment of the polarized localization of PIN1 could be the result of a feedback loop involving auxin, as suggested in the canalization hypothesis (Berleth and Sachs, 2001). The redistribution of PIN1 in early embryo cells and its differential localization during the globular to heart stage transition presage the specification of vascular cells, precisely correlating with the remodeling of the pattern of expression of the *MONOPTEROS* gene, an auxin-responsive transcriptional regulator (Hardtke and Berleth, 1998).

gnom embyros have an abnormal distribution of PIN1; however, in *monopteros* the distribution of PIN1 is similar to wild type. The deregulation of *PIN1* expression in *gnom* embryos suggests that the GNOM protein actively participates in the subcellular transport and localization of PIN1 in early embryo cells that is required for the differentiation of the apical basal pattern of embryo morphogenesis (Steinmann *et al.*, 1999). *monopteros* may fail to interpret or relay a canonical auxin signal.

5.6.3   An auxin maximum marks the position of the quiescent center in the embryo axis

The visualization of auxin concentration levels using the DR5 *in vivo* reporter assay shows the accumulation of auxin in the basal region of the embryo and reveals the presence of a highly localized auxin concentration maximum in the cells just above the quiescent center. The auxin concentration maximum is required for the differentiation of the quiescent center. This observation shows that in early embryogenesis auxin is actively transported in a polarized flux from apical to basal embryo poles across embryo domains, as suggested by the distribution of PIN1 (Steimann *et al.*, 1999).

The auxin maximum formed during embryogenesis is maintained in the postembryonic root meristem, suggesting that the cellular organization of the root meristem is continuously regulated by the distribution of auxin in the root tip. Mutations that affect root development and auxin homeostasis also affect the position of the auxin maximum, indicating that the precise position of the auxin maximum results from active communication processes within the root tip (Sabatini *et al.*, 1999). It is not currently known whether auxin is synthesized in the embryo, transported from maternal tissues, or inherited from the megagametophyte during early embryogenesis.

5.6.4   Auxin gradients in the embryo modulate cell-specific processes

An interesting aspect of auxin signaling is that the auxin signal appears to be interpreted contextually, eliciting different responses depending on cellular differentiation state, cell position and cellular history. The auxin hormone also has multiple mechanisms of action by regulating different cellular processes. One of the molecular mechanisms of auxin action takes place directly at the gene regulation level. Auxin directly activates and represses multiple genes that are responsible for different processes during plant growth and development (Abel and Theologis, 1996; Guilfoyle *et al.*, 1998). This aspect of auxin action appears to be integrated into the cellular differentiation programs taking place during early embryogenesis. In early embryos the distribution of the auxin efflux carrier protein PIN1 is similar to the pattern of expression of the monopteros gene. The expression of the *MONOPTEROS* gene is required to form the embryo axis and also for the differentiation and correct alignment of vascular cells in the embryo and postembryonically. The *MONOPTEROS* gene encodes a transcription factor with DNA-binding affinity to auxin-responsive gene promoters and probably directly regulates gene expression programs mediating the morphogenetic effects of auxin associated with the *monopteros* mutant phenotype (Berleth and Jürgens, 1993; Ulmasov *et al.*, 1997, 1999; Hardtke and Berleth, 1998). The epistatic and synergistic interaction of *monopteros* with other embryo morphogenesis genes suggests that additional processes of embryo morphogenesis may also be coordinated by auxin-regulated programs of gene expression. Auxin is also involved in the regulation of protein stability and turnover in the plant cell, suggesting a posttranscriptional regulatory role of the hormone.

Auxin is also involved in signal transduction processes mediated by the ubiquitously conserved cascade of mitogen-activated protein kinases. This cascade of protein kinases also relays other molecular signals and broadly coordinates cell cycle progression and cell division. The experimental interference and suppression of auxin signal transduction via the mitogen-activated protein kinase cascade results in repression of early auxin response genes and most notably in the arrest of embryogenesis at heart stage in transgenic tobacco plants. These results suggest that auxin signaling may be directly regulating cell cycle progression in embryo cells and differential cell division during embryo morphogenesis. It is worth noticing that the initial stages of embryo development do not appear to be affected (Kovtun *et al.*, 1998). It has been suggested that some of the components of the mitogen-activated protein kinase cascade also control the directional extension of the cell plate (Nishihama *et al.*, 2001).

Recently the serine-threonine protein kinase encoded by the *PINOID* gene has been proposed to negatively regulate auxin signaling. Interestingly, the pattern of expression of the *PINOID* and *MONOPTEROS* genes are somewhat exclusive and complementary in early embryos (Christensen *et al.*, 2000).

Although embryo development is not solely regulated by auxin, the multifaceted aspects of auxin action at the molecular and cellular levels together with the system of multiple positive and negative regulators of auxin signaling provide a conceptual framework for approaching the processes coordinating regional specification programs during embryogenesis.

5.7 Prospects

In recent years a tremendous advance in the molecular characterization of embryo development mutants has taken place. These genetic and molecular studies are beginning to provide a molecular understanding of plant embryogenesis. The identification of novel genes, molecules and processes involved in the morphogenesis of the plant embryo should continue at an increased pace with the availability of the entire sequence of several plant genomes. The observations that 'housekeeping' genes correspond to several embryo developmental mutations, although somewhat disappointing in the search for global regulators of embryogenesis, have been quite informative about the coordination of basic cellular processes and developmental programs. This coordination appears to be particularly relevant in plant development since many plant cells can be considered autonomous entities at the metabolic as well as developmental levels. Somewhat unexpectedly, plant embryogenesis has benefited in recent years from the molecular and genetic study of developmental processes classically considered to be characteristic of postembryonic development, such as the establishment and differentiation of the shoot and root meristem. These studies have revealed a rich system of short-range cellular interactions in the early embryo and have facilitated the identification of genes and molecules that assign long-term cell differentiation fates and regulate tissue formation in the embryo. One aspect that will probably be emphasized in further research is the understanding of the molecular basis for cell differentiation and for the coordination of cellular roles and functions during embryogenesis. What makes an epidermis cell different from a vascular cell, and how are cell specific metabolic pathways differentiated in different cell types early on during embryogenesis? Also, the search for key regulatory genes of embryogenesis will continue given their utility in plant reproduction and seed production.

Acknowledgements

I am grateful to the members of The Seed Institute for their support and particularly to Professor John J. Harada for graciously hosting me in his laboratory. I also thank Professor Bob Goldberg for his continued support and exemplary encouragement. I thank Professor Roy Brown's laboratory for the illustration in figure 5.1.

References

Abel, S. and Theologis, A. (1996) Early genes and auxin action. *Plant Physiol.*, **111**, 9-17.
Aida, M., Ishida, T. and Tasaka M. (1999) Shoot apical meristem and cotyledon formation during *Arabidopsis* embryogenesis: interaction among the CUP-SHAPED COTYLEDON and SHOOT MERISTEMLESS genes. *Development*, **126**, 1563-1570.
Assaad, F.F., Mayer, U., Wanner, G. and Jürgens, J. (1996) The KEULE gene is involved in cytokinesis in *Arabidopsis. Mol. Gen. Genet.*, **253**, 267-277.
Barton, M.K. and Poethig, R.S. (1993) Formation of the shoot apical meristem in *Arabidopsis thaliana*: an analysis of development in the wild type and in the shoot meristemless mutant. *Development*, **119**, 823-831.

Benfey, P., Linstead, P., Roberts, K., Schiefelbein, J., Hauser, M.-T. and Aeschbacher, R. (1993) Root development in *Arabidopsis*: four mutants with dramatically altered root morphogenesis. *Development*, **119**, 57-70.

Bennett, M.J., Marchant, A., May, S.T. and Swarup, R. (1998) Going the distance with auxin: unravelling the molecular basis of auxin transport. *Phil. Trans. R. Soc. London*, **353**, 1511-1515.

Berleth, T. and Jürgens, G. (1993) The role of the monopteros gene in organising the basal body region of the *Arabidopsis* embryo. *Development*, **118**, 575-587.

Berleth, T. and Sachs, T. (2001) Plant morphogenesis: long-distance coordination and local patterning. *Curr. Opin. Plant Biol.*, **4**, 57-62.

Bohmert, K., Camus, I., Bellini, C., Bouchez, D., Caboche, M. and Benning, C. (1997) AGO1 defines a novel locus of *Arabidopsis* controlling leaf development. *EMBO J.*, **17**, 170-180.

Bowman, J.L. and Eshed, Y. (2000) Formation and maintenance of the shoot apical meristem. *Trends Plant Sci.*, **5**, 110-115.

Brown J.D. and Beggs J.D. (1992) Roles of the PRP8 protein in the assembly of splicing complexes. *EMBO J.*, **11**, 3721-3729.

Brownlee, C. and Berger, F. (1995) Extracellular matrix and pattern in plant embryos: on the lookout for developmental information. *Trends Genet.*, **11**, 344-348.

Busch, M., Mayer, U. and Jürgens, G. (1996) Molecular analysis of the *Arabidopsis* pattern formation gene GNOM: gene structure and intragenic complementation. *Mol. Gen. Genet.*, **250**, 681-691.

Christensen, S., Dagenais, N., Chory, J. and Weigel, D. (2000) Regulation of auxin response by the protein kinase PINOID. *Cell*, **100**, 4769-4787.

Clouse, S.D. (2000) A role for sterols in embryogenesis. *Curr. Biol.*, **10**, R601-604.

Davis, P.J. (1995) *Plant Hormones: Physiology, Biochemistry and Molecular Biology*, Kluwer Academic, Dordrecht.

Di Laurencio, L., Wysocka-Diller, J., Malamy, J.E., *et al.* (1996) The SCARECROW gene regulates an asymmetric cell division that is essential for generating the radial organization of the *Arabidopsis* root. *Cell*, **86**, 423-433.

Diener, A., Li, H., Zhou, W.H., Whoriskey, W., Nes, W. and Fink, G. (2000) STEROL METHYLTRANSFERASE1 controls the level of cholesterol in plants. *Plant Cell*, **12**, 853-870.

Dolan, L., Janmaat, K., Willemsen, V., *et al.* (1993) Cellular organization of the *Arabidopsis thaliana* root. *Development*, **119**, 71-84.

Endrizzi, K., Moussian, B., Haecker, A., Levin, J. and Laux, T. (1996) The SHOOT MERISTEMLESS gene is required for maintenance of undifferentiated cells in *Arabidopsis* shoot and floral meristem and acts at a different regulatory level than the meristem genes WUSCHEL and ZWILLE. *Plant J.*, **10**, 967-979.

Essau, K. (1977) *Anatomy of Seed Plants*. Wiley, New York.

Evans, M.S.M. and Barton, M.K. (1997) Genetics of angiosperm shoot apical meristem development. *Annu. Rev. Plant Physiol. Plant Mol. Biol.*, **48**, 673-701.

Galweiler, L., Guan, C., Muller, A., *et al.* (1998) Regulation of polar auxin transport by AtPIN1 in *Arabidopsis* vascular tissue. *Science*, **282**, 2226-2230.

Guilfoyle, T., Hagen, G., Ulmasov, T. and Murfett, J. (1998) How does auxin turn on genes? *Plant Physiol.*, **118**, 341-347.

Goldberg, R.B., Barker, S.J. and Perez-Grau, L. (1989) Regulation of gene expression during plant embryogenesis. *Cell*, **56**, 149-160.

Goldberg, R.B., De Paiva, G. and Yadegari, R. (1994) Plant embryogenesis: zygote to seed. *Science*, **266**, 605-614.

Goodner, B. and Quatrano, R.S. (1993) Fucus embryogenesis: a model to study the establishment of polarity. *Plant Cell*, **5**, 1471-1481.

Grebe, M., Gadea, J., Steinmann, T., *et al.* (2000) A conserved domain of the *Arabidopsis* GNOM protein mediates subunit interaction and cyclophilin 5 binding. *Plant Cell*, **12**, 343-356.

Hadfi, K., Speth, V. and Neuhaus, G. (1998) Auxin-induced developmental patterns in *Brassica juncea* embryos. *Development*, **125**, 879-887.

Hamann, T., Mayer, U. and Jürgens, G. (1999) The auxin-insensitive bodenlos mutation affects primary root formation and apical-basal patterning in the *Arabidopsis* embryo. *Development*, **126**, 1387-1395.

Harada, J.J. (1997) Seed maturation and control of germination, in *Cellular and Molecular Biology of Plant Seed Development* (eds. B.A. Larkins and I.K. Vasil), Kluwer Academic, Dordrecht, pp. 545-592.

Hardtke, C.S. and Berleth, T. (1998) The *Arabidopsis* gene MONOPTEROS encodes a transcription factor mediating embryo axis formation and vascular development. *EMBO J.*, **17**, 1405-1411.

Helariutta, Y., Fukaki, H., Wysocka-Diller, J., *et al.* (2000) The SHORT-ROOT gene controls radial patterning of the *Arabidopsis* root through radial signaling. *Cell*, **101**, 555-567.

Hobbie, L.J., McGovern, M., Hurwitz, L.R., *et al.* (2000) The axr6 mutants of *Arabidopsis thaliana* define a gene involved in auxin response and early development. *Development*, **127**, 23-32.

Inoue, T., Higuchi, M., Hashimoto, Y., *et al.* (2001) Identification of CRE1 as a cytokinin receptor from *Arabidopsis*. *Nature*, **409**, 1060-1063.

Irish, V.F. and Sussex, I.M. (1992) A fate map of the *Arabidopsis* embryonic shoot apical meristem. *Development*, **115**, 745-753.

Jang, J., Fujioka, S., Tasaka, M., *et al.* (2000) A critical role of sterols in embryonic patterning and meristem programming revealed by the fackel mutants of *Arabidopsis thaliana*. *Genes Dev.*, **14**, 1485-1497.

Jürgens, G. (1995) Axis formation in plant embryogenesis: cues and clues. *Cell*, **81**, 467-470.

Jürgens, G. and Mayer, U. (1994) *Arabidopsis*, in *Embryos. Color Atlas of Development* (ed. J.B.L. Bard), Wolfe Publishing, London, pp. 7-21.

Kaplan, D.R. and Cooke, J.T. (1997) Fundamental concepts in the embryogenesis of dicotyledons: a morphological interpretation of embryo mutants. *Plant Cell*, **9**, 1903-1919.

Koornneef, M. and Karsen, C.M. (1994) Seed dormancy and germination, in *Arabidpsis* (eds. E.M. Meyerowitz and C. Somerville), pp. 313-334.

Kovtun, Y., Chiu, W., Zeng, W. and Sheen, J. (1998) Suppression of auxin signal transduction by a MAPK cascade in higher plants. *Nature*, **395**, 716-720.

Lauber, M.H., Waizenegger, I., Steinmann, T., *et al.* (1997) The *Arabidopsis* KNOLLE protein is a cytokinesis soecific syntaxin. *J. Cell Biol.*, **139**, 1485-1493.

Laux, T. and Jürgens, G. (1997) Embryogenesis: a new start in life. *Plant Cell*, **9**, 989-1000.

Laux, T. and Mayer, K.F.X. (1998) Cell fate regulation in the shoot meristem. *Semin. Cell Dev. Biol.*, **9**, 195-200.

Laux, T., Mayer, K.F.X., Berger, J. and Jürgens, G. (1996) The wuschel gene is required for shoot and floral meristem integrity in *Arabidopsis*. *Development*, **122**, 87-96.

Liu, C. and Meinke, M.W. (1998) The titan mutants of *Arabidopsis* are disrupted in mitosis and cell cycle control during seed development. *Plant J.*, **16**, 21-31.

Liu, C., Xu, Z. and Chua, N.-H. (1993) Auxin polar transport is essential for the establishment of bilateral symmetry during early plant embryogenesis. *Plant Cell*, **5**, 621-630.

Lomax, T.L., Muday, G.K. and Rubery, P.H. (1995) Auxin transport, in *Plant Hormones; Physiology, Biochemistry and Molecular Biology* (ed. P.J. Davis), Kluwer Dordrecht, pp. 509-530.

Long, J.A. and Barton M.K. (1998) The development of apical embryonic pattern in *Arabidopsis*. *Development*, **125**, 3027-3035.

Long, J.A., Moan, E.I., Medford, J.I. and Barton, M.K. (1996) A member of the knotted class of homeodomain proteins encoded by the stm gene of *Arabidopsis*. *Nature*, **379**, 66-69.

Lotan, T., Ohto, M.-A., Yee, K.M., *et al.* (1999) *Arabidopsis* leafy cotyledon1 is sufficient to induce embryo development in vegetative cells. *Cell*, **93**, 1195-1205.

Lukowitz, W., Mayer, U. and Jürgens, G. (1996) Cytokinesis in the *Arabidopsis* embryo involves the synthaxin-related KNOLLE gene product. *Cell*, **84**, 61-71.

Lynn, K., A. Fernandez, M. Aida, J., *et al.* (1999) The PINHEAD/ZWILLE gene acts pleiotropically in *Arabidopsis* development and has overlapping functions with the ARGONAUTE gene. *Development*, **126**, 469-481.

Mahonen, A.P., Bonke, M., Kauppinen, L., Riikonen, M., Benfey, P.N. and Helariutta, Y. (2000) A novel two-component hybrid molecule regulates vascular morphogenesis of the *Arabidopsis* root. *Genes Dev.*, **14**, 2938-2943.

Mansfield, S.G. and Briarty, L.G. (1991) Early embryogenesis in *Arabidopsis thaliana*. II. The developing embryo. *Can. J. Bot.*, **69**, 461-476.

Mansfield, S.G., Briarty, L.G. and Erni, S. (1991) Early embryogenesis in *Arabidopsis thaliana*. I. The mature embryo sac. *Can. J. Bot.*, **69**, 447-460.

Mayer, K.F.X., Schoof, H., Haecker, A., Lenhard, M., Jürgens, G. and Laux, T. (2000) Role of WUSCHEL in regulating stem cell fate in the *Arabidopsis* shoot meristem. *Cell*, **95**, 805-815.

Mayer, U. and Jürgens, G. (1998) Pattern formation in plant embryogensis: a reassessment. *Semin. Cell Dev. Biol.*, **9**, 187-193.

Mayer, U., Torres Ruiz, R.A., Berleth, T., Misera, S. and Jürgens, G. (1991) Mutations affecting body organization in the *Arabidopsis* embryo. *Nature*, **353**, 402-407.

Mayer, U., Buttner, G. and Jürgens, G. (1993) Apical–basal pattern formation in the *Arabidopsis* embryo: studies on the role of the gnom gene. *Development*, **117**, 149-162.

Mayer, U., Herzog, U., Berger, F., Inze, D. and Jürgens, J. (1999) Mutations in the PILZ group genes disrupt the microtubule cytoskeleton and uncouple cell cycle progression from cell division in *Arabidopsis* embryo and endosperm. *Eur. J. Cell Biol.*, **78**, 100-108.

McDonell, J.R. and Barton, M.K. (1995) Effects of mutations in the PINHEAD gene of *Arabidopsis* on the formation of shoot apical meristems. *Dev. Genet.*, **16**, 358-366.

McElver, J., Patton, D., Rumbaugh, M., Liu, C., Yang, L. and Meinke, D. (2000) The TITAN5 gene of *Arabidopsis* encodes a protein related to the ADP ribosylation factor family of GTP binding proteins. *Plant Cell*, **12**, 3179-1392.

Meinke, D.W. (1985) Embryo-lethal mutants of *Arabidopsis thaliana*: analysis of mutants wide range of lethal phases. *Theor. Appl. Genet.*, **69**, 543-552.

Meinke, D.W. (1995) Molecular genetics of plant embryogenesis. *Annu. Rev. Plant Physiol. Plant Mol. Biol.*, **46**, 369-394.

Morton, R.L., Quiggin, D. and Higgins, T.J.V. (1995) Regulation of seed storage protein gene expression, in *Seed Development and Germination* (eds. J.K. Kigel and G. Galili), New York, Marcel Dekker, pp. 103-138.

Moussian, B., Scoof, A.H., Haecker, A., Jürgens, G. and Laux, T. (1998) Role of the ZWILLE gene in the regulation of central shoot meristem cell fate during *Arabidopsis* embryogenesis. *EMBO J.*, **17**, 1799-1809.

Nishihama, R., Ishikawa, M., Araki, S., Soyano, T., Asada, T. and Machida, Y. (2001) The NPK1 mitogen-activated protein kinase kinase kinase is a regulator of cell-plate formation in plant cytokinesis. *Genes Dev.*, **15**, 352-363.

Ogas, J., Cheng, J., Sung, Z.R. and Somerville, C. (1997) Cellular differentiation regulated by gibbrellin in the *Arabidopsis thaliana* pickle mutant. *Science*, **277**, 91-94.

Palmer, K. and Galweiler, L. (1999) PIN-pointing the molecular basis of auxin transport. *Curr. Opin. Plant Biol.*, **2**, 375-381.

Perez-Grau, L. and Goldberg, R.B. (1989) Soybean seed protein genes are regulated spatially during embryogenesis. *Plant Cell*, **1**, 1095-1109.

Przemeck, G.K.H., Mattson, J., Hardtke, C.S., Sung, Z.R. and Berleth, T. (1996) Studies on the role of the *Arabidopsis* gene MONOPTEROS in vascular development and plant cell axialization. *Planta*, **200**, 229-237.

Pysh, L.D., Wysocka-Diller, J.W., Camilleri, C. and Benfey, P.N. (1999) The GRAS gene family in *Arabidopsis*: sequence characterization and basic expression analysis of the SCARECROW-LIKE genes. *Plant J.*, **18**, 111-119.

Quatrano, R.S. and Shaw, S.L. (1997) Role of the cell wall I the determination of cell polarity and the plane of cell division in Fucus embryos. *Trends Plant Sci.*, **2**, 15-21.

Quatrano, R.S., Marcotte, W.R.J. and Guiltinan, M. (1993) Regulation of gene expression by abscisic acid, in *Control of Plant Gene Expression* (ed. D.P.S. Verma), CRC Press, Boca Raton, FL, pp. 69-90.

Raghavan, V. (1986) *Embryogenesis in Angiosperms*, Cambridge University Press, Cambridge.

Raz, V., Bergervoet, J.H.W. and Koorneef, M. (2001) Sequential steps for developmental arrest in *Arabidopsis* seeds. *Development*, **128**, 243-252.

Sabatini, S., Beis, D., Wolkenfelt, H., *et al.* (1999) An auxin-dependent distal organizer of pattern and polarity in the *Arabidopsis* root. *Cell*, **99**, 463-472.

Scheres, B., Kenfelt, H., Willemsen, V., *et al.* (1994) Embryonic origin of the *Arabidopsis* primary root and root meristem initials. *Development*, **120**, 2475-2487.

Scheres, B., Di Laurenzio, L., Willemsen, V., *et al.* (1995) Mutations affecting the radial organization of the *Arabidopsis* root display specific defects throughout the embryonic axis. *Development*, **121**, 53-62.

Schoof, H., Lenhard, M., Haecker, A., Mayer, K.F.X., Jürgens, G. and Laux, T. (2000) The stem cell population of *Arabidopsis* shoot meristems is maintained by a regulatory loop between the CLAVATA and WUSCHEL genes. *Cell*, **100**, 635-644.

Schrick, K., Mayer, U., Horrichs, A., *et al.* (2000) FACKEL is a sterol C-14 reductase required for organized cell division and expansion in *Arabidopsis* embryogensis. *Genes Dev.*, **14**, 1471-1484.

Schwarz, B.W., Yeung, E.C. and Meinke, D.W. (1994) Disruption of morphogenesis and transformation of the suspensor in abnormal suspensor mutants of *Arabidopsis. Development*, **120**, 3235-3245.

Schwartz, B.W., Vernon, D.M. and Meinke D.W. (1997) Development of the suspensor: differentiation, communication, and programmed cell death during plant embryogenesis, in *Cellular and Molecular Biology of Plant Seed Development* (eds. B.A. Larkins and I.K. Vasil), Kluwer Academic, pp. 53-72.

Shevell, D.E., Leu, W.M., Gillmor, C.S., Xia, G.X., Feldmann, K.A. and Chua, N.H. (1994) emb30 is essential for normal cell division, cell expansion, and cell adhesion in arabidopsis and encodes a protein that has similarity to sec7. *Cell*, **77**, 1051-1062.

Shevel, D.E., Kunkel, T. and Chua, N. (2000) Cell wall alterations in the *Arabidopsis* emb30 mutant. *Plant Cell.*, **12**, 2047-2059.

Soueges, M.R. (1914) Nouvelles reserches sur le development de l'embrion chez les crucifers. *Ann. Sci. Nat. Bot. Ser. 9*, **19**, 311-339.

Souter, M. and Lindsey, K. (2000) Polarity and signalling in plant embryogenesis. *J. Exp. Bot.*, **51**, 971-983.

Steeves T.A. and Sussex, I.A. (1989) *Patterns in Plant Development*, Cambridge University Press, Cambridge.

Steinmann, T., Geldner, N., Grebe, M., *et al.* (1999) Coordinated polar localization of auxin efflux carrier PIN1 by GNOM ARF GE. *Science*, **286**, 316-318.

Thimann, K.V. (1977) *Hormone Action in the Life of the Whole Plant*, University of Massachusetts Press, Amhrest, MA.

Topping, J.F., May, V.J. Muskeet, P.R. and Lindsey, K. (1997) Mutations in the HYDRA1 gene of *Arabidopsis* perturb cell shape and disrupt embryonic and seedling morphogenesis. *Development*, **124**, 4415-4424.

Torres-Ruiz, R.A. and Jürgens., G. (1994) Mutations in the FASS gene uncouple pattern formation and morphogenesis in *Arabidopsis* development. *Development*, **120**, 2967-2978.

Trass, J., Bellini, C., Nacry, P., Kronenberger, J., Bouchez, D. and Caboche, M. (1995) Normal differentiation patterns in plants lacking the microtubular preprophase bands. *Nature*, **375**, 676-677.

Ulmasov, T., Hagen, G. and Guilfoyle. T.J. (1997) ARF1, a transcription factor that binds to auxin response elements. *Science*, **276**, 1865-1868.

Ulmasov, T., Hagen, G. and Guilfoyle. T.J. (1999) Activation and repression of transcription by auxin-response factors. *Proc. Natl. Acad. Sci. USA*, **96**, 5844-5849.

Uwer, U., Willmitzer, L. and Altmann, T. (1998) Inactivation of a glycyl-tRNA synthase leads to an arrest in plant embryo development. *Plant Cell*, **10**, 1277-1294.

Van den Berg, C., Willemsen, V., Hendriks, G., Weisbeek, P. and Scheres, B. (1997) Short-range control of cell differentiation in the *Arabidopsis* root meristem. *Nature*, **390**, 287-289.

Vernon, D.M. and Meinke, D.W. (1994) Embryogenic transformation of the suspensor in twin, a polyembryonic mutant of arabidopsis. *Dev. Biol.*, **165**, 566-573.

Vroemen, C.W., Langeveld, S., Mayer, U., *et al.* (1996) Pattern formation in the arabidopsis embryo revealed by position-specific lipid transfer protein gene expression. *Plant Cell*, **8**, 783-791.

Waizenegger, I., Lukowitz, W., Assaad, F., Schwarz, H., Jürgens, G. and Mayer, U. (2000) The *Arabidopsis* KNOLLE and KEULE genes interact to promote vesicle fusion during cytokinesis. *Curr. Biol.*, **10**, 1371-1374.

Wardlaw, C.C. (1995) *Embryogenesis in Plants*. Methuen, London.

West, M. and Harada, J.J. (1993) Embryogenesis in higher plants: an overview. *Plant Cell*, **5**, 1361-1369.

West, M.A.L., Matsudaira-Yee, K.L., Danao, J., *et al.* (1994) Leafy cotyledon is an essential regulator of late embryogenesis and cotyledon identity in arabidopsis. *Plant Cell*, **7**, 1731-1745.

Willemsen, V., Wolkenfelt, H., de Vrieze, G., Weisbeek, P. and Scheres, B. (1998) The HOBBIT gene is required for formation of the root meristem in the *Arabidopsis* embryo. *Development*, **125**, 521-531.

Wobus, U. and Weber, H. (1999) Seed maturation: genetic programs and control signals. *Curr. Opin. Plant Biol.*, **2**, 33-38.

Woodrich, R., Martin, P.R., Birman, I. and Pickett, F.B. (2000) The *Arabidopsis* embryonic shoot fate map. *Development*, **127**, 813-820.

Wysocka-Diller, J., Helariutta, Y., Fukaki, H., Malamy, J.E. and Benfey, P.N. (2000) Molecular analysis of SCARECROW function reveals a radial patterning mechanism common to root and shoot. *Development*, **127**, 595-603.

Yadegari, R. and Goldberg, B.G. (1997) Embryogenesis in dicotyledoneous plants, in *Cellular and Molecular Biology of Plant Seed Development* (eds. B.A. Larkins and I.K. Vasil), Kluwer Academic, pp. 3-52.

Yadegari, R., De Paiva, G.R., Laux, T., *et al.* (1994) Cell differentiation and morphogenesis are uncoupled in arabidopsis raspberry embryos. *Plant Cell*, **7**, 1713-1729.

Zhang, J.Z. and Somerville, C.R. (1997) Suspensor-derived polyembryony by altered expression of valyl-tRNA synthase in the twn2 mutant of *Arabidopsis*. *Proc. Natl. Acad. Sci. USA*, **94**, 7349-7355.

6 Endosperm development

Roy C. Brown, Betty E. Lemmon and Hong Nguyen

6.1 Introduction

The evolution of sexual reproduction in plants culminated in the seed of angiosperms, in which the embryo is provisioned with a dedicated nutrient source, the endosperm. Traditionally the development of endosperm has been treated as a component of embryology and, so far as we know, the term endospermology does not exist. This may be unfortunate, since endosperm develops along a unique pathway that has little in common with that of the embryo. Because of the importance of endosperm in sexual reproduction and of angiosperm seeds in agriculture and industry, there has been a resurgence of interest in endosperm biology. Recent reviews have emphasized development (Olsen *et al.*, 1995, 1999, 2001; DeMason 1997; Berger, 1999; Becraft *et al.*, 2000), nutrient storage (Lopes and Larkins, 1993), evolution (Floyd and Friedman, 2000; Friedman, 2001), and the genetic programs that initiate development (e.g. Vinkenoog *et al.*, 2000; Luo *et al.*, 2000; Sorensen *et al.*, 2001). This review will deal principally with the development of endosperm in model systems (the cereals and *Arabidopsis*) with emphasis on genetic controls and developmental mechanisms involving the cytoskeleton. The cytoskeleton plays a fundamental role in plant morphogenesis at the cellular level, including cell division and control of the placement and subsequent expansion of walls. The microtubule cycle that drives development of the endosperm is different from that of the embryo (Pickett-Heaps *et al.*, 1999; Brown and Lemmon, 2001).

6.2 Background

Reproduction in flowering plants is characterized by a highly reduced female gametophyte and endosperm initiated by double fertilization. The majority of flowering plants, about 70% including some basal angiosperms (Floyd and Friedman, 2000), cereals (Russell, 1979; Huang and Sheridan, 1994), and *Arabidopsis* (Webb and Gunning, 1990; Mansfield *et al.*, 1991; Reiser and Fischer, 1993; Schneitz *et al.*, 1995) have the 8-nucleate monosporic *Polygonum* type of embryo sac. In this type of female gametophyte development, three of the four megaspores resulting from meiosis abort (Russell, 1979; Webb and Gunning, 1990; Bajon *et al.*, 1999) and the surviving megaspore undergoes three rounds of mitosis without cytokinesis to produce two quartets of nuclei within the original megaspore. Three of the four nuclei at the chalazal pole are walled off to become antipodals, with one nucleus remaining in the central cell. Three of the four nuclei at the micropylar pole become walled off into one egg and two flanking synergids, with the fourth nucleus remaining in the central cell. The two nuclei in the central cell are termed polar nuclei because they move from the polar regions into the center of the central cell where they fuse either before fertilization or at the time of fertilization.

The male gametophyte is likewise reduced in angiosperms and consists of the germinated pollen grain that delivers two sperm to the micropyle of the ovule. One sperm fertilizes the egg to produce the zygote, and the other fertilizes the diploid central cell nucleus or joins with the two polar nuclei in triple fusion (Engell, 1989). Thus, the primary endosperm nucleus resulting from two haploid polar nuclei and one haploid sperm nucleus is triploid. Although the two fusion products are alike genetically and in a like environment, they develop into totally different structures. The zygote enters the developmental pattern typical of plant meristems and gives rise to the embryo sporophyte. The primary endosperm nucleus enters an unusual pathway leading to the highly specialized nutritive endosperm. From the first division of zygote and primary endosperm nucleus, the microtubule cycles that drive the developmental pathways are different (Brown and Lemmon, 2001). The microtubule cycle of endosperm differs from that in the co-developing embryo in that hoop-like cortical microtubules of interphase and preprophase bands of microtubules (PPBs) are absent. Instead, cytoplasmic organization and wall placement is a function of nuclear-based radial microtubule systems.

Identity of the plant generation represented by the endosperm has long been controversial. While the embryo is clearly the offspring sporophyte plant, the endosperm surrounding it is an enigma with respect to its origin, level of ploidy, and specialized form and function. Historically, two hypotheses have been tendered to account for the evolutionary origin of the endosperm (Friedman, 1994). Endosperm could either be a delayed continuation of female gameto-phyte development (Thomas, 1907) and therefore homologous to the nutritive gametophyte in gymnosperms, or it could be a highly modified supernumerary embryo (Sargant, 1900). Whereas the embryo quickly organizes meristems, the endosperm is essentially without meristems and matures into a short-lived terminal structure specialized for food storage. Thus, the final form of the endosperm as well as its function is totally different from that of the sporophyte and its origin as a sexual product clearly precludes it from being a gametophyte. Endosperm may be viewed as a third plant generation, a 'trophophyte' that has evolved in angiosperms (Battaglia, 1989). The term trophophyte, meaning nourishment-producing plant, was coined by Pearson (1909) in recognition of the nature of this third generation as distinct from both gametophyte (gamete-producing plant) and sporophyte (spore-producing plant). Regardless of its designation, the endosperm remains one of evolution's most intriguing and ultimately important accomplishments.

An emerging consensus about the extant plants constituting the basal angiosperms has set the stage for determination of the plesiomorphic features of endosperm. The primitive endosperm is thought to be cellular, copious, and the sole nutritive tissue in the seed (Friedman, 2001). The typical endosperm is triploid, stemming from a megagametophyte of the 8-nucleate, 7-celled *Polygonum* type. Three major patterns of endosperm development—cellular, helobial and nuclear—have long been recognized. In the cellular and helobial types, the first division of the primary endosperm nucleus is followed by cytokinesis. In the cellular type, all subsequent mitoses are accompanied by wall formation. In the helobial type, the micropylar and chalazal chambers may develop independently, either with or without cytokinesis following mitosis. The nuclear type is characterized by a long period of nuclear divisions without cytokinesis, resulting in a multinucleate cytoplasm (syncytium) before the eventual cellularization. It is interesting that, in spite of the developmental type, the endosperms of all basal eudicots exhibit bipolarity resulting in distinct micropylar and chalazal developmental domains (Floyd and Friedman, 2000). When differences in development in the two domains are considered, at least six patterns can be recognized in the basalmost angiosperms (Floyd and Friedman, 2000). Endosperm bipolarity may be considered an embryo-like attribute (Floyd and Friedman, 2000) or as having been inherited from developmental programs of two-chambered female gametophytes such as *Gnetum* (Carmichael and Friedman, 1996) and *Welwitschia* (Chamberlain, 1935).

Whatever its origin, the endosperm is functionally nutritive. Bipolarity leading to different developmental domains establishes conditions for differentiation of the endosperm for nutrient uptake, storage and release. This is especially apparent in the primitive angiosperm *Cabomba*, where the first division of the primary endosperm nucleus is transverse, dividing the central cell into micropylar and chalazal cells (Floyd and Friedman, 2000). Further development is helobial; the micropylar cell undergoes nuclear-type development with subsequent cellularization occurring first around the embryo. The chalazal cell does not divide again but instead elongates greatly into a thread-like unicellular haustorium. There are numerous reports of highly specialized modifications of both micropylar and chalazal endosperm into haustoria that penetrate the maternal tissues (reviewed by Vijayraghavan and Prahabkar, 1984; Nguyen *et al.*, 2000).

6.3 Nuclear endosperm development

Although little is known about developmental mechanisms in cellular and helobial types, there is considerable information on the nuclear type. It is by far the most common type of endosperm development (Dahlgren, 1991; Johri *et al.*, 1992; Floyd *et al.*, 1999) and occurs in *Arabidopsis* and cereals, as well as other major groups of crop plants such as legumes (Yeung and Cavey, 1988; Dute and Peterson, 1992; Chamberlin *et al.*, 1994; XuHan and van Lammeren, 1994) and cotton (Schulz and Jensen, 1977). The nuclear type is thought to have evolved independently in the monocots and several times in the dicots (Floyd and Friedman, 2000). In all of the taxa studied, including the basal eudicots *Papaver* (Olson, 1981; Bhandari *et al.*, 1986), *Ranunculus* (Chitralaka and Bhandari, 1993; XuHan and van Lammeren, 1993) and *Platanus* (Floyd *et al.*, 1999), the overall pattern of development appears to be remarkably similar.

The mechanism of placement and growth of walls in nuclear endosperm remained enigmatic until the advent of appropriate techniques for three-dimensional imaging of developing systems *in situ*. Of special importance are techniques of immunolocalization that provide global views of cytoskeletal arrays and wall development (van Lammeren, 1988; Webb and Gunning, 1991; Brown *et al.*, 1994, 1999; Nguyen *et al.*, 2001). Initial wall formation in the syncytial endosperm results in alveolation, a unique and ancient developmental pathway that occurs in female gametophytes of extant gymnosperms (Chamberlain, 1935; Singh, 1978) and is suspected to have occurred as far back as the Devonian (Taylor and Taylor, 1993).

Discovery of genes involved in the initiation of endosperm (Sorensen *et al.*, 2001) provides evidence for a distinctive developmental pathway under genetic control. Two important and interrelated components of endosperm developmental regulation are currently under intense investigation. These are (1) the genetic mechanisms that initiate and regulate endosperm development, and (2) the role of parental imprinting and its relationship to polyploidy in endosperm evolution.

Mutations causing autonomous development of the endosperm without fertilization have been identified in three gene loci of the *Arabidopsis* genome. These genes have been variously designated as *MEA* (*MEDEA*), *FIS1* (*FERTILIZATION-INDEPENDENT SEED 1*)/*F644*/*EMB173*; *FIS2*, and *FIE* (*FERTILIZATION-INDEPENDENT ENDOSPERM*)/*FIS3* (Vinkenoog *et al.*, 2000). Mutants in *FIE*, *MEA*, and *FIS2* cause proliferation of the unfertilized central cell nucleus, suggesting that the wild-type genes function in the normal suppression of central cell proliferation until release by fertilization. *MEA* and *FIE* encode proteins of the Polycomb group that are known to maintain transcriptional repression of homeotic genes in animal cells (Grossniklaus *et al.*, 1998; Kiyosue *et al.*, 1999; Ohad *et al.*, 1999). In *Drosophila*, the Polycomb proteins are thought to form complexes that modulate chromatin configuration or prevent access of transcriptional factors through many rounds of cell division (Pirotta, 1997, 1998). It is hypothesized that the wild-type gene products of *MEA* and *FIE* form complexes that prevent endosperm-specific gene transcription in the unfertilized central cell (Kiyouse *et al.*, 1999; Luo *et al.*, 2000; Ohad *et al.*, 1999). These genes all are maternally inherited and phenotypes are not altered by paternal wild-type alleles. Although endosperm proliferates autonomously in mutant female gametophytes, downstream events in both programs are affected; the endosperm does not complete development and, when fertilization does occur, the endosperm overproduces and embryos abort.

The three gene loci known to be involved in suppression of endosperm until fertilization all show parent-of-origin effect. The mutant phenotype (autonomous endosperm proliferation without fertilization) occurs when the mutant gene is inherited maternally. To study normal expression patterns of the *MEA, FIS2,* and *FIE* genes, their promoters were fused to β-glucuronidase (GUS). Although the patterns of expression varied, all three maternally derived genes showed activity in the central cell of the female gametophyte before pollination and in the nuclear endosperm after pollination. The early endosperm shows expression of only maternally derived genes, indicating that the paternally derived genes are imprinted. Paternally inherited wild-type genes introduced at fertilization fail to rescue developing seeds with the maternally inherited mutant gene, suggesting that silencing prevents production of active protein. Since DNA methylation is known to be an essential component of the imprinting mechanism in animals, considerable effort has been made to identify its role in silencing the genes that suppress endosperm development. When the paternal wild-type *MEA* gene is inherited together with a mutation (*ddm1*) that causes a decrease in DNA methylation, seeds with the maternally inherited lethal *mea* gene can be rescued (Vielle-Calzada *et al.,* 1999). Vinkenoog *et al.* (2000) used transgenic plants with the *METHYL TRANSFERASE 1* antisense (*MET1* a/s) to investigate the role of methylation in *FIE* expression. It has been shown that unlike the situation in normally methylated male plants, pollen from *MET1* a/s plants carrying the wild-type *FIE* can rescue *fie-1* mutant seeds (Vinkenoog *et al.,* 2000). Both of these studies show evidence of paternal gene silencing and provide insight into the parent-of-origin effect of genes controlling endosperm development. This is particularly interesting in view of the strong evidence that parental conflict has influenced the evolution of seeds.

The concept of parental conflict is based upon the assumption that maternal and paternal genomes have conflicting interests in allocation of resources to offspring. It is more advantageous for the female genome to conserve resources for many offspring, whereas the male genome encourages allocation of extra resources to its own offspring. Thus, genes that promote increased amounts of endosperm, which control nourishment of the embryo, would be preferentially expressed by the male allele, whereas the maternal allele would favor restricted nutrient flow to the embryo via a limited development of endosperm. The balance established in wild-type seed development involves not only silencing of genes but levels of ploidy as well. The endosperm of wild-type *Arabidopsis* seeds has a 2:1 ratio of maternal to paternal genomes, as would be expected in plants with a *Polygonum* type megagametophyte. Experimental crossing of plants with different levels of ploidy resulting in altered ratios of maternal to paternal genes showed that an excess of paternal genes leads to increased production of endosperm, whereas an excess of maternal genes has the opposite effect (Scott et al., 1998). In seeds with paternal excess, the chalazal endosperm showed overproliferation, supporting the idea that the specialized chalazal endosperm cyst is instrumental in facilitating seed loading (see section 6.4.2).

Four stages of free nuclear endosperm development are recognized (Bosnes et al., 1992): (1) syncytial, a period during which nuclear divisions without cytokinesis result in a large syncytium that lines the periphery of the central cell; (2) cellularization, a period of cytokinesis resulting in discrete cells; (3) differentiation, which results in distinct tissues; and (4) maturation, which (particularly in albuminous seeds) is characterized by engorgement of tissues with storage reserves and the development of desiccation tolerance. The first three developmental stages will be emphasized in this review.

6.3.1 Syncytial

The behavior of nuclei during early development of the endosperm in maize has been deduced from analyses of color sectors in mature kernels (for summary see Walbot, 1994). Anthocyanin pigmentation in the aleurone and iodine staining of starchy endosperm in the interior of the grain provide clonal evidence that sectors stem from the first three mitoses occurring in relatively predictable planes. When viewed from the top, kernels with two half-sector hemispheres result from the first mitosis, placing a nucleus to the right and left of the zygote. The next mitosis results in bisection of the hemispheres into quarter-sectors and the third mitosis results in eight sectors. At the 8-nucleate stage, the nuclei are in the peripheral layer of syncytial cytoplasm surrounding the central vacuole. It is generally assumed that nuclei migrate to the specific positions that give rise to the characteristic sectors in the mature kernel. The nuclei continue to proliferate synchronously until there are about 256–512 nuclei in the syncytial cytoplasm before the onset of cellularization.

The early events in the development of nucleate endosperm in Arabidopsis (Vandendries, 1909; Webb and Gunning, 1991; Mansfield, 1994; Schneitz et al., 1995; Brown et al., 1999; Nguyen et al., 2001; and unpublished observations) are summarized in the following account. The early mitoses are synchronous. Following double fertilization, the primary endosperm nucleus begins to divide almost immediately so that there are usually several endosperm nuclei before the first division of the zygote (figure 6.1). The primary endosperm nucleus divides to produce a pair of nuclei. Each develops a radial microtubule system and the sister nuclei are displaced toward the poles with a large vacuole between them. The next round of division is oblique to the long axis of the central cell and places a nucleus near the zygote and one near the chalazal pole, the other two occupying the central portion of the central cell. Following the third round of division, two nuclei reside in the 'chalazal pocket'. Subsequently, the

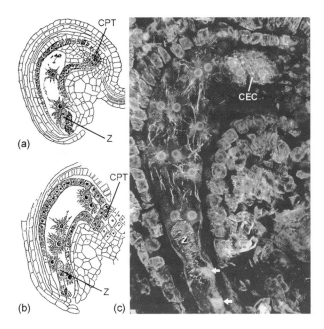

Figure 6.1 Early development of the syncytial endosperm in the mustard *Arabidopsis thaliana* showing the establishment of three developmental chambers. The first two figures (a and b) are taken from the exceptionally detailed drawings of Vandendries (1909). The filaments radiating from the nuclei in drawings, now known to be microtubules, are shown by immunofluorescence staining in (c). (a) Binucleate stage, showing the micropylar-chalazal positioning of two nuclei with vacuole between. CPT, chalazal proliferative tissue. (b) Six endosperm nuclei shown; two surround the zygote (Z) in the micropylar region, two occupy the 'chalazal pocket', and two are in the developing central chamber. (c) Sixteen-nucleate stage. Endosperm nuclei around the zygote (Z) are irregular in shape (arrows) and those in the chalazal region are tightly packed in dense cytoplasm (CEC). Microtubules of the endosperm are nuclear-based radial systems, while those of the still undivided zygote (Z) are organized into a hoop-like cortical system typical of meristems.

early syncytium is subtly differentiated into three developmental domains as nuclei tend to cluster in dense cytoplasm at the micropylar and chalazal poles, a general feature of the nuclear type of endosperm development recognized by Maheshwari (1950). Three distinct developmental domains—micropylar, central and chalazal—result (figure 6.2). The three domains exhibit differences in patterns of development leading to specializations in mature structure and presumed function.

During the syncytial stage, unusual fusiform endosperm nuclei are sheathed with parallel microtubules that are connected to a reticulate network of microtubules in the cytoplasm (figure 6.3a). With the cessation of proliferative nuclear division, the reticulate cytoskeleton is reorganized into radial microtubule systems (RMSs) emanating from the nuclei (Nguyen *et al.*, 2001). This

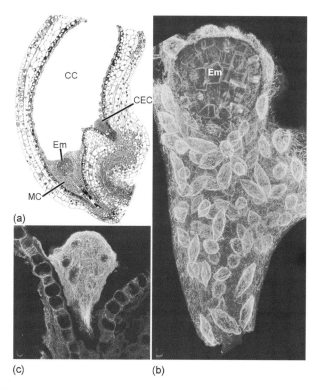

Figure 6.2 The three developmental domains in mustards. (a) Sagittal section showing micropylar chamber (MC), central chamber (CC), and chalazal endosperm cyst (CEC). (b and c) Isolated endosperm with microtubules stained by immunofluorescence. (b) Micropylar chamber with endosperm surrounding the globe stage embryo (Em). Irregularly shaped endosperm nuclei are sheathed with parallel microtubules that focus at tips and interconnect with a reticulate network in the cytoplasm. (c) Chalazal chamber. Microtubules form an irregular network in the common cytoplasm of the cyst and are parallel to each other in the narrow stalk-like portion where they extend toward the region of transfer walls in the tip adjacent to the maternal tissues.

organizes the common cytoplasm into nuclear cytoplasmic domains (NCDs) at the boundaries of which the initial walls will be deposited. The RMSs serve to position nuclei equidistantly into a regular polygonal (usually hexagonal) pattern (figure 6.3b). NCDs first develop in the central chamber followed by the micropylar chamber. In the mustards, nuclei in the chalazal region never develop RMSs and nuclei remain unevenly spaced. This region differentiates into a distinct multinucleate cyst (figure 6.2c) that does not cellularize or cellularizes late in development and remains distinct from the bulk of the endosperm (Brown *et al.*, 1999; Nguyen *et al.*, 2000, 2001). A transitional zone with 'nodules' occurs between the cyst and evenly spaced nuclei in the thin peripheral cytoplasm of the central chamber.

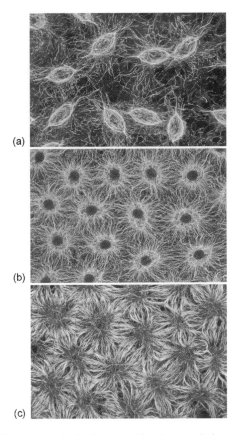

Figure 6.3 Microtubule reorganization in the syncytial endosperm during establishment of nuclear cytoplasmic domains (NCDs) and adventitious phragmoplasts associated with initial wall formation. All shown in face view. (a) Syncytium before organization into NCDs. (b) Nuclear-based radial microtubule systems (RMSs) organize syncytium into NCDs. (c) Adventitious phragmoplasts formed at interfaces of opposing RMSs are associated with wall deposition among NCDs.

6.3.2 Initial anticlinal walls

Cellularization begins with anticlinal wall formation between adjacent NCDs. These initial walls are unusual in several respects. They can form independently of mitosis among all nuclei and in the absence of distinct phragmoplasts. These unusual anticlinal walls have been termed 'free-growing' in the literature (see Olsen *et al.*, 1995). The term 'free-growing' has led to much confusion, particularly as some workers (e.g. Morrison and O'Brien, 1976; Bhandari *et al.*, 1986; Mansfield and Briarty, 1990; Chitraleka and Bhandari, 1992; Chamberlin *et al.*, 1994) interpreted the origin of these walls as ingrowths of the central cell wall. Other investigators (Chitraleka and Bhandari, 1993; Brown *et al.*, 1994,

1999; Olsen *et al.*, 1995) were unable to find evidence that the initial anticlinal walls begin as ingrowths. The explanation for anticlinal walls as ingrowths was discounted in favor of the NCD model in a series of endosperm studies using modern methods of light and electron microscopy (Brown *et al.*, 1994, 1996a, 1999; Olsen *et al.*, 1995; Otegui and Staehelin, 2000; Nguyen *et al.*, 2001). Likewise, an earlier model proposed by Fineran *et al.* (1982) for initial anticlinal wall formation in association with interzonal phragmoplasts between telophase nuclei has been discounted by the recent studies.

The NCD model (Brown and Lemmon, 1992; Pickett-Heaps *et al.*, 1999) provides the most plausible explanation for the unusual phenomenon of cell plate initiation in the absence of prominent phragmoplasts. This model holds that the pattern of wall placement is determined by nuclear-based RMSs and that formation of adventitious phragmoplasts at the boundaries of opposing RMSs (figure 6.3c) is responsible for wall deposition. A summary diagram (figure 2a in Olsen *et al.*, 1995) illustrated the 'free growing anticlinal wall' starting at sites along shared boundaries of NCDs and not yet fused with each other or with the central cell wall. Details of the events in wall initiation in the micropylar chamber have been documented using high-pressure freezing fixation and transmission electron microscopy (TEM) (Otegui and Staehelin, 2000). Cell plates are initiated in individual phragmoplast units of 4–12 microtubules ('mini-phragmoplasts') at the perimeter of NCDs. Thus, the initial anticlinal walls originate in a patchwork fashion among the interacting RMSs defining adjacent NCDs and spread to merge with each other laterally to establish an interconnected network of expanding walls. Cell plate maturation after fusion with the central cell wall is similar to final cell plate assembly in somatic cells described by Samuels *et al.* (1995). However, the endospermic cell walls remain callosic for a prolonged period (Brown *et al.*, 1994, 1997; Otegui and Staehelin, 2000). Peg-like ingrowths frequently reported in earlier TEM studies may not be inconsistent with this model since in some cases patches could be initiated immediately adjacent to the central cell wall and soon fuse with it. In the narrow micropylar chamber of mustards, where the syncytium is not forced to the periphery by a large central vacuole, initial anticlinal walls may complete cellularization by joining with central cell and embryo walls (Otegui and Staehelin, 2000) or completely surrounding NCDs (Nguyen *et al.*, 2001). Similar events may account for the precocious cellularization surrounding the embryo of cereals (Engell, 1989).

The same basic process of wall placement via the NCD mechanism also occurs in the large central chamber, but with variations that result in a unique pattern of cellularization termed alveolation. Alveolation begins with the delimitation of NCDs by nuclear-based RMSs in the thin layer of peripheral syncytium between the central cell wall and the large central vacuole. The NCDs become polarized in axes perpendicular to the central cell wall and anticlinal wall formation results in open-ended compartments termed alveoli. Numerous works (e.g. Olsen *et al.*, 1995; XuHan, 1995; Brown *et al.*, 1996a,b, 1999; Nguyen *et al.*, 2001) have demonstrated that alveolation is the typical mechanism for cellularization of a syncytium peripheral to a large central vacuole. In some developing seeds such as the elongate cereal grains, this is a very large region and the process of alveolation dominates endosperm development. It is even more prominent in the large megagametophytes of gymnosperms, where in some cases the entire process of initial cellularization is attributable to a single peripheral layer of alveoli that elongates centripetally until meeting in the center to effect closure (Singh, 1978).

Two phenomena contribute to alveolation. Each NCD becomes polarized in an axis perpendicular to the central cell wall and a ring of vacuoles accumulates around the nucleate column of cytoplasm (Brown *et al.*, 1996a,b; Nguyen *et al.*, 2001). The vacuoles serve to isolate a thin layer of shared cytoplasm between adjacent NCDs in which the anticlinal walls are formed. Little is known of either process, but it seems likely that vacuolation is important in driving the elongation of NCDs. Certainly vacuolation is conspicuous as anticlinal walls continue to grow. A recently isolated mutant of *Arabidopsis* (*mangled*) that causes defects in endosperm development shows sequence and structural similarity to a yeast protein required for vacuole protein sorting and vacuole biogenesis (Rojo *et al.*, 2000).

Concomitant with elongation of alveoli is a dramatic reorganization of the nuclear-based radial microtubules into highly polar systems. As seen from the side (figure 6.4a), the microtubule system in an elongating alveolus resembles a tree with axially aligned microtubules in the trunk of cytoplasm containing the nucleus, root-like processes at the central cell wall, and a canopy extending into the syncytial cytoplasm adjacent to the central vacuole and overtopping the anticlinal walls. Phragmoplasts form adventitiously (not in interzones between telophase nuclei) at the interfaces of opposing microtubule systems emanating from adjacent NCDs. In face view (figure 6.3c), as seen from the central vacuole, the alveoli are arranged in a honeycomb-like pattern. The arboreal configuration, which is a consistent and conspicuous feature in both cereals and mustards, is unlike any other cytoskeletal configuration known in plant cells.

The adventitious phragmoplasts that form in the syncytial cytoplasm at boundaries of NCDs mediate continued growth of the initial anticlinal walls. Centrifugal growth of the initial phase of anticlinal wall formation results in the development of a junction with the nearby central cell wall on one side and a blind edge in the cytoplasm on the other. The subsequent stage of anticlinal wall growth is centripetal relative to the entire central cell as walls grow unidirectionally toward the center in association with adventitious phragmoplasts. This second phase of wall formation is unusual in a number of respects. The adventitious phragmoplasts guiding wall deposition in the syncytial cytoplasm remain tethered to adjacent nuclei that themselves advance centripetally in the vacuolate cytoplasm within the alveoli. The complex of

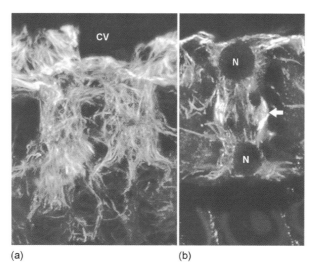

(a)   (b)

Figure 6.4 Microtubules in alveoli of the central chamber of mustards as viewed from the side. (a) As anticlinal walls form among NCDs and grow toward the central vacuole (CV), microtubules are arranged in a tree-like configuration; they are parallel in the elongated trunk-like cytoplasm containing the nucleus, extend into root-like cytoplasmic strands at the periphery, and flare into a canopy in the syncytial front of cytoplasm adjacent to the central vacuole. Adventitious phragmoplasts (shown in face view in figure 6.3) guide anticlinal wall deposition. (b) Periclinal division in an alveolus. An interzonal phragmoplast (arrow) guides cell plate deposition, which cuts off the first complete layer of cells at the periphery and displaces the inner alveolar portion inward. After periclinal walls complete alveolar division, adventitious phragmoplasts reform between the non-sister nuclei in adjacent alveoli and anticlinal walls resume centripetal growth.

merged alveolar phragmoplasts and the syncytial front of cytoplasm in which they are formed is continuously elevated as the six walls surrounding each alveolar NCD grow into the center of the central cell.

Descriptions of centripetally growing alveolar walls based on TEM are remarkably consistent (Olsen *et al.*, 1995; Brown *et al.*, 1996a, 1997). The phragmoplast microtubules are at right angles to the forming wall at its leading edge. The leading edge, which ends blindly in the cytoplasm, is often enlarged and bulb-shaped, and associated with ER (endoplasmic reticulum) cisternae, dictyosomes, and vesicles. Flattened ER cisternae are appressed to the sides of the developing walls. The abundant ER cisternae have been suggested to function in the trapping of dictyosome vesicles (Gori, 1987) or the formation of plasmodesmata (Dute and Peterson, 1992). However, the contribution of this distinctive ER system to the process of alveolar wall construction remains to be resolved. The forming walls exhibit a distinct gradation with a terminus of fusing vesicles, a lamellar zone, and a continuous zone (Brown *et al.*, 1997). Callose is first detected when walls become lamellar. In *Arabidopsis*, xyloglucans are present at all stages, including the earliest ones (Otegui and Staehelin, 2000).

An important aspect of development in nuclear endosperm is the temporal relationship of alveolar wall formation to mitosis. Although it seems certain that the wave of anticlinal wall deposition that initiates alveoli is independent of the mitotic division cycle, the two may be widely separated or overlapping. In barley, a long interval (up to 48 hours) occurs between cessation of mitosis and the formation of anticlinal walls (Brown et al., 1994). However, it is more usual for the wave of simultaneous wall deposition to occur quickly after cessation of mitosis, as it does in rice (Brown et al., 1996a) and maize (Randolph, 1936; Walbot, 1994). The two phenomena may even overlap. After alveoli are initiated in Ranunculus, the nuclei in them undergo another round of mitosis (Chitralekha and Bhandari, 1993). Unlike subsequent divisions in alveoli, the mitotic spindles are parallel to the central cell wall and the new cell plates are anticlinal. Thus, the number of alveoli is doubled and the merged alveolar walls are of mixed origin. The new bisecting wall is the result of the interzonal phragmoplast following mitosis, while the others are the anticlinal walls deposited earlier in the absence of mitosis. In addition, a fast-moving wave of anticlinal wall formation can overtake the final wave of mitosis in the chalazal region of the central cell and both types of phragmoplasts (interzonal and adventitious) contribute simultaneously to alveolar wall formation (Chitralekha and Bhandari, 1993). In this case, a six-sided alveolus would have one wall formed in association with an interzonal phragmoplast between sister nuclei and the other five walls formed adventitiously between non-sister nuclei. Thus it appears that NCDs can divide anticlinally after their formation and even after being partially encased by alveolar walls. Although anticlinal division in alveoli are not prominent in cereals and Arabidopsis, scattered reports are found in the literature (e.g. Fineran et al., 1982; Brown et al., 1999).

There also appears to be variation in the polarity of developmental waves in nuclear endosperm. In most taxa, the gradient of cellularization proceeds from micropylar to chalazal chambers (Bhatnagar and Sawhney, 1981; Vijayraghavan and Prahabkar, 1984; Engell, 1989; Bosnes et al., 1992; Chitralekha and Bhandari, 1993; Brown et al., 1999), but the opposite has been reported for the basal eudicot Platanus (Floyd et al., 1999), where the chalazal chamber appears to cellularize by a different pattern from the micropylar chamber. Whereas the large micropylar chamber undergoes alveolation, the chalazal syncytium is irregularly divided into multinucleate compartments and finally cells with a single nucleus. In the small grains of barley (Brown et al., 1994) and rice (Brown et al., 1996a), where the developing seed has vascular tissue adjacent to the long central chamber, developmental waves start at the vascular crease and move in both directions around the central chamber. These observations would seem to argue for factors in addition to the embryo as a source of signals in endosperm development.

6.3.3   Periclinal division

Each alveolus, starting with only one wall (the central cell wall) and growing usually six merged anticlinal walls, remains open-ended until it is divided periclinally by cytokinesis following mitosis. Prior to the wave of periclinal divisions, the alveolar phragmoplasts are disassembled and nuclei become more centrally located in the alveoli. This has been observed in both cereals (e.g. Brown *et al.*, 1994, 1996a) and mustards (e.g. Nguyen *et al.*, 2001). The factors involved in positioning nuclei are not known. Alveoli are of nearly uniform length and the prophase nuclei are suspended in rafts of cytoplasm (phragmosomes) nearly equidistant from the central cell wall. Phragmosomes predict the plane of the future division as is typical of vacuolate cells (Sinnott and Bloch, 1941; Lloyd and Traas, 1988; Lloyd *et al.*, 1992) but no PPBs are formed in alveoli (Brown *et al.*, 1994, 1999) and no actin pattern indicates division site selection (Nguyen *et al.*, 2001). Interzonal phragmoplasts/cell plates expand to form a junction with the anticlinal walls of the alveoli (figure 6.4b). In this manner, the peripheral portion of each alveolus receives its final wall and becomes a cell, while the inner portion remains an alveolus. Following the wave of periclinal divisions, microtubules again emanate from the tips of interphase nuclei (all of which are non-sisters) in the alveolar layer, interact at their interfaces and organize adventitious phragmoplasts that direct renewed growth of the anticlinal walls between non-sister nuclei.

The second layer of alveolar growth in endosperm is identical to the first and again is followed by a periclinal division. This repeated cycle of anticlinal wall formation between non-sister nuclei followed by periclinal wall formation between sister nuclei completes centripetal cellularization of the endosperm. The alternating cycles are conspicuous in rice, where orderly files of several cells are produced (Brown *et al.*, 1996a). Nuclei of any given file are sister nuclei and the walls between them are the result of interzonal phragmoplast activity following mitosis. The anticlinal walls between adjacent cell files were formed in association with adventitious phragmoplasts between non-sister nuclei. Some variability exists in the process of centripetal cellularization. In many plants including the cereals (Brown *et al.*, 1996a) and *Arabidopsis* (Brown *et al.*, 1999) files of cells are produced in a regular pattern (figure 6.5a) whereas in others, e.g. *Platanus* (Floyd *et al.*, 1999), cells of variable size with no distinct layering results in an irregular pattern. In female gametophyte development of many gymnosperms a single layer of alveoli may continue to grow until meeting in the center before they are subdivided by periclinal and oblique divisions (Singh, 1978).

Following initial cellularization, cell divisions occur in both the starchy endosperm and peripheral aleurone. These later divisions depart from the strictly RMS-driven NCD developmental pathway of initial cellularization. Cells of the starchy endosperm develop cortical microtubules but no PPBs (Brown *et al.*, 1994; Clore *et al.*, 1996). Cells of the multilayered peripheral aleurone of barley

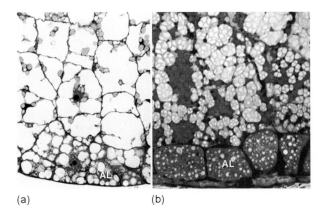

(a)   (b)

Figure 6.5 Tissue differentiation in the endosperm of rice. (a) Immediately following cellularization, the peripheral layer of aleurone initials (AL) are characterized by numerous small vacuoles and dense cytoplasm. Files of cells in the starchy endosperm (reflecting the regular pattern of cellularization) are highly vacuolate and are beginning to accumulate starch. (b) In the mature grain, the peripheral layer of thick-walled aleurone (AL) containing aleurone grains surrounds the central region of thin-walled storage endosperm packed with compound starch granules.

develop both hoop-like cortical microtubules and PPBs (Brown *et al.*, 1994). It is significant that the switch to the PPB microtubule cycle typical of histogenesis occurs only in later stages of endosperm development when cells are added in an orderly fashion to the growing aleurone. It is not known whether PPBs develop in the single-layered aleurone (endosperm epidermis) that occurs in many seed types (see section 6.4.3).

6.4 Differentiation

The fully differentiated endosperm of grasses and mustards is structurally simple, consisting of three principal cell types: an inner mass of storage endosperm cells, an epidermis (aleurone) from one to three cell layers thick, and a region of basal transfer cells adjacent to the main maternal vasculature.

6.4.1 Central or storage endosperm

The storage endosperm constitutes the central bulk of the endosperm. It is frequently known as starchy endosperm because of abundant starch granules produced in amyloplasts (figure 6.5). Whereas starch storage is a characteristic of cereals, other albuminous seeds may store reserves either as proteins or as oils (Jacobsen *et al.*, 1976; Lopes and Larkins, 1993; Floyd and Friedman, 2000), or as polysaccharides such as mannans and xyloglucans in the very thick walls of living endosperm (DeMason *et al.*, 1992; Lopes and Larkins, 1993; Otegui *et al.*, 1999; Nonogaki *et al.*, 2000). The primitive endosperm stores protein and oil, with starch producing and sequestering ability appearing in higher eudicots and monocots (Floyd and Friedman, 2000). According to Jacobsen *et al.* (1976), about two-thirds of the dicotyledons store reserves as aleurone grains (protein bodies) and lipids rather than starch.

The storage tissue is depleted during development of exalbuminous seeds such as mustards and legumes but remains copious in cereal grains and other albuminous seeds such as celery. The starchy endosperm of cereals also contains storage proteins (Bosnes *et al.*, 1992; Lopes and Larkins, 1993), which are deposited either in protein bodies sequestered from the ER or in protein storage vacuoles (Müntz, 1998). A small amount of oil bodies similar to those present in high concentration in the aleurone layer are found in the starchy endosperm of wheat (Hargin and Morrison, 1980), and probably other cereals as well.

The starchy endosperm of cereals is generally thought to be dead at maturity and recent studies have shown that programmed cell death is linked to the ethylene pathway. In the mutant *shrunken2* (*sh2*) that overproduces ethylene, cell death is accelerated (Young *et al.*, 1997). During seed germination in cereals, a cocktail of enzymes produced by the aleurone degrades the starchy endosperm. In albuminous seeds with living endosperm, such as celery (Jacobsen *et al.*, 1976) and tomato (Nonogaki *et al.*, 2000), the breakdown of endosperm is due to hydrolases produced in the endosperm itself as a response to embryo-generated gibberellin.

The prevalence of polyploidy in endosperm of diverse plants (Chamberlin *et al.*, 1993) suggests that it may play an important role. An amazing level of 3072C has been reported to occur in the haustorium of the curcubit *Echinocystis lobata* (Chamberlin *et al.*, 1993). Extremely high levels of DNA resulting from endoreduplication is an important developmental feature in cereals (Lopes and Larkins, 1993). In the starchy endosperm of maize, the DNA content may reach 96C (Grafi and Larkins 1995). Endoreduplication begins during the differentiation stage (10–12 days after pollination (DAP)) after cellularization is complete. It requires the induction of S-phase kinase and the inhibition of M-phase promoting factors.

Rapidly accumulating evidence indicates that the embryo-surrounding region (ESR), a pocket within the starchy endosperm in which the embryo develops, is a distinct developmental domain. In maize, the ESR is a restricted region of dense endosperm that surrounds the entire embryo at 5 DAP, the suspensor at 7 DAP, and only the lower part of the suspensor at 9 DAP (Schel *et al.*, 1984; Opsahl-Ferstad *et al.*, 1997). Three genes (*Esr1–3*) are specifically expressed in the ESR (Opsahl-Ferstad *et al.*, 1997). Expression of these genes can be detected from 4 DAP and is maintained throughout seed development, the expression being most abundant at the adaxial side of the embryo close to the pericarp. The absence of gene expression in spontaneously occurring embryoless endosperms suggests a dependence on signaling from the embryo for *Esr* gene expression (Opsahl-Ferstad *et al.*, 1997). The tightly controlled gene expression pattern, as demonstrated by GUS expression (Bonello *et al.*, 2000), suggests that the gene products may play a role in embryo nutrition and/or the establishment of an interface between the embryo and the endosperm leading to the creation of the embryonic cavity.

In *Arabidopsis* and other mustards, the ESR is easily recognizable as a distinct micropylar domain early in development. It probably stems from early events in syncytial proliferation that result in a population of nuclei in the region around the zygote (Pacini *et al.*, 1975). In the globe embryo stage seeds of *Coronopus didymus* (figure 6.2b), the endosperm in the micropylar chamber consists of a system of fusiform to multangular nuclei interconnected by a reticulum of microtubules and F-actin (Nguyen *et al.*, 2001). Microtubules but not F-actin closely ensheathe the nuclei. This contrasts dramatically with the evenly spaced NCDs with spherical nuclei and RMSs in the central chamber. At the time of cellularization, coinciding with the heart stage embryo, NCDs are established in the ESR as well as in the central chamber and cellularization occurs via a similar basic mechanism (see section 6.3.2). Unlike the case in cereals, there is evidence of nutrient uptake for support of the young embryo in the micropylar chamber of mustards. In the exalbuminous seeds of mustards (e.g. Schulz and Jensen, 1974; Simoncioli, 1974) and legumes (Yeung and Clutter, 1979; Dute and Peterson, 1992; XuHan, 1995), transfer type wall invaginations occur in the ESR both from the central cell wall and from the embryo suspensor during the syncytial stage of endosperm development. The suspensors of dicotyledons are in general much larger and more complex than those of monocotyledons and have been implicated in the uptake of solutes and in the production of specific substrates for the young embryo (Yeung and Meinke, 1993). However, as pointed out by Lopes and Larkins (1993), the process of endosperm degradation and utilization by the embryo is one of the least-understood aspects of seed development.

6.4.2 Transfer tissues

All metabolites must enter the endosperm via symplastic or apoplastic transport as there are no direct connections to the maternal vascular tissue. Wall protuberances and labyrinthines are generally accepted as evidence of transfer cells in which amplification of the plasma membrane is associated with an increased capacity for uptake of assimilate into the cell (Gunning and Pate, 1969). The transfer cells of endosperm, which are characterized by heavy secondary wall ingrowths, tend to be located adjacent to the maternal vasculature. In the small cereal grains, the vascular tissue extends the length of the ventral surface of the elongated grains. Nucellar cells in the ventral crease develop into the nucellar projection, a specialized region of elongate cells with extensive wall infoldings between the vascular bundle and the endosperm. The adjacent endosperm exhibits transfer cell type walls grading from the aleurone to the subaleurone (Wang *et al.*, 1994; Linnestad *et al.*, 1998). In maize, aleurone transfer cells are located at the broad region overlying the hilar pad (Davis *et al.*, 1990). In addition to the basalmost cell layer, two or three adjacent layers of endosperm cells possess wall ingrowths in a gradient decreasing toward the interior of the endosperm (Schel *et al.*, 1984).

Several recent studies have reported differentiation of the chalazal endosperm in the Brassicaceae into a coenocytic haustorium that invades a specialized region of the nucellus known as the chalazal proliferative tissue lying between the endosperm and the vasculature of the pedicel (figure 6.1 and 6.2). The chalazal cyst remains a distinct syncytial structure after cellularization has been completed in the micropylar and central chambers. In some species such as *Diplotaxis erucoides* (Pacini *et al.*, 1975) and *Arabidopsis* (Mansfield and Briarty, 1990; Brown *et al.*, 1999), the cyst is the last region of the endosperm to cellularize and is absent in the mature seed. In *Iberis amara* (Vijayraghavan and Prahabkar, 1984) and *Lepidium virginicum* (Nguyen *et al.*, 2000) the cyst remains syncytial and persists in the mature seed.

Ultrastructural studies of the early chalazal endosperm in *Capsella bursa-pastoris* (Schulz and Jensen, 1971, 1974), *Diplotaxis erucoides* (Pacini *et al.*, 1975), *Alyssum maritimum* (Vijayraghavan *et al.*, 1984), *Iberis amara* (Vijayraghavan and Prahabkar, 1984) and *Arabidopsis thaliana* (Mansfield and Briarty, 1990; Mansfield, 1994; Nguyen *et al.*, 2000) show a multinucleate cytoplasm with lobed nuclei, abundant rough endoplasmic reticulum, dictyosomes active in vesicle production and profuse wall invaginations. Polyploid nuclei have been reported in the persistent cyst of *Iberis* (Vijayraghavan and Prahabkar, 1984). In the shallow chalazal depression of *Arabidopsis*, the cyst is mushroom-shaped with short tentacle-like processes penetrating the maternal tissues. The narrow chalazal chamber of *Lepidium virginicum* (Nguyen *et al.*, 2000) and *Coronopus didymus* (Nguyen *et al.*, 2001) is filled by an elongate stalk-like portion of the cyst. The basal portions, which lack nuclei and plastids, contain parallel arrays of microtubules and F-actin and terminate in a region of extensive wall ingrowths (Nguyen *et al.*, 2001). Mitochondria are concentrated in the cytoplasm adjacent to the wall ingrowths. The apparently active cytoplasm, extensive transfer cell walls and positioning of the endosperm cyst above the chalazal proliferative tissue strongly suggest an involvement in its degradation (Schulz and Jensen, 1971) and a role in nutrient transfer (Pacini *et al.*, 1975; Vijayraghavan and Prahabkar, 1984; Scott *et al.*, 1998; Nguyen *et al.*, 2000).

The importance of nutrient loading to endosperm development is supported by evidence that seed abortion can be traced to the failure of transfer cells or haustoria to develop. In certain interspecific crosses of *Trifolium*, seed abortion was found to be related to abnormal development of the haustorium (Williams, 1987). Deviation from the normal endospermic ratio of 2 maternal to 1 paternal genomes can result in seed abortion in maize. Ultrastructural studies of seeds with an unbalanced genomic ratio of 2 maternal to 2 paternal revealed that development of the transfer cell layer was almost completely suppressed (Charlton *et al.*, 1995). The efficient uptake of nutrients appears to be a factor essential to proper differentiation of the transfer layer. An invertase-deficient mutant (*miniature-1*) of maize results in aberrant endosperm development. Development is normal until 9 DAP, but soon thereafter a gap occurs between the endosperm and the maternal pedicel tissues. The placental pad disintegrates and aleurone transfer cells do not differentiate (Miller and Chourey, 1992). Thus, the wild-type allele is required for normal development of both endosperm and maternal cells, emphasizing the complex interplay of various tissues in seed development (Cheng *et al.*, 1996).

Molecular markers for the basal endosperm transfer layer include *BETL1–4* in maize and *END1* in barley. The genes *BETL1–4*, are specifically expressed in differentiating transfer cells of maize (Hueros *et al.*, 1995, 1999a). Analysis of mRNA steady-state levels show that *BETL2* and *BETL3* are detectable in 6 DAP endosperm just prior to transfer cell differentiation, whereas *BETL1* and *BETL4* appear around 8–9 DAP, coincident with transfer cell differentiation. *BETL1* encodes a 7 kDa cell wall polypeptide proposed to play a role in the structural specialization of the transfer cells (Hueros *et al.*, 1995). A weak similarity of BETL1 and 3 to plant defensins (Hueros *et al.*, 1999a) suggests a possible role for these proteins in plant defense. The *BETL1* upstream promoter sequence is capable of conferring transfer cell-specific expression of GUS in transgenic maize (Hueros *et al.*, 1999b). In barley, the *END1* transcript is first expressed in the syncytial endosperm adjacent to the nucellar projection. Following cellularization, expression remains restricted to the basal endosperm transfer layer (Doan *et al.*, 1996). The function of the *END1* gene product is unknown, but the fact that it is expressed in the syncytium prior to transfer wall development suggests a role in differentiation of the transfer cells.

6.4.3 Aleurone

The peripheral layer of endosperm is an epidermis generally known as the aleurone (figure 6.5). In the cereal grains, the persistent aleurone has evolved important function in the mobilization of storage reserves in germination by secreting hydrolases into the central endosperm. Genes encoding amylases, hydrolases and proteases are induced by the phytohormone gibberellin released by the germinating embryo. Following gene expression, the aleurone cells undergo programmed cell death involving DNA fragmentation and wall degradation (Wang *et al.*, 1998). Starchy endosperm is degraded into sugars and other simple molecules that support embryo growth during germination.

In the cereal grains, the aleurone layer is situated outside the starchy endosperm, and is the only layer of the endosperm that may be pigmented, as in maize and rice. The aleurone consists of a single layer of cells in maize

(Walbot, 1994) and wheat (Morrison *et al.*, 1975), from one to several layers in rice depending upon location and growing conditions (Bechtel and Pomeranz, 1980), and three layers in barley (Jones, 1969). The subaleurone is a transitional zone between the peripheral aleurone and the starchy endosperm. It is a site of cell proliferation in late development, where repeated periclinal divisions result in files of cells that increase the seed size (Randolph, 1936; Olsen *et al.*, 1992). Subaleurone may contain much of the storage protein in the mature grain (Olsen *et al.*, 1992).

Mature aleurone cells are characterized by thick, highly autofluorescent walls, prominent nuclei, and aleurone grains (protein storage bodies). The aleurone grains produced in protein-storing vacuoles (PSVs) are very complex, consisting of a protein matrix in which globoid inclusions (phytin, niacin and phenolics) or carbohydrates are embedded (Bethke *et al.*, 1998). Both types may develop in the same PSV (Jacobsen *et al.*, 1971). Lipid droplets surround the aleurone grains (Jones, 1969; Morrison *et al.*, 1975; Kyle and Styles, 1977). In addition to the PSV, a second kind of vacuole, the lytic vacuole, occurs in the aleurone cells of barley (Swanson *et al.*, 1998).

Although aleurone is characteristic of cereals, it has been reported in *Arabidopsis* (Brown *et al.*, 1999; Beeckman *et al.*, 2000) and other mustards (Groot and van Caeseele, 1993), and in certain other dicots (e.g. Neubauer, 1971). In *Arabidopsis* the aleurone layer is first recognizable at the torpedo embryo stage (Beeckman *et al.*, 2000). The aleurone persists and sticks to the seed coat in mature seeds after desiccation. The aleurone cells are tabular and thick-walled, and contain numerous protein bodies and few chloroplasts. A specific role for the persistent aleurone in dicot seeds is unknown. It may function as an epidermis in maintaining water balance and/or protecting the germinating embryo from invasion by microorganisms.

The first sign of aleurone formation in maize (Randolph, 1936) and wheat (Morrison *et al.*, 1975) is a preponderance of anticlinal cell divisions in the peripheral layer, resulting in cuboidal cells. After the establishment of aleurone cell shape, small vacuoles accumulate and the cytoplasm becomes dense. Differentiation begins at 4 DAP in rice (Juliano and Bechtel, 1985), 6–8 DAP in wheat (Morrison *et al.*, 1975), 8 DAP in barley (Bosnes *et al.*, 1992), and 10–15 DAP in maize (Kyle and Styles, 1977). This variation may be due in part to different criteria for the determination of the onset of differentiation (e.g. anticlinal divisions versus aleurone grain accumulation). In the mutilayered aleurone of barley, the peripheral cells are the first to differentiate, with the second and third layers differentiating sequentially afterward. The earliest aleurone-specific transcript known in barley is *Ltp2*, which is detectable at about 8 DAP, i.e. simultaneously with the first morphological signs of aleurone cell differentiation (Kalla *et al.*, 1994). The aleurone specificity of this gene has been confirmed in transgenic rice expressing the *Ltp2* promoter::GUS construct (Kalla *et al.*, 1994).

Previous literature described the aleurone layer of maize as a separate lineage from the starchy endosperm (Randolph, 1936; Walbot, 1994; Becraft et al., 1996). Recent investigation of the cytoskeleton and division plane in small grains suggested that the first periclinal division sets off a peripheral layer of cells destined to become aleurone (Brown et al., 1996b). Although divisions in the peripheral layer were thought to be exclusively anticlinal, periclinal divisions have been reported to occur in both wheat (Morrison et al., 1975) and maize (Becraft and Ascuncion-Crabb, 2000). In wheat aleurone the interior daughter cell undergoes a process of dedifferentiation before converting to starchy endosperm, while the outer cell retains aleurone identity (Morrison et al., 1975). Likewise in maize, the interior cells never form aleurone.

Aleurone differentiation appears to require both a cue from its peripheral position and a hierarchy of genetic regulation (Becraft and Ascuncion-Crabb, 2000). At least two genes of maize are known to be involved in determining aleurone fate. The mutants crinkly4 (cr4) and defective kernel1 (dek1) can have starchy endosperm cells at the periphery, indicating that the wild-type genes are essential to normal differentiation of aleurone (Becraft et al., 1996). The cr4 gene encodes a receptor-like kinase, suggesting a model for endosperm development in which CR4 might be responsible for the reception of signals that specify aleurone cell fate (Olsen et al., 1998). Studies of the dek1 mutant, which can completely lack aleurone, suggest that positional cues occur throughout endosperm development and that cells require continual input of positional cues to maintain aleurone fate (Becraft and Ascuncion-Crabb, 2000). Revertant sectors from a mutable dek1 mutant allele show that starchy endosperm cells at the periphery can acquire aleurone fate late in development, indicating that the cues that specify aleurone fate are present throughout development and that the Dek1+ gene product is required to perceive these signals and maintain aleurone identity. Conversely, sectors with loss of the Dek1+ gene late in development cause peripheral cells to switch fate from aleurone to starchy endosperm. Thus, neither aleurone nor starchy endosperm cells possess a stably determined state.

Following perception of position and the acquisition of aleurone cell fate, several more genes are involved in differentiation of aleurone characteristics. Although Mosaic1 (Msc1) and Dappled (Dap) mutants show abnormalities in aleurone cell morphology and anthocyanin pigmentation, the peripheral cells do not differentiate as starchy endosperm suggesting that they block a late step in aleurone differentiation (Becraft and Ascuncion-Crabb, 2000). Other Dap mutants have a similar effect on aleurone phenotype and inhibit anthocyanin gene expression (Gavazzi et al., 1997). Genetic variants of maize with multiple layers of aleurone indicate that the number of aleurone layers is also under genetic control (Wolf et al., 1972). Finally, a specialized developmental program involving late aleurone genes confers desiccation tolerance to the aleurone cells, allowing them to survive the maturation process.

Similar studies are needed on *Arabidopsis* where the persistent epidermis is remarkably aleurone-like with thickened walls and aleurone grains and should provide a good model for studies of cell identity and differentiation. The homeobox-containing gene *meristem layer1* (*ATML1*), a marker for protoderm development in the embryo of *Arabidopsis*, is not expressed in cellularized endosperm or the persistent endosperm epidermis (Lu *et al.*, 1996). This case suggests caution in assuming homology between embryo and endosperm tissues. Even if the endosperm is a reprogrammed embryo and the initial developmental pathway is similar to that of female gametophytes of certain gymnosperms, many innovative features have appeared in angiosperms. The endosperm seems less likely to be a failed embryo than a highly evolved plant generation with a complex developmental pathway initiated by double fertilization. In our search for the fundamental processes underlying endosperm development, we have stressed commonalities among mustards and cereals. The radial microtubule systems, nuclear cytoplasmic domains, and specialized microtubule arrays responsible for wall deposition during alveolation are strikingly similar. Variations in the cell cycle associated with specific events in endosperm development indicate precise regulation. Much is to be gained by pursuit of the underlying genetic controls and developmental processes of this unique and important plant system.

Acknowledgments

Supported in part by NSF MCB-9726968 to R.C.B and B.E.L. and Louisiana Board of Regents Fellowship LEQSF-1998-03-GF-28 to H.N.

References

Bajon, C., Horlow, C., Motamayor, J.C., Sauvanet, A. and Robert, D. (1999) Megasporogenesis in *Arabidopsis thaliana* L.: an ultrastructural study. *Sex. Plant Reprod.*, **12**, 99-109.

Battaglia, E. (1989) The evolution of the female gametophyte of angiosperms: an interpretative key. (Embriological questions: 14). *Ann. Bot. (Roma)*, **47**, 7-144.

Bechtel, D.B. and Pomeranz, Y. (1980) The rice kernel, in *Advances in Cereal Science and Technology*, vol. 3 (ed. Y. Pomeranz), American Association of Cereal Chemistry, St. Paul, MN, pp. 73-113.

Becraft, P.W. and Asuncion-Crabb, Y. (2000) Positional cues specify and maintain aleurone cell fate in maize endosperm development. *Development*, **127**, 4039-4048.

Becraft, P.W., Stinard, P.S. and McCarty, D.R. (1996) CRINKLY4: A TNFR-like receptor kinase involved in maize epidermal differentiation. *Science*, **273**, 1406-1409.

Becraft, P.W., Brown, R.C., Lemmon, B.E., Olsen, O.-A. and Opsahl-Ferstad, H.-G. (2001) Endosperm development, in *Current Trends in the Embryology of Angiosperms* (ed. S.S. Bhojwani and W.-Y. Soh), Kluwer Academic, Dordrecht, pp. 353-374.

Beeckman, T., De Rycke, R., Viane, R. and Inzé, D. (2000) Histological study of seed coat development in *Arabidopsis thaliana. J. Plant Res.*, **113**, 139-148.

Bethke, P.C., Swanson, S.J., Hillmer, S. and Jones, R.L. (1998) From storage compartment to lytic organelle: the metamorphosis of the aleurone storage vacuole. *Ann. Bot.*, **82**, 399-412.

Berger, F. (1999) Endosperm development. *Curr. Opin. Plant Biol.*, **2**, 28-32.

Bhandari, N.N., Bhargava, M. and Chitralekha, P. (1986) Cellularization of free-nuclear endosperm of *Papaver somniferum* L. *Phytomorphology*, **36**, 357-366.

Bhatnagar, S.P. and Sawhney, V. (1981) Endosperm—its morphology, ultrastructure, and histochemistry. *Int. Rev. Cytol.*, **73**, 55-102.

Bonello, J.-F., Opsahl-Ferstad, H.-G., Perez, P., Dumas, C. and Rogowsky, P.M. (2000) *Esr* genes show different levels of expression in the same region of maize endosperm. *Gene*, **246**, 219-227.

Bosnes, M., Weideman, F. and Olsen, O.-A. (1992) Endosperm differentiation in barley wild-type and *sex* mutants. *Plant J.*, **2**, 661-674.

Brown, R.C. and Lemmon, B.E. (1992) Cytoplasmic domain: a model for spatial control of cytokinesis in reproductive cells of plants. *EMSA Bull.*, **22**, 48-53.

Brown R.C. and Lemmon, B.E. (2001) The cytoskeleton and spatial control of cytokinesis in the plant life cycle. *Protoplasma*, **215**, 35-49.

Brown, R.C., Lemmon, B.E. and Olsen, O.-A. (1994) Endosperm development in barley: microtubule involvement in the morphogenetic pathway. *Plant Cell*, **6**, 1241-1252.

Brown, R.C., Lemmon, B.E. and Olsen, O.-A. (1996a) Development of the endosperm in rice (*Oryza sativa* L.): cellularization. *J. Plant Res.*, **109**, 301-313.

Brown, R.C., Lemmon, B.E. and Olsen, O.-A. (1996b) Polarization predicts the pattern of cellularization in cereal endosperm. *Protoplasma*, **192**, 168-177.

Brown, R.C., Lemmon, B.E., Stone, B.A. and Olsen, O.-A. (1997) Cell wall (1-3) and (1-3, 1-4) β-glucans during early grain development in rice (*Oryza sativa* L.). *Planta*, **202**, 414-426.

Brown, R.C., Lemmon, B.E., Nguyen, H. and Olsen, O.-A. (1999) Development of endosperm in *Arabidopsis thaliana*. *Sex. Plant Reprod.*, **12**, 32-42.

Carmichael, J.S. and Friedman, W.E. (1996) Double fertilization in *Gnetum gnemon* (Gnetaceae): its bearing on the evolution of sexual reproduction within the Gnetales and the anthophyte clade. *Am. J. Bot.*, **83**, 767-780.

Chamberlain, C.J. (1935) *Gymnosperms: Structure and Function*, University of Chicago Press, Chicago.

Chamberlin, M.A., Horner, H.T. and Palmer, R.G. (1993) Nuclear size and DNA content of the embryo and endosperm during their initial stages of development in *Glycine max* (Fabaceae). *Am. J. Bot.*, **80**, 1209-1215.

Chamberlin, M.A., Horner, H.T. and Palmer, R.G. (1994) Early endosperm, embryo, and ovule development in *Glycine max* (L.) Merr. *Int. J. Plant Sci.*, **155**, 421-436.

Charlton, W.L., Keen, C.L., Merriman, C., Lynch, P., Greenland, A.J. and Dickinson, H.G. (1995) Endosperm development in *Zea mays*: implication of gametic imprinting and paternal excess in regulation of transfer layer development. *Development*, **121**, 3089-3097.

Cheng, W.-H., Taliercio, E.W. and Chourey, P.S. (1996) The *minature1* seed locus of maize encodes a cell wall invertase required for normal development of endosperm and maternal cells in the pedicel. *Plant Cell*, **8**, 971-983.

Chitralekha, P. and Bhandari, N.N. (1992) Cellularization of endosperm in *Asphodelus tenuifolius* Cav. *Phytomorphology*, **42**, 185-193.

Chitralekha, P. and Bhandari, N.N. (1993) Cellularization of free-nuclear endosperm in *Ranunculus sceleratus* Linn. *Phytomorphology*, **43**, 165-183.

Clore, A.M., Dannenhoffer, J.M. and Larkins, B.A. (1996) EF-1alpha is associated with a cytoskeletal network surrounding protein bodies in maize endosperm cells. *Plant Cell*, **8**, 2003-2014.

Dahlgren, G. (1991) Steps toward a natural system of the dicotyledons: embryological characters. *El Aliso*, **13**, 107-165.

Davis, R.W., Smith, J.D. and Cobb, B.G. (1990) A light and electron microscope investigation of the transfer cell region of maize caryopses. *Can. J. Bot.*, **68**, 471-479.

DeMason, D.A. (1997) Endosperm structure and development, in *Cellular and Molecular Biology of Plant Seed Development* (ed. B.A. Larkins and I.K. Vasil), Kluwer Academic, Dordrecht, pp. 73-115.

DeMason, D.A., Madore, M.A., Sekhar, C. and Harris, M.J. (1992) Role of α-galactosidase in cell wall metabolism of date palm (*Phoenix dactylifera*) endosperm. *Protoplasma*, **166**, 177-184.

Doan, D.N., Linnestad, C. and Olsen, O.-A. (1996) Isolation of molecular markers from the barley endosperm coenocyte and the surrounding nucellus cell layers. *Plant Mol. Biol.*, **31**, 877-886.

Dute, R.R. and Peterson, C.M. (1992) Early endosperm development in ovules of soybean, *Glycine max* (L.) Merr. (Fabaceae). *Ann. Bot.*, **69**, 263-271.

Engell, K. (1989) Embryology of barley: time course and analysis of controlled fertilization and early embryo formation based on serial sections. *Nord. J. Bot.*, **9**, 265-280.

Fineran, B.A., Wild, D.J.C. and Ingerfeld, M. (1982) Initial wall formation in the endosperm of wheat *Tritichum aestivum*: a reevaluation. *Can. J. Bot.*, **60**, 1776-1795.

Floyd, S.K. and Friedman, W.E. (2000) Evolution of endosperm developmental patterns among basal flowering plants. *Int. J. Plant Sci.*, **161**, S57-S81.

Floyd, S.K., Lerner, V.T. and Friedman, W.E. (1999) A developmental and evolutionary analysis of embryology in *Platanus* (Platanaceae), a basal eudicot. *Am. J. Bot.*, **86**, 1523-1537.

Friedman, W.E. (1994) The evolution of embryogeny in seed plants and the developmental origin and early history of endosperm. *Am. J. Bot.*, **81**, 1468-1486.

Friedman, W.E. (2001) Comparative embryology of basal angiosperms. *Curr. Opin. Plant Biol.*, **4**, 14-20.

Gavazzi, G., Dolfini, S., Allegra, D., Castiglioni, P., Todesco, G. and Hoxha, M. (1997) *Dap* (defective aleurone pigmentation) mutations affect maize aleurone development. *Mol. Gen. Genet.*, **256**, 223-230.

Gori, P. (1987) The fine structure of the developing *Euphorbia dulcis* endosperm. *Ann. Bot.*, **60**, 563-569.

Grafi, G. and Larkins, B.A. (1995) Endoreduplication in maize endosperm: involvement of M phase-promoting factor inhibition and induction of S phase-related kinases. *Science*, **269**, 1262-1264.

Groot, E.P. and van Caeseele, L.A. (1993) The development of the aleurone layer in canola (*Brassica napus*). *Can. J. Bot.*, **71**, 1193-1201.

Grossniklaus, U., Vielle-Calzada, J.-P., Hoeppner, M.A. and Gagliano, W.B. (1998) Maternal control of embryogenesis by *MEDEA*, a polycomb-group gene in *Arabidopsis*. *Science*, **280**, 446-450.

Gunning, B.E.S. and Pate, J.S. (1969) 'Transfer cells'; plant cells with wall ingrowths, specialized in relation to short distance transport of solutes-their occurrence, structure, and development. *Protoplasma*, **68**, 107-133.

Hargin, K.D. and Morrison, W.R. (1980) The distribution of acyl lipids in the germ, aleurone, starch and nonstarch endosperm of 4 wheat varieties. *J. Sci. Food Agric.*, **31**, 877-888.

Huang, B.-Q. and Sheridan, W.F. (1994) Female gametophyte development in maize: microtubular organization and embryo sac polarity. *Plant Cell*, **6**, 845-861.

Hueros, G., Varotto, S., Salamini, F. and Thompson, R.D. (1995) Molecular characterization of *BET1*, a gene expressed in the endosperm transfer cells of maize. *Plant Cell*, **7**, 747-757.

Hueros, G., Gomez, E., Cheikh, N., *et al.* (1999b) Identification of a promotor sequence from the *BETL1* gene cluster able to confer transfer-cell-specific expression in transgenic maize. *Plant Physiol.*, **121**, 1143-1152.

Hueros, G., Royo, J., Maitz, M., Salamini, F. and Thompson, R.D. (1999a) Evidence for factors regulating transfer cell-specific expression in maize endosperm. *Plant Mol. Biol.*, **41**, 403-414.

Jacobsen, J.V., Knox, R.B. and Pyliotis, N.A. (1971) The structure and composition of aleurone grains in the barley aleurone layer. *Planta*, **101**, 189-209.

Jacobsen, J.V., Pressman, E. and Pyliotis, N.A. (1976) Gibberellin-induced separation of cells in isolated endosperm of celery seed. *Planta*, **129**, 113-122.

Johri, B.M., Ambegaokar, K.B. and Srivastava, P.S. (eds.) (1992) *Comparative Embryology of Angiosperms*, Springer-Verlag, Berlin.

Jones, R.L. (1969) The fine structure of barley aleurone cells. *Planta*, **85**, 359-375.

Juliano, B.O. and Bechtel, D.B. (1985) The rice grain and its gross composition, in *Rice: Chemistry and Technology*, 2nd edn (ed. B.O. Juliano) American Association of Cereal Chemistry St. Paul, MN, pp. 17-57.

Kalla, R., Shimamoto, K., Potter, R., Nielsen P.S., Linnestad, C. and Olsen, O.-A. (1994) The promoter of the barley aleurone-specific gene encoding a putative 7 kDa lipid transfer protein confers aleurone cell-specific expression in transgenic rice. *Plant J.*, **6**, 849-860.

Kiyosue, T., Ohad, N., Yadegari, R., *et al.* (1999) Control of fertilization-independent endosperm development by the *MEDEA* polycomb gene in *Arabidopsis*. *Proc. Natl. Acad. Sci. USA*, **96**, 4186-4191.

Kyle, D.J. and Styles, E.D. (1977) Development of aleurone and sub-aleurone layers in maize. *Planta*, **137**, 185-193.

Linnestad, C., Doan, D.N.P., Brown, R.C., *et al.* (1998) Nucellain, a barley homolog of the dicot vacuolar-processing protease, is localized in nucellar cell walls. *Plant Physiol.*, **118**, 1169-1180.

Lloyd, C.W. and Traas, J.A. (1988) The role of F-actin in determining the division plane of carrot suspension cells. Drug studies. *J. Cell Sci.*, **102**, 211-221.

Lloyd, C.W., Venverloo C.J., Goodbody, K.C. and Shaw, P.J. (1992) Confocal laser microscopy and three-dimensional reconstruction of the nucleus-associated microtubules in the division plane of vacuolated plant cells. *J. Microsc.*, **166**, 99-109.

Lopes, M.A. and Larkins, B.A. (1993) Endosperm origin, development, and function. *Plant Cell*, **5**, 1383-1399.

Lu, P., Porat, R., Nadeau, J.A. and O'Neill, S.D. (1996) Identification of a meristem L1 Layer-specific gene in *Arabidopsis* that is expressed during embryonic pattern formation and defines a new class of homeobox genes. *Plant Cell*, **8**, 2155-2168.

Luo, M., Bilodeau, P., Dennis, E.S., Peacock, W.J. and Chaudhury, A. (2000) Expression and parent-of-origin effects for *FIS2*, *MEA* and *FIE* in the endosperm and embryo of developing *Arabidopsis* seeds. *Proc. Natl. Acad. Sci. USA*, **97**, 10637-10642.

Maheshwari, P. (1950) *An Introduction to the Embryology of Angiosperms*, McGraw-Hill, New York.

Mansfield, S.G. (1994) Endosperm development, in *Arabidopsis: An Atlas of Morphology and Development* (ed. J. Bowman), Springer-Verlag, New York, pp. 385-397.

Mansfield, S.G. and Briarty, L.G. (1990) Endosperm cellularization in *Arabidopsis thaliana* (L.) *Ara. Info. Serv.*, **27**, 65-72.

Mansfield, S.G., Briarty, L.G. and Erni, S. (1991) Early embryogenesis in *Arabidopsis thaliana*. I. The mature embryo sac. *Can. J. Bot.*, **69**, 447-460.

Miller, M.E. and Chourey, P.S. (1992) The maize invertase-deficient *minature-1* seed mutation is associated with aberrant pedicel and endosperm development. *Plant Cell*, **4**, 297-305.

Morrison, I.N. and O'Brien, T.P. (1976) Cytokinesis in the developing wheat grain; division with and without a phragmoplast. *Planta*, **130**, 57-67.

Morrison, I.N., Kuo, J. and O'Brien, T.P. (1975) Histochemistry and fine structure of developing wheat aleurone cells. *Planta*, **123**, 105-116.

Müntz, K. (1998) Deposition of storage proteins. *Plant Mol. Biol.*, **38**, 77-99.

Neubauer, B.F. (1971) The development of the achene of *Polygonum pensylvanicum*: embryo, endosperm, and pericarp. *Am. J. Bot.*, **58**, 655-664.

Nguyen, H., Brown, R.C. and Lemmon, B.E. (2000) The specialized chalazal endosperm in *Arabidopsis thaliana* and *Lepidium virginicum* (Brassicaceae). *Protoplasma*, **212**, 99-110.

Nguyen, H., Brown, R.C. and Lemmon, B.E. (2001) Patterns of cytoskeletal organization reflect distinct developmental domains in endosperm of *Coronopus didymus* (Brassicaceae). *Int. J. Plant Sci.*, **162**, 1-14.

Nonogaki, H., Gee, O.H. and Bradford, K.J. (2000) A germination-specific endo-β-mannanase gene is expressed in the micropylar endosperm cap of tomato seeds. *Plant Physiol.*, **123**, 1235-1245.

Ohad, N., Yadegari, R., Margossian, L., *et al.* (1999) Mutations in FIE, a WD polycomb group gene, allow endosperm development without fertilization. *Plant Cell*, **11**, 407-415.

Olsen, O.-A., Potter, R.H. and Kalla, R. (1992) Histo-differentiation and molecular biology of developing cereal endosperm. *Seed Sci. Res.*, **2**, 117-131.

Olsen, O.-A., Brown, R.C. and Lemmon, B.E. (1995) Pattern and process of wall formation in developing endosperm. *BioEssay*, **17**, 803-812.

Olsen, O.-A., Lemmon, B. and Brown, R. (1998) A model for aleurone development. *Trends Plant Sci.*, **3**, 168-169.

Olsen, O.-A., Linnestad C. and Nichols, S.E. (1999) Developmental biology of the cereal endosperm. *Trends Plant Sci.*, **4**, 253-257.

Olsen, O.-A. (2001) Endosperm development: cellularization and cell fate specification. *Ann. Rev. Plant Phys. Plant Mol. Biol.*, **52**, 233-267.

Olson, A.R. (1981) Embryo and endosperm development in ovules of *Papaver nudicaule* after *in vitro* placental pollination. *Can. J. Bot.*, **59**, 1738-1748.

Opsahl-Ferstad, H.-G., Le Deunff, E., Dumas, C. and Rogowsky, P.M. (1997) *ZmEsr*, a novel endosperm-specific gene expressed in a restricted region around the maize embryo. *Plant J.*, **12**, 235-246.

Otegui, M. and Staehelin, L.A. (2000) Syncytial-type cell plates: a novel kind of cell plate involved in endosperm cellularization of *Arabidopsis*. *Plant Cell*, **12**, 933-947.

Otegui, M., Lima, C., Maldonado, S. and de Lederkremer, R.M. (1999) Development of the endosperm of *Myrsine laetevirens* (Myrsinaceae). I. Cellularization and deposition of cell-wall storage carbohydrates. *Int. J. Plant Sci.*, **160**, 491-500.

Pacini, E., Simoncioli, C. and Cresti, M. (1975) Ultrastructure of nucellus and endosperm of *Diplotaxis erucoides* during embryogenesis. *Caryologia*, **28**, 525-538.

Pickett-Heaps, J.D., Gunning, B.E.S., Brown, R.C., Lemmon, B.E. and Cleary, A.L. (1999) The cytoplast concept in dividing plant cells: cytoplasmic domains and the evolution of spatially organized cell division. *Am. J. Bot.*, **86**, 153-172.

Pearson, H. (1909) Further observations on *Welwitschia*. *Phil. Trans. R. Soc. London*, B, **200**, 331-402.

Pirrotta, V. (1997) PcG complexes and chromatin silencing. *Curr. Opin. Genet. Dev.*, **7**, 249-258.

Pirrotta, V. (1998) Polycombing the genome: PcG, trxG, and chromatin silencing. *Cell*, **93**, 333-336.

Randolph, L.F. (1936) Developmental morphology of the caryopsis in maize. *J. Agric. Res.*, **53**, 881-916.

Reiser, L. and Fischer, R.L. (1993) The ovule and the embryo sac. *Plant Cell*, **5**, 1291-1301.

Rojo, E., Gilmore, C.S., Somerville, C.R. and Raikel, N.K. (2000) Characterization of *mangled*, a gene required for normal mitosis and cell growth in *Arabidopsis* ovules. *American Society for Cell Biology Late Abstract Poster Session*, L51.

Russell, S.D. (1979) Fine structure of megagametophyte development in *Zea mays*. *Can. J. Bot.*, **57**, 1093-1110.

Samuels, A.L., Giddings, T.H. Jr and Staehelin, L.A. (1995) Cytokinesis in tobacco BY-2 and root tip cells: a new model of cell plate formation in higher plants. *J. Cell Biol.*, **130**, 1345-1357.

Sargant, E. (1900) Recent work on the results of fertilization in angiosperms. *Ann. Bot.*, **14**, 689-712.

Schel, J.H.N., Kieft, H. and van Lammeren, A.A.M. (1984) Interactions between embryo and endosperm during early developmental stages of maize caryopses (*Zea mays*). *Can. J. Bot.*, **62**, 2842-2853.

Schneitz, K., Hulskamp, M. and Pruitt, R.E. (1995) Wild-type ovule development in *Arabidopsis thaliana*: a light microscope study of cleared whole-mount tissue. *Plant J.*, **7**, 731-749.

Schulz, P. and Jensen, W.A. (1971) *Capsella* embryogenesis: the chalazal proliferating tissue. *J. Cell Sci.*, **8**, 201-227.

Schulz, P. and Jensen, W.A. (1974) *Capsella* embryogenesis: the development of the free nuclear endosperm. *Protoplasma*, **80**, 183-205.

Schulz, P. and Jensen, W.A. (1977) Cotton embryogenesis: the early development of the free nuclear endosperm. *Am. J. Bot.*, **64**, 384-394.

Scott, R.J., Spielman, M., Bailey, J. and Dickinson, H.G. (1998) Parent-of-origin effects on seed development in *Arabidopsis thaliana*. *Development*, **125**, 3329-3341.

Simoncioli, C. (1974) Ultrastructural characteristics of *Diplotaxis erucoides* (L.) DC. Suspensor. *Giorn. Bot. Ital.*, **108**, 175-189.

Singh, H. (1978) Embryology of gymnosperms, in *Handbuch der Pflanzenanatomie* 10. Borntraeger, Berlin, pp. 110-287.

Sinnot, E.W. and Bloch, R. (1941) Division in vacuolate plant cells. *Am. J. Bot.*, **28**, 225-232.

Sorensen M.B., Chaudhury A.M., Robert H., Bancharel E. and Berger, F. (2001) Polycomb group genes control pattern formation in plant seed. *Curr. Biol.*, **11**, 277-281.

Swanson, S.J., Bethke, P.C. and Jones, R.L. (1998) Barley aleurone cells contain two types of vacuoles: characterization of lytic organelles by use of fluorescent probes. *Plant Cell*, **10**, 685-698.

Taylor, T.N. and Taylor, E.L. (1993) *The Biology and Evolution of Fossil Plants*, Prentice Hall, Englewood Cliffs, N.J.

Thomas, E.N. (1907) Some aspects of 'double fertilization' in plants. *Sci. Prog.*, **1**, 420-426.

Vandendries, R. (1909) Contribution à l'histoire du développement des crucifères. *Cellule*, **25**, 415-459.

Van Lammeren, A.A.M. (1988) Structure and function of the microtubular cytoskeleton during endosperm development in wheat: an immunofluorescence study. *Protoplasma*, **146**, 18-27.

Vielle-Calzada, J.-P., Thomas, J., Spillane, C., Collucio, A., Hoeppner, M.A. and Grossniklaus, U. (1999) Maintenance of genomic imprinting at the *Arabidopsis medea* locus requires zygotic *DDM1* activity. *Genes Dev.*, **13**, 2971-2982.

Vijayraghavan, M.R. and Prahabkar, K. (1984) The endosperm, in *Embryology of Angiosperms* (ed. B.M. Johri), Springer-Verlag, Berlin, pp. 319-376.

Vijayraghavan, M.R., Prabhakar, K. and Puri, B.K. (1984) Histochemical, structural and ultrastructural features of endosperm in *Alyssum maritimum* Lam. *Acta Bot. Neerl.*, **33**, 111-122.

Vinkenoog, R., Spielman, M., Adams, S., Fischer, R.L., Dickinson, H.G. and Scott, R.J. (2000) Hypomethylation promotes autonomous endosperm development and rescues postfertilization lethality in *fie* mutants. *Plant Cell*, **12**, 2271-2282.

Walbot, V. (1994) Overview of key steps in aleurone development, in *The Maize Handbook* (ed. M. Freeling and V. Walbot), Springer-Verlag, New York, pp. 78-80.

Wang, H.L., Offler C.E. and Patrick J.W. (1994) Nucellar projection transfer cells in the developing wheat grain. *Protoplasma*, **182**, 39-52.

Wang, M., Oppedijk, B.J., Caspers, M.P.M., *et al.* (1998) Spatial and temporal regulation of DNA fragmentation in the aleurone of germinating barley. *J. Exp. Bot.*, **49**, 1293-1301.

Webb, M.C. and Gunning, B.E.S. (1990) Embryo sac development in *Arabidopsis thaliana* I. Megasporogenesis, including the microtubular cytoskeleton. *Sex. Plant Reprod.*, **3**, 244-256.

Webb, M.C. and Gunning, B.E.S. (1991) The microtubular cytoskeleton during development of the zygote, proembryo and free-nuclear endosperm in *Arabidopsis thaliana* (L.) Heynh. *Planta*, **184**, 187-195.

Williams, E.G. (1987) Interspecific hybridization in pasture legumes, in *Plant Breeding Reviews* (ed. J. Janik), VanNostrand Reinhold, New York, pp. 237-305.

Wolf, M.J., Cutler, H.C., Zuber, M.S. and Khoo, U. (1972) Maize with multilayer aleurone of high protein content. *Crop. Sci.*, **12**, 440-442.

XuHan, X. (1995) Seed development in *Phaseolus vulgaris* L., *Populus nigra* L., and *Ranunculus scleratus* L. with Special Reference to the Microtubular Cytoskeleton, CIP-Gegevens Koninklijke Bibliotheek, Den Haag, Netherlands

XuHan, X. and van Lammeren, A.A.M. (1993) Microtubular configurations during the cellularization of coenocytic endosperm in *Ranunculus scleratus* L. *Sex. Plant Reprod.*, **6**, 127-132.

XuHan, X. and van Lammeren, A.A.M. (1994) Microtubular configurations during endosperm development in *Phaseolus vulgaris*. *Can. J. Bot.*, **72**, 1489-1495.

Yeung, E.C. and Cavey, M.J. (1988) Cellular endosperm formation in *Phaseolus vulgaris*. I. Light and scanning electron microscopy. *Can. J. Bot.*, **66**, 1209-1216.

Yeung, E.C. and Clutter, M.E. (1979) Embryogeny of *Phaseolus coccineus*: the ultrastructure and development of the suspensor. *Can. J. Bot.*, **57**: 120-136.

Yeung, E.C. and Meinke, D.W. (1993) Embryogenesis in angiosperms: development of the suspensor. *Plant Cell*, **5**, 1371-1381.

Young, T.E., Gallie, D.R. and DeMason, D.A. (1997) Ethylene-mediated programmed cell death during maize endosperm development for wild-type and *shrunken2* genotypes. *Plant Physiol.*, **115**, 737-751.

7 The central role of the ovule in apomixis and parthenocarpy

Anna M. Koltunow, Adam Vivian-Smith,
Matthew R. Tucker and Nicholas Paech

*But for the ovule, the chain of events which initiate the development of
a fruit [and seed] would not take place*

J.P. Nitsch, 1952

7.1 Introduction

In flowering plants, the reproductive lifecycle initiates with a change in meristem growth from an indeterminate vegetative mode to a determinate flowering mode and ends with the production of seed and fruit. Once the flower opens and growth of the floral organs has diminished, further information is necessary to enable the initiation of fruit and seed development. Pollination and subsequent double fertilization provide cues that enable this transition, and the development of fruit and seed is coordinated as the fruit facilitates seed dispersal (figure 7.1).

Double fertilization occurs in the ovule (chapter 5). It involves the fusion of male sperm cells with two specific cells of the female gametophyte or embryo sac, the egg cell and the central cell, which then give rise to the embryo and endosperm of the seed respectively. If fertilization does not occur within a specific window of time referred to as female receptivity, the entire flower undergoes senescence (figure 7.1). This contrasts with the selective abscission of floral structures that do not contribute to the fruit after fertilization.

Cues for seed initiation are considered to emanate from the ovule after fertilization, and stimuli for fruit development may come from three sources; the vegetative part of the plant, pollen or the ovules (Nitsch, 1952). Critical questions pertain to the identity of the signals generated by pollination and fertilization that directly promote seed and fruit development, and how post-fertilization information is integrated to enable the selective abscission of floral structures while promoting the growth and further differentiation of others. These questions can be addressed by analysing plants where seed and fruit development is naturally uncoupled from fertilization, as occurs in apomictic and parthenocarpic species, in addition to mutants that exhibit fertilization-independent growth characteristics.

Both apomixis and parthenocarpy are genetically controlled traits that have agronomic importance. Apomicts initiate embryogenesis without fertilization.

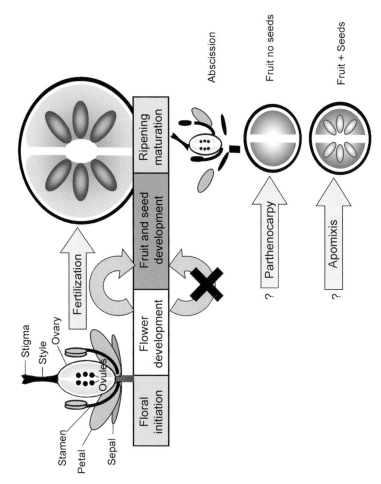

Figure 7.1 Fertilization-dependent and fertilization-independent reproductive events in plants.

The embryo is of a maternal genotype because it originates from somatic cells of the ovule or from cells within an embryo sac that forms without meiosis. Fruit development is coordinated with apomictic seed development. Apomixis is potentially a useful trait because it maintains and fixes a genotype, but it is not prevalent in agricultural crops. Engineering of this trait in crop species could be used to fix hybrid vigour, which would enable the propagation of hybrids over many generations by seed (Koltunow, 1993; Koltunow *et al.*, 1995a). By contrast, plants that form fruit in the absence of fertilization and seed initiation are termed parthenocarpic. As parthenocarpy results in seedless fruit, it is a sought-after trait in fruit and vegetable breeding programmes (Varoquaux *et al.*, 2000).

A number of reviews have discussed the agronomic benefits of engineering apomixis and parthenocarpy in crops (Hanna and Bashaw, 1987; Koltunow *et al.*, 1995a; Grossniklaus *et al.*, 1998a, 2001; Van Baarlen *et al.*, 2000; Varoquaux *et al.*, 2000). In this chapter, we consider the processes of apomixis and partheno-carpy, and examine mutants in which seed and fruit development is uncoupled from fertilization to understand the factors regulating the initiation of these processes. The data we discuss uncover three instrumental factors that regulate the transition between flower maturity and fruit and seed set. First, there are molecular mechanisms in place that actively prevent the development of the carpel into a fruit and the ovule into a seed in the absence of fertilization. Sec-ondly, plant hormones are required for seed and fruit development, and they are also involved in long-range signalling processes within the flower to orchestrate the events of floral abscission. Finally, signalling pathways involving the ovules are essential for fertilization-dependent and also fertilization-independent fruit set. The latter point supports the early hypothesis of Nitsch (1952), suggesting that events occurring in the ovule are critical for both seed and fruit initiation.

7.2   Development of floral, seed and fruit structures

The genetic control of flower development has been described in preceding chapters. Angiosperm flowers are diverse in form but have common structural elements that originate from floral organ primordia with specified floral organ identity. Similarly, the components of a seed are derived from common ovule tissues when different species are compared. For example, the ovule integuments give rise to the seed coat and the embryo and endosperm are generally formed within a gametophytic structure as described in chapters 8 and 9. By contrast, the structures that make up the diverse array of fruits do not always have a common origin and they differ extensively between species.

Fruits do not necessarily form entirely from the gynoecial tissues of the carpel. Other floral and seed structures such as bracts, receptacles, the aril of the seed and even the complete inflorescence, can contribute to the structure that

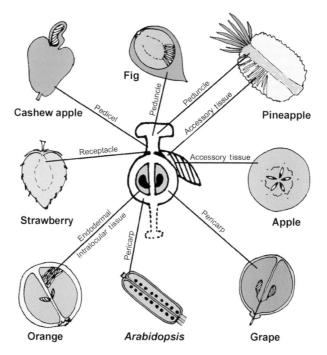

Figure 7.2 Diversity of floral structures that contribute to fruit form. The diagram was redrawn from Coombe (1975).

we call a fruit (Hulme, 1970, 1971; Coombe, 1975; figure 7.2). The form of an angiosperm fruit is dependent upon the contributing number and type of floral organ components, the position of the contributing organs and how the different tissues within them grow and differentiate (Coombe, 1975; figure 7.2). The number and type of cell divisions and the relationship between cell division and expansion across a range of different tissues are therefore critical in determining the final shape and size of the fruit (Lyndon, 1990).

7.3   Evolutionary origin of ovules, seeds and fruits

A deeper understanding of angiosperm fruit and seed development may be attained through an examination of the evolutionary origins of the various structures contributing to the reproductive programme. Angiosperms are defined by their reproductive biology. Unlike other groups of existing seed plants, they develop ovules that are completely enclosed within a carpel, and during seed development they produce an embryo-nourishing endosperm (Friedman, 2001; Sun *et al.*, 1998; Theissen *et al.*, 2000).

Plants first developed seed some 350 million years ago (Mya), well before the advent of angiosperms. Flowers and fruits developed later as a way to shield seed from predation, protect against self-fertilization, attract pollinators and aid in seed dispersal (Theissen *et al.*, 2000; Dilcher, 2000). Although the ovule is one of the last structures to form during the temporal development of the modern angiosperm flower, it was probably the first floral organ to arise in evolution. Unlike the flower, which is generally considered homologous to a compressed, determinate shoot, the ovule does not have an obvious direct counterpart with any vegetative organ (Pennel and Roberts, 1990; Bowman *et al.*, 1999).

Current concepts of the evolutionary origin of the ovule are based on the telomic theory of megasporophyll development (Zimmermann, 1952). Telomes are simple vascularized axes that composed the most basic units of early vascular plants. The fertile telomes consisted of a spore-forming structure (sporangium) borne on a vascular axis: the sporangiophore (a relic stem). Ovules developed as a result of the reduction and fusion of a series of telome axes, some of which bore sporangia at their tips. Reduction of fertile telomes placed the sporangium at the base of accessory telomes and fusion of these produced the integument (Herr, 1995).

Analysis of ovules of Palaeozoic plants, existing seed plants and *Arabidopsis* ovule mutants by Herr (1995) suggests that the nucellus of the ovule, where the progenitor cells of the female gametophyte arise, represents a complex of the sporangiophore and its terminal megasporangium. In addition, genetic evidence suggests that those telomes that fused to form the integument were at least partially fertile, as the modern angiosperm integument still retains a measure of reproductive fertility. Evolution of a megagametophyte enclosed within the sporophytic tissue of the ovule involved changes enabling nutrient transfer to the developing gametophyte. These changes probably occurred in a single phylogenetic line beginning in a population of progymnosperms from the late Devonian period (370 Mya) (Herr, 1995). Unanswered questions relate to the origin of the endosperm and to whether a 4-nucleate embryo sac or an 8-nucleate embryo sac is the ancestral condition in angiosperms (Friedman, 2001). Two main views have been advanced concerning the former. In one hypothesis, it has been postulated that the endosperm arose from an altruistic twin embryo that assumed a growth support role. In the other, the endosperm is thought to result from the extended development of the megagametophyte, which is thought to be promoted when the central cell is fertilized by the second male gamete (Friedman, 1994, 2001).

At some stage in evolution, telomes adjacent to those that went on to form the ovule became fused and flattened to form a leaf-like structure; the predecessor of the angiosperm carpel. The earliest fossil evidence of a closed carpel dates to 142 million years ago and is that of *Archaefructus* from China (Sun *et al.*, 1998). Carpels from these primitive angiosperms resemble longitudinally folded leaves with unfused margins bearing two to four ovules. According to Dilcher (2000), evolution of a closed carpel coincided with the evolution of the bisexual flower; as male and female organs were brought into proximity, the need to protect against self-fertilization necessitated the evolution of this mechanical barrier.

The MADS-Box gene *AGAMOUS* (*AG*) is required for carpel and ovule identity and is also expressed in developing fruits (Yanofsky *et al.*, 1990; Bowman *et al.*, 1991, 1999; Western and Haughn, 1999). It is present in the most primitive angiosperms and in gymnosperms, but is absent from ferns (Theissen *et al.*, 2000). Conifer homologues of *Arabidopsis AG* also have conserved gene function (Theissen *et al.*, 2000), indicating that the ancestral function of *AG* was probably to specify the primary identity of ovule-bearing organs. During the evolution of angiosperm fruits, *AG* may have been co-opted into new functions to regulate carpel structure and then, subsequently, in the downstream events of fruit development (Thiessen *et al.*, 2000; Bowman *et al.*, 1999). Alvarez and Smyth (1999) provided evidence from *Arabidopsis* to support the theory that genetic networks were recruited during evolution to convert primitive leaf-like sporophyllous organs into the closed angiosperm carpel. Loss-of-function mutations in the *Arabidopsis* genes *AG*, *PISTILLATA*, *APETALA2*, *CRABS CLAW* and *SPATULA* together resulted in the conversion of the carpel into leaf-like organs containing vestigial ovule primordia at their margins.

A large radiation of angiosperm fruit types occurred during the Paleocene and Eocene (34–65 Mya), and this event is associated with coincident radiation in rodents and birds. Fruit evolution accelerated during this period as angiosperms adopted biotic dispersal of their seed (Dilcher, 2000). The range of fleshy, nutritious fruits consumed by megafauna today is possibly a product of these earlier co-evolutionary events. However, the molecular processes responsible for generating the diversity of fruit forms seen in modern angiosperms have not yet been elucidated. Nitsch (1952) argued that the common factor underlying the varied structures that we call fruit is a physiological one with a hormonal basis. Plant hormones do play a significant role in seed and fruit development (Gillaspy *et al.*, 1993) and this will be discussed further in the next section. Considering the evolutionary evidence, we propose that the developmental transition from a flower to a fruit might be marked by linkage between genes like *AG* and others regulating flower development, with genes involved in the biosynthesis, perception and transduction of hormone signals.

7.4 Plant hormones facilitate long-range communication within and between floral organs during fruit and seed initiation

In the majority of angiosperm species, ovule and female gametophyte development is completed prior to pollination. Auxin appears to be important for pistil development in a number of plants (Nitsch, 1952) including orchid (O'Neill,

1997) and *Arabidopsis* (Nemhauser *et al.*, 2000). De Martinis and Mariani (1999) have shown that ethylene is required for the prepollination development of the ovule and female gametophyte in tobacco. Silencing of the ethylene-forming enzyme ACC (1-aminocyclopropane-1-carboxylate) oxidase or the inhibition of its action by silver thiosulfate causes ovule arrest, and mature embryo sacs do not form because megasporocytes are unable to initiate and complete meiosis. Ethylene might also be required for the prepollination development of ovules and gametophytes in *Arabidopsis*, because the *constitutive triple response 1-1* (*ctr1-1*) ethylene-response mutant is partially defective in female gametophyte development (Drews *et al.*, 1998; Wang and Sundaresan, 2000). Pollination is required to directly induce growth and differentiation of the ovule and female gametophyte in orchid. Ethylene is induced in orchid ovaries soon after pollination; however, a precise role for ethylene in orchid ovule development has yet to be established (Zhang and O'Neill, 1993; Wang and Sundaresan, 2000).

At flower maturity, pollination of the stigma also induces a series of signals that are amplified and transmitted to the other floral organs to aid in the senescence of structures not required for seed and fruit development (O'Neill, 1997; O'Neill and Nadeau, 1997). As such, seed and fruit become the major sinks for photosynthates translocated from the leaves. The primary cues for selective floral abscission are derived from the pollen, but the pollen signal molecule is not known for many plants. In orchid, auxin induces a rapid increase in ethylene production in floral organs to effect perianth abscission, but the molecular details of this long-range signal transduction chain remain unknown. Ethylene is a known causal agent for organ abscission in a range of species (O'Neill, 1997; Nadeau and O'Neill, 1997). Cues for seed and fruit initiation must be independent of, or negate, events leading to senescence in the other floral organs, so that fruit and seed development can be sustained in a coordinated manner.

Significant increases and alterations in plant hormone levels occur in developing seeds and in the ovary soon after fertilization. The seeds are sites of phytohormone synthesis, and the endosperm in particular contains high levels of auxin, gibberellins and cytokinin (Van Overbeek *et al.*, 1941; Nitsch, 1952; Jensen and Bandurski, 1994; Swain *et al.*, 1995a). Fruit growth, shape and form is also known to be modified by differences in seed genotype (Denny, 1992) and seed number (Sedgley and Griffin, 1989). Gustafson (1936, 1939, 1942) and Nitsch (1950, 1952) were among the first to observe that exogenous hormone application to pistil tissues induced seedless fruit growth and circumvented the requirement for fertilization. This has since been demonstrated in a wide range of species (Schwabe and Mills, 1981). Gustafson and Nitsch both advocated that increased phytohormone levels in seed and ovary tissues after fertilization represented the physiological basis for fruit formation. Subsequent research approaches consequently focused upon identifying the types and levels of endogenous plant hormones involved in fruit growth and development, (Garcia-Martinez *et al.*, 1987, 1991, 1997; Fos *et al.*, 2000; Mehouachi *et al.*, 2000).

The identification of genes involved in plant hormone biosynthesis and perception, and in some cases the genetic and molecular analysis of their function, is providing a deeper understanding of their role in fruit and seed development. Such studies have confirmed the functional requirement for gibberellin biosynthesis in fruit growth in some species (Barendse *et al.*, 1986; Swain *et al.*, 1995b; Vivian-Smith and Koltunow, 1999) and defined the role of ethylene biosynthesis and perception in fruit maturation and ripening with useful agronomic consequences (Gillaspy *et al.*, 1993; Zarembinski and Theologis, 1994; Johnson and Ecker, 1998).

The elevation of auxin in the placenta and ovules of tobacco, tomato and eggplant by a transgenic approach directly induced seedless fruit development without fertilization (Rotino *et al.*, 1997; Ficcadenti *et al.*, 1999). Apart from providing a means for parthenocarpic fruit formation, these data are developmentally informative. They identify the ovule and placental tissues as being important for fruit initiation, quite independently from stimuli arising from pollination. They also imply that auxin might be involved in long-range signalling processes between the ovule, placenta and the other ovary tissues to stimulate fruit development. It is not certain from the available data whether auxin itself or a secondary molecule, involved in auxin signal transduction, is transported to the ovary. The role of auxin in long-distance signal transduction and patterning in plant development has been reviewed recently (Berleth and Mattsson, 2000; Geldner *et al.*, 2000; Berleth and Sachs, 2001).

Collectively, the data summarized in this section lead to the conclusion that once a flower is formed, the balance between floral organ senescence and fertilization-induced seed and fruit development involves different plant hormones and long-range communication events between different floral organs. In parthenocarpic and also apomictic species, the regulation of floral organ senescence and fruit and seed development can be independent of pollination and fertilization. What does the analysis of these processes tell us about molecular factors regulating the transition between the mature flower and the initiation of fruit and seed development? Apomixis and parthenocarpy will be considered separately in the sections below.

7.5   Apomixis—a seed by any means?

Apomixis is an asexual mode of reproduction that results in the formation of seeds containing embryos with a maternal genotype. However, the general association of the term 'seeds without sex' with apomixis is not strictly correct, because the majority of apomicts require fertilization of the central cell (pseudogamy) to initiate endosperm development and form viable seeds. One striking feature of apomixis is the enormous variability found in this reproductive process. Because of this, apomictic processes have been classified in a number of ways. There is continued debate concerning the validity of particular classification schemes and confusion arises because of the often complex nomenclature. The types of plants that apomixis naturally occurs in have been reviewed by Carman (1997) and here we primarily focus on the developmental aspects of apomixis.

In contrast to apomixis, the events of sexual reproduction in angiosperms occur in an ordered sequence (figure 7.3). Female gametophyte development requires meiosis of the megaspore mother cell (MMC), spore selection, mitosis and cellular differentiation (figure 7.3). Double fertilization of the egg cell and central cell of the embryo sac is then required to give rise to the embryo and endosperm. Apomicts omit critical events from this normally ordered pathway because they avoid meiosis, embryo development is autonomous, and endosperm formation may or may not require fertilization (figure 7.3). These three features are often referred to as the components of apomixis. Polyploidy is very common among natural apomicts but is not an essential requirement for apomictic seed production (Bicknell, 1997; Savidan, 2000).

Sexual reproduction and apomixis are not mutually exclusive reproductive modes. The majority of natural apomicts are termed facultative because they produce seeds from both sexual and apomictic processes. In apomictic plants, sexual and apomictic processes can occur simultaneously and, depending on the mechanism, functionally coexist in an ovule (Koltunow, 1993). Interestingly, apomixis does not initiate in all ovules of facultative plants, and the events of apomixis may or may not influence the events of sexual reproduction occurring in adjacent cells. Environmental parameters also influence the degree of apomictic seed set (Koltunow, 1993; Savidan, 2000). Thus apomixis is not a predictable process in facultative species and is highly stochastic in nature (Koltunow *et al.*, 1998, 2000).

7.5.1   Types of apomixis and their temporal initiation during ovule development

A commonly accepted classification scheme divides apomictic processes into two types. The first, termed gametophytic apomixis, is defined by the mitotic formation of an embryo sac. Embryogenesis occurs autonomously inside the unreduced embryo sac and endosperm develops with or without fertilization of the central cell. The second type of apomixis in this simple scheme is called adventitious embryony, and it is defined by the direct formation of an embryo inside the ovule but external to an embryo sac structure.

Two forms of gametophytic apomixis are also defined, dependent upon the origin of the cell initiating unreduced embryo sac formation (figure 7.3). The

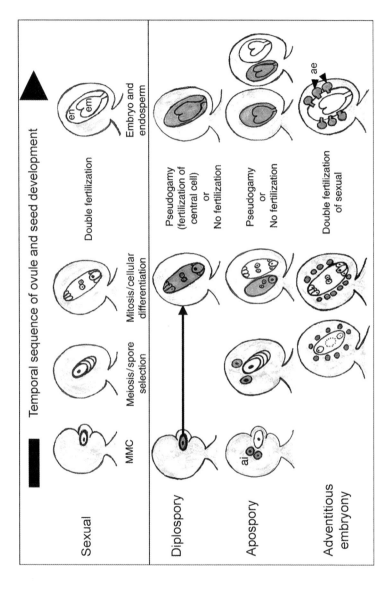

Figure 7.3 A summary of apomictic mechanisms relative to the temporal sequence of sexual reproduction in flowering plants. Sexual reproductive events are unshaded and the apomictic reproductive events are shaded grey. ae, adventitious embryo(s); ai, aposporous initial(s); em, embryo; en, endosperm; MMC, megaspore mother cell.

cell initiating diplospory has the positional location of the MMC. In some species the MMC initiates meiosis, then aborts the process and proceeds to initiate the mitotic events of gametophyte development. Other diplosporous species show no histological evidence of entry into meiosis. The absence of a functional MMC in ovules undergoing diplospory means that by default the sexual process is excluded from that ovule. However, some MMCs in plants with facultative diplospory retain the capacity for sexual reproduction. This implies that either the MMC changes fate to a diplosporous pathway, or it is predetermined to do so as it differentiates in the same location as the MMC. In either case, it is possible that positional signals specifying cell identity in this region are altered in diplosporous plants. Analysis of callose in the walls of cells initiating diplospory shows reduced deposition in the cell wall, consistent with the observation that callose is absent from cells undergoing the mitotic events of embryo sac formation in sexual plants (Carman *et al.*, 1991). Other molecular markers are not yet available for examining cell identity and specification events in diplosporous plants.

The initiation of unreduced embryo sac development from cells within the ovule other than the MMC is defined as apospory (figure 7.3). The retention of a functional MMC means that sexual and aposporous processes can initiate and coexist in an individual ovule. Apospory initiates from a field of cells close to those undergoing sexual development. Apospory has been observed to initiate from cells in the nucellus, the nucellar epidermis and the integument. Numerous cells can enlarge and immediately begin the mitotic events of aposporous embryo sac formation. The cells initiating apospory, like those undergoing megagametogenesis in sexual plants, lack callose in their walls (Peel *et al.*, 1997; Tucker *et al.*, 2001).

The timing of aposporous embryo sac initiation, although stochastic, can preferentially occur at particular times relative to sexual gametophyte development. Initiation prior to MMC meiosis can lead to an early arrest of sexual development. By contrast, the later apospory initiates, the greater is the chance that both types of embryo sacs will coexist. There are, however, exceptions where aposporous and sexual embryo sacs coexist in the ovule irrespective of when apospory begins. Some of the factors leading to the displacement of the sexual process have been examined in two aposporous *Hieracium* species that predominantly initiate apospory at distinct times relative to the sexual pathway; before and soon after meiosis (Tucker *et al.*, 2001). The mode of displacement differs in the two types of plants. Therefore, it was proposed that displacement may relate to the timing of aposporous initial formation and the developmental events occurring within the ovule at that point in time (Tucker *et al.*, 2001).

Diplosporous and also aposporous species can form the 8-nucleate Polygonum-type of embryo sac that is predominantly found in sexual species (Willemse and van Wendt, 1984). In some accessions of aposporous *Hieracium*, the mode of embryo sac formation can involve an amalgamation of multiple embryo sacs containing few nuclei into a single disorganized structure with a crude egg apparatus, quite distinct from the Polygonum-type embryo sac observed in the sexual plant (Koltunow *et al.*, 1998). Aposporous grasses often form a diagnostic 4-nucleate (Panicum-type) embryo sac (Dujardin and Hanna, 1984). Thus, the type of embryo sac formed in gametophytic apomicts is apparently not crucial for the subsequent events of seed development. Variant embryo sacs formed during apomixis do, however, maintain a structural polarity as autonomous embryo development is often favoured at the pole containing cells resembling an egg apparatus.

Embryo formation is not restricted to one cell type in gametophytic apomicts and has been observed to initiate from egg cells, synergids, antipodals and combinations of all these cell types. Multiple embryo formation (polyembryony) is common but not universal in plants exhibiting gametophytic apomixis. The development of adventitious embryos, like the initiation of autonomous embryos inside the embryo sacs of gametophytic apomicts, generally occurs later in the temporal events of ovule development. Adventitious embryos have been observed to initiate from the nucellus, and within the integuments adjacent to a mature fertilized embryo sac (Koltunow, 1993; Savidan, 2000; figure 7.3). They can be quite numerous in species such as citrus (Koltunow *et al.*, 1995b). If the reduced embryo sac is not fertilized, the embryos may consume the nucellus or integument cells, or the embryos can arrest during development. These embryos can be rescued in tissue culture, indicating that their potential to develop to maturity may be limited only by nutrient supply (Koltunow *et al.*, 1995b).

Having made the above generalizations about diplospory, apospory and adventitious embryony, it is important to note that these modes of apomixis are not mutually exclusive. These processes can occur in various combinations, together with sexual reproduction, within an ovule or among the ovules of an individual apomictic plant. All three processes can occur in *Beta* species (Jassem, 1990) and in members of the Rosaceae (Nybom, 1988). In aposporous grasses and *Hieracium*, adventitious embryony can occur in addition to apospory (Koltunow *et al.*, 2000). In citrus, adventitious embryony is the common mode of apomixis, yet the formation of unreduced gametophytes has also been documented (Naumova, 1993).

Given that apomixis is genetically controlled, two possibilities might explain these observations. Different genes might confer diplospory, apospory and adventitious embryony, and plants with multiple apomictic modes might contain the composite mix of requisite genes. Alternatively, a common gene or set of genes might confer the potential for cells to initiate and undergo apomixis. In this second possibility, other genetic factors might influence the final mode of apomixis that we recognize and classify as a specific apomictic mechanism.

7.5.2   Genetic regulation of apomixis

The genetic regulation of apomixis has been examined in apomictic species exhibiting simple 'classical' modes of diplospory, apospory and adventitious embryony. Such genetic studies require that the apomict be crossed as the pollen donor to a sexual female recipient, because reciprocal crosses in which the female recipient is apomictic favour the recovery of maternal apomictic progeny. Genetic studies have been problematic because of the polyploid nature of the plants involved, the requirement for compatible pollination in pseudogamous species, and the methods employed during the assessment of apomixis in progeny (Koltunow *et al.*, 2000; Nogler, 1984a).

Genetic analyses have shown that apomixis is dominant over sexuality. Apomixis is conferred by a monogenic locus in diplosporous *Tripsacum* (Grimanelli *et al.*, 1998), aposporous *Ranunculus* (Nogler, 1984b, 1995), *Panicum* (Savidan, 1982), *Pennisetum* (Ozias-Akins *et al.*, 1998) and *Brachiaria* (do Valle and Savidan, 1996; Pessino *et al.*, 1997). Adventitious embryony (polyembryony) in citrus is under monogenic control and is also dominant over sexual monoembryony (Parlevliet and Cameron, 1959; Iwamasa *et al.*, 1967). The three main components of apomixis in *Hieracium*—apospory, autonomous embryo and endosperm development—are conferred by a monogenic dominant locus (Bicknell *et al.*, 2000). However, in *Taraxacum* independent loci are required for the three components of apomixis: diplospory, autonomous embryo and endosperm development (de Jong *et al.*, 2001).

The observation that components of apomixis have been observed in sexual systems, including the formation of unreduced spores in potato (Jongedijk, 1985) and the formation of autonomous embryos in barley from reduced egg cells (Hagberg and Hagberg, 1980), has led to the suggestion that apomixis and sexuality might share regulatory components (Koltunow, 1993; Grossniklaus *et al.*, 1998a). The dominant nature of apomixis further suggests that it might be caused by a gain of function mutation(s) in a gene(s) involved in sexual reproduction. Alternatively, genes involved in apomixis may induce stochastic misregulation of genes involved in the sexual process.

Do the observed apomixis loci contain one gene or many genes, and do common genes control all apomictic mechanisms? Soon after it was demonstrated in *Panicum* that apospory, autonomous embryogenesis and the rate of facultativeness were transmitted by a single locus, Savidan (1982) proposed that a linkat controlled apomixis within this species, explaining the complexity of the phenotype. A linkat is a chromosomal segment containing genes that remain tightly linked and that contribute to a single phenotypic function (Savidan, 1982). Molecular mapping has since shown that the apomixis locus in diplosporous *Tripsacum* (Grimanelli *et al.*, 2001) and aposporous *Pennisetum* (Ozias-Akins *et al.*, 1998) is located on a chromosomal segment where recombination appears absent. Surprisingly, the current level of marker analysis of the apomixis locus in *Pennisetum* suggests that it has no known counterpart in sexual *Pennisetum* species, and implies the activity of a novel apomixis-specific genomic region. This contradicts the hypothesis of a gain-of-function mutation within a sexual gene, but does not rule out the possibility that the locus directly misregulates genes involved in sexual reproduction.

Synteny among grass species is well known and it has been suggested that apospory may be located in the same area of the genome in aposporous *Pennisetum, Cenchrus* and *Brachiaria* (Savidan, 2000). Further comparative molecular mapping work using probes from other apomictic grasses such as diplosporous *Tripsacum* will provide information concerning the genetic relatedness of various types of apomixis (Savidan, 2000).

The apomixis locus has been shown to be allelic in two *Hieracium* species with quite distinct modes of embryo sac formation (Bicknell *et al.*, 2000). Several unlinked loci also moderate the expression level of apomixis by influencing the timing, the number of cells initiating apomixis and other aspects relating to the mode of its progression (Bicknell *et al.*, 2000; Koltunow *et al.*, 2000). The action of modifier loci has also been implicated from genetic studies of apospory and autonomous embryo development in *Ranunculus* (Nogler, 1995). It has been proposed that the apomixis locus in *Hieracium* enables a simple competency for apomixis and that the mode of apomixis is dependent upon the modifier loci in various genetic backgrounds (Bicknell *et al.*, 2000; Koltunow *et al.*, 2000). Molecular mapping of the apomixis locus in *Hieracium* is currently under way (Bicknell *et al.*, 2001).

The activities of genes conferring apomixis need not only to account for the diversity in apomictic phenotypes, but they must also explain why sexual process can be left completely intact and occur to completion in the same plant. Collectively, apomixis is clearly stochastic in nature, a quality that implies that factors required for initiation oscillate around a critical threshold level. Changes in this level may influence the cellular potential for apomixis. Therefore, if apomixis and sexuality share regulatory components, particular levels of the factor might promote sexual events and others apomixis, enabling the coexistence of both processes.

7.5.3 Overlaps exist between sexual and apomictic processes in facultative plants and lead to alterations in progeny ploidy level

In facultative gametophytic apomictic plants where sexual and apomictic processes coexist, and where the plant is self-fertile, it is possible to obtain seeds containing embryos originating from four different reproductive pathways. This results in four genotypic seed classes. Such seed heterogeneity is common in pseudogamous species (Nogler, 1984a; Savidan, 2000) and also in apomicts where endosperm formation is autonomous (Skalinska, 1970, 1971, 1973;

Koltunow *et al.*, 1998; R.A. Bicknell, personal communication). Embryos formed during apomixis retain the maternal genotype, and embryos derived from sexual reproduction that combine both maternal and paternal genomes are termed B_{II} hybrids.

The other two rarer classes of embryos are called 'off-types'. These result from overlaps in the sexual and apomictic pathways. B_{III} hybrid embryos have a higher ploidy level than those formed by apomixis or sexual reproduction because they result from the fusion of an unreduced egg cell with a male sperm cell. The polyhaploid class of embryos have decreased ploidy because they are derived from the autonomous development of a reduced egg cell. Polyhaploids are the rarest embryo class, but their frequency in apomicts is higher than that observed in sexually reproducing plants.

The development of B_{III} hybrids is decreased in apomicts where precocious embryo formation occurs prior to flower opening and pollination or if the unreduced egg cell develops a thick wall to prevent pollen tube penetration and fertilization (Savidan, 2000). If the sexual pathway terminates soon after the initiation of apomixis the production of sexual hybrids and polyhaploids is reduced.

The capacity to form B_{III} hybrids suggests that the unreduced egg cell formed during gametophytic apomixis is comparable, functionally, to an egg derived from meiotic reduction because it is able to attract the pollen tube and complete fertilization. The increased potential for the reduced egg to undergo autonomous embryogenesis in an apomictic background supports the concept of shared regulation of reproductive pathways and patterning during apomictic and sexual reproduction. A model in which a factor is limiting might explain the inability of all reduced egg cells to undergo autonomous embryogenesis in the apomictic background.

7.5.4   Cell fate decisions and patterns of gene expression in ovules of apomicts

The study of mutants defective in ovule and seed development has shown that signals from sporophytic tissues influence gametophyte and embryo development (Gasser *et al.*, 1998). The initiation and coexistence of sexual and apomictic reproductive modes within the ovule raises questions concerning the dependence of apomixis on cell specification events within the ovule, and signals from sporophytic tissues during the temporal events in ovule and seed development.

Apomixis usually initiates in ovules following the specification of ovule primordia and initiation of the sexual process. The events of gametophytic apomixis often initiate close to cells undergoing sexual reproduction. Adventitious embryos form later in the temporal sequence of ovule development adjacent to mature embryo sacs. In the case of citrus, embryos initiate in the chalazal domain of the ovule if the embryo sac aborts during development, rather than primarily forming in the nucellus adjacent to the micropylar region of the embryo sac (Koltunow *et al.*, 1995b). Factors emanating from or associated with a field of cells adjacent to those undergoing sexual reproduction in the ovule may therefore be critical for apomixis.

Guerin *et al.* (2000) have shown that mRNA levels of a *DEFICIENS* homologue, *DEFH*, are downregulated in a broad but localized zone of ovule cells in the chalaza of apomictic *Hieracium* adjacent to cells undergoing sexual events. The zone of depleted *DEFH* mRNA is not observed in sexual plants and in apomicts it overlaps with the area where apospory initiates in *Hieracium*. The downregulation of *DEFH* has been postulated as a prerequisite for the subsequent differentiation of aposporous initial cells because the number of cells in the zone is greater than the number of aposporous initial cells formed. However, there is no functional proof as yet for a direct role of *DEFH* in apomixis (Guerin *et al.*, 2000).

Other studies have focused on the comparison of genes expressed in flowers of sexual and apomictic plants. This has met with mixed success even though a range of different methods have been used, including differential display, differential screening, virtual subtraction and cold plaque screening (Vielle-Calzada *et al.*, 1996; Leblanc *et al.*, 1997; Guerin *et al.*, 2000; Tucker *et al.*, 2001). In most cases the spatial and temporal pattern of gene expression by *in situ* hybridization (if it has been carried out) does not obviously correlate with particular aspects of apomixis or sexual reproduction. This might relate to problems in sample collection and staging and possibly to genetic differences between the species being compared. Future screening procedures with higher-throughput chip technologies might prove informative for analysing the temporal expression patterns of many genes in a comparative manner between sexual and apomictic plants.

The cells that differentiate to initiate apomixis may not be immediately committed to a particular fate. Reproductive analyses in transgenic aposporous *Hieracium* plants containing genes that alter ovule development show an increased frequency of 'aposporous initial cells', and these appear to develop into a range of structures in the one ovule (Koltunow *et al.*, 2001). Isolated embryos, developing embryo sacs and embryo sacs with cellularizing endosperm can be observed in the deformed ovules (Koltunow *et al.*, 2000; A.M. Koltunow and S.D. Johnson, unpublished observations). The correlation between the alteration in ovule form, the increased frequency of apomixis and the diverse resultant structures suggests that positional information and signalling events play a critical role in determining the fate of cells initiating apomixis during the temporal course of ovule development in *Hieracium*.

These observations support the conclusions of inheritance studies that suggest that the apomictic locus in *Hieracium* creates a competency for apomixis within a particular group of cells (Koltunow *et al.*, 2000). If their fate is also determined by external signals during the course of ovule and seed development, then some of the modifier loci observed in the inheritance studies may well include genes involved in signal transduction. The *Hieracium* model can also explain the other classical modes of apomixis because the final mode of apomixis would be dependent upon modifiers present in a particular genetic background. However, this would imply that the causal locus for apomixis is common to all apomictic species.

7.5.5   Mutagenesis in sexual and apomictic plants

Mutagenesis strategies in sexual and apomictic plants provide the means to directly identify the genes involved in apomixis in addition to uncovering information about the genetic regulation of ovule and female gametophyte development.

7.5.5.1   Mutagenesis in apomictic Hieracium

Mutagenesis in apomicts could possibly enable reversion to a sexual phenotype if a gene essential for apomictic competency is made nonfunctional. Alternatively, hyperactivation of apomixis might also be observed if a mutation promotes the activity of the apomixis gene. Another important outcome expected from the mutagenesis of facultative apomicts is that information on the interrelationships between sexual and apomictic processes and the kinds of genes required for the two reproductive modes might be obtained. Mutant screens using γ-irradiation and insertional mutagenesis (T-DNA tagging and transposons) are being carried out in *Hieracium* with plants being scored for their capacity to form sexual progeny. Over 5000 M1 plants have been screened. Within this screened population, some plants have increased sexuality to compensate for decreased apomixis, some do not, and one mutant appears pseudogamous (Bicknell *et al.*, 2001).

7.5.5.2   Mutagenesis in sexual pearl millet (Pennisetum)
induces apospory in ovules

The hypothesis that genes providing a competency for apomixis are gain-of-function mutations in genes normally involved in sexual reproduction or that the apomixis locus misregulates genes involved in sexual reproduction implies that it should be possible to induce either apomixis or components of the apomictic process in sexual plants. Two mutants have been induced in sexual pearl millet that form aposporous embryo sacs with similar structure to those of the apomictic wild species. *female sterile* (*fs*; Hanna and Powell, 1974; Arthur *et al.*, 1993) and *stubby head* (Hanna and Powell, 1973) were induced using thermal neutrons and a combination of irradiation by thermal neutrons followed by exposure to diethyl sulfate, respectively.

Both mutants exhibit significant alterations in inflorescence, floral and ovule structure. *stubby head* displays multiple embryo sacs per ovule, and both meiotic and aposporous embryo sacs have been documented. The functionality of the unreduced embryo sacs was inferred because of the lack of transmission of a dominant colour marker in 23% of the progeny. In the *fs* mutant, multiple embryo sacs are observed in 35% of the ovules, but it does not set seed and is maintained by out-crossing viable pollen. A genetic characterization of *stubby head* and *fs* showed that, despite the similarities in phenotypes, these mutations represent different loci (Morgan *et al.*, 1997). As both mutations affect gynoecium and particularly ovule structure, leading to aposporous embryo sac formation, it was suggested that alterations in positional information and cellular signalling within the ovule may have resulted in the observed sporadic failure of meiosis, and the activation of normally quiescent somatic cells during embryo sac development (Morgan *et al.*, 1997). The capacity to induce aposporous events in these sexual *Pennisetum* plants contradicts observations from *Pennisetum* mapping studies where the apomixis locus does not have a counterpart in the sexual genome (Ozias-Akins *et al.*, 1998). Together these studies predict the possibility that there may be functionally redundant pathways for apospory.

7.5.6 *Autonomous endosperm mutants in* Arabidopsis *and the role of* FIS *genes in apomixis*

The recent isolation of a group of mutants in sexually reproducing *Arabidopsis*, and the identification and characterization of the corresponding gene products has provided significant insight regarding endosperm formation and the processes regulating the postfertilization expression of genes derived from maternal and paternal genomes in plants. Mutations in one of three different genes of the *FERTILIZATION INDEPENDENT SEED* (*FIS*) class, *FIS1/MEA*, *FIS2* and *FIS3/FIE*, can activate endosperm development, seed coat maturation and fruit development in the absence of fertilization (Grossniklaus *et al.*, 2001). Reproductive development in *fis*-class mutants differs from apomixis because embryo sacs form by meiotic reduction, fertilization-independent embryo formation does not occur and the seed aborts.

Two of the *FIS*-class genes, *MEDEA* (*MEA*) and *FERTILIZATION INDEPENDENT ENDOSPERM* (*FIE*), encode proteins with homology to the polycomb group (PcG) chromatin remodelling proteins of *Drosophila* that form interactive complexes and regulate gene expression by packaging chromatin into transcriptionally permissive or nonpermissive structures. This higher-order transcriptional regulation by PcG proteins appears to be generated and maintained by methylation and covalent histone modification (Strahl and Allis, 2000). *MEA* and *FIE* are similar in sequence to *Drosophila ENHANCER OF ZESTE* (*EZ*) and *EXTRA SEX COMBS* and they also interact in two hybrid analysis (Luo *et al.*, 2000; Spillane *et al.*, 2000). The third *FIS* gene, *FIS2*, encodes a zinc-finger protein that has no known counterpart in *Drosophila* or mammals. *FIS2* protein does not interact directly with either *FIE* or *MEA* in two hybrid analysis, but the similarity of the mutant phenotypes for all three genes and their overlapping promoter::GUS expression patterns suggest that the proteins encoded by these different genes operate together at the chromatin level, in an integrated manner, to control seed development (Luo *et al.*, 2000).

It has been proposed that one of the early developmental functions of a complex involving FIS proteins is to stably maintain transcriptional repression of the genes required for seed development (*SDGs*) until fertilization occurs (Luo *et al.*, 1999; Ohad *et al.*, 1999).The three known members of the *FIS* class are all co-expressed in the female gametophyte immediately prior to fertilization and during early endosperm divisions (Luo *et al.*, 2000). RT-PCR and promoter::GUS analysis shows that these genes display only maternal expression during the early stages of seed development (Kinoshita *et al.*, 1999; Luo *et al.*, 2000). Such selective allelic expression is known as parental imprinting, and it is likely that all three *FIS* genes are imprinted. Methylation in the genome is known to play a role in imprinting in plants, animals and insects (Alleman and Doctor, 2000; Mann *et al.*, 2000). Recent experiments that monitor seed viability and selectively lower the methylation status of maternal and paternal genomes have shown that imprinting and the expression of the *FIS* genes can be altered, leading to the rescue of viable seeds in the appropriate combination (Vielle-Calzada *et al.*, 1999; Adams *et al.*, 2000; Luo *et al.*, 2000; Vinkenoog *et al.*, 2000; Yadegari *et al.*, 2000). Genomic imprinting is therefore an important component regulating the expression of the downstream *SDGs* targeted by the FIS class of proteins.

The FIS-class proteins, however, are pleiotropic in their function, like their chromatin remodelling counterparts in *Drosophila* and mammals. FIS proteins also control the proliferation of endosperm nuclei and appear to regulate the establishment of the anterior–posterior axis in the endosperm following fertilization, while also regulating seed size at later stages (Luo *et al.*, 2000; Sørenson *et al.*, 2001). The FIS member MEDEA contains a putative retinoblastoma-binding motif (our observations), indicating potential functions that are directly linked to cell cycle events (Grossniklaus *et al.*, 1998b; Williams and Grafi, 2000). *FIE* may also play a role in vegetative development because it is also expressed in vegetative tissues (Kiyosue *et al.*, 1999; Luo *et al.*, 2000).

In sexually reproducing plants such as cereals and also *Arabidopsis*, a balanced ratio of two maternal to one paternal genomes in the endosperm of the developing seed is critical for viability. It appears that genomic imprinting, maintained by methylation, is important for seed development and endosperm balance (Scott *et al.*, 1998). Although rare, autonomous apomicts are insensitive to endosperm imprinting requirements and yet they produce viable seed. The functional relevance of imprinting in endosperm development is therefore challenged in apomicts that produce autonomous endosperm where the maternal genome is the sole contributor. Even pseudogamous apomictic grasses like *Tripsacum* and *Paspalum* (Quarin *et al.*, 1998; Quarin, 1999) exhibit unbalanced parental genome ratios in their endosperm following fertilization of the central cell and manage to produce viable seeds.

Grossniklaus *et al.* (2001) have suggested that the high degree of seed sterility observed in apomictic hybrids in introgression programmes and genetic studies where crosses were made between sexual and apomictic plants might reflect an unbalanced endosperm ratio in the progeny. They suggest that this may pose a barrier for the transfer of apomixis into sexual crops. An investigation of the role and activity of *FIS* genes in autonomous and pseudogamous apomicts may provide clues towards understanding how apomicts avoid imprinting barriers to produce viable seed.

The fundamental observation that fruit development occurs in *fis* mutants when endosperm formation is initiated in the absence of fertilization suggests that processes specifically relating to endosperm formation are sufficient to trigger the spatially removed events of fruit development in *Arabidopsis*. It remains to be determined whether these are the same cues triggered during fertilization-dependent seed set, so that fruit development can occur.

7.6   Parthenocarpic fruit development and the analysis of fruit initiation

Parthenocarpy is defined as the formation of fruit without fertilization. In a restricted sense, parthenocarpy has similarities to apomixis because both processes result in fertilization-independent fruit development. However, unlike apomixis, parthenocarpy restricts reproduction because it does not enable seed formation. Parthenocarpy exists in some species as a form of mimicry, preventing seed loss to herbivores and as such can offer some selective advantage (Zangerl *et al.*, 1991; Traveset, 1993).

Parthenocarpy can occur naturally and have a genetic basis or it can be induced artificially. There are two types of natural parthenocarpy, autonomous and stimulatory, with the latter requiring pollination for the induction of fruit development (Yasuda, 1930, 1935). In each type, parthenocarpy is only observed if the fertility of the plant is compromised so that fertilization does not occur; for example, if the plant is sterile or self-incompatible, or insect and pollination vectors are excluded, or if the flowers are surgically manipulated. Parthenocarpy is considered facultative if the plant can also produce seeded fruit, which occurs because the plant retains fertility as opposed to obligate parthenocarpic species that are reproductively sterile and always set seedless fruit.

Artificial induction of parthenocarpy involves the manual application of dead (mentor) pollen, pollen extracts and growth regulators such as auxins, gibberellins and cytokinins to floral tissues (Gustafson, 1942; Schwabe and Mills, 1981; Knox *et al.*, 1987; Gillaspy *et al.*, 1993). As pollen contains phytohormones and developing seeds produce significant amounts of phytohormones, and because artificially elevated auxin biosynthesis within ovules and placenta of the carpel can induce fruit development (Rotino *et al.*, 1997; Ficcadenti *et al.*, 1999), it is believed that the transition from a flower to a fruit has a hormonal basis that is dependent on the level, distribution and perception of hormones within the ovary and other structures contributing to fruit form (Gustafson, 1942; Nitsch, 1952, 1970). However, it remains unclear how changes in hormone level and perception generate the diverse array of fruits present in today's angiosperms (figure 7.2).

Another overlooked consideration relates to the processes regulating growth during flower maturation. Growth of floral organs ceases at flower opening and most angiosperms then require pollination and fertilization for fruit growth. Early studies on fruit development in cucurbits suggested that mechanisms exist to stop fruit growth at the time of flower opening and that without these mechanisms the program of ovary to fruit development would be continuous (Nitsch, 1952).

Recent analysis of fertilization-induced and parthenocarpic fruit formation in crops and model plants has shed light on the factors arresting the initiation of fruit development. Furthermore, the identity of genes required for cell division and expansion in developing fruit layers, and the interaction of particular ovule tissues required for signal transduction during parthenocarpic fruit growth are also being elucidated.

7.6.1 Arabidopsis *as a model system for understanding fruit development*

Although tomato has been the primary model for molecular studies of parthenocarpic and fertilization-induced fruit development (George *et al.*, 1984; Gillaspy *et al.*, 1993; Mazzucato *et al.*, 1999), recent attention has also focused on fruit development in *Arabidopsis*. Gynoecium development in *Arabidopsis* has been reviewed recently (Bowman *et al.*, 1999) and the identity of many essential genes contributing to ovary pattern has been established.

Growth of the *Arabidopsis* carpel continues until the flower opens and then, as in other species, carpel growth abruptly diminishes unless the ovules are fertilized. If fertilization does not occur, floral organs surrounding the *Arabidopsis* carpel abscise and cells of the carpel expand slightly, causing a small increase in carpel length prior to senescence (Vivian-Smith and Koltunow, 1999).

Arabidopsis produces a simple fruit called a silique following fertilization (Meinke and Sussex, 1979). Silique growth involves cell division, cell expansion and differentiation to produce four anatomical layers; an external exocarp, a mesocarp, a thin supportive schlerenchyma and an inner endocarp cell layer. After fertilization, longitudinal cell expansion is primarily observed in all carpel layers except those forming the mesocarp, where a high level of cell division is observed in the plane of silique elongation. The amount of cell division that occurs during mesocarp development together with the coordinated expansion in other layers affects final silique length (Vivian-Smith and Koltunow, 1999). The girth of the *Arabidopsis* silique is established by cell expansion in all tissue layers. Silique maturity is indicated by the yellowing of silique valves, followed by their dehiscence to release seeds (Ferrándiz *et al.*, 1999; Vivian-Smith and Koltunow, 1999, 2001). In unpollinated pistils, dehiscence is characteristically absent.

The synthesis and metabolism of GAs is essential for silique development in *Arabidopsis* (Barendse *et al.*, 1986). Parthenocarpy can be induced by the exogenous application of auxin, GA and cytokinin directly to carpels where all of the surrounding floral organs have been removed. Growth regulator application to such naked carpels results in siliques of different lengths and widths, suggesting that these applied regulators are differentially perceived, transported or metabolized in the various layers of the carpel to influence silique development. Ecotype differences also contribute to a variation in parthenocarpic silique growth following the exogenous application of growth regulators (Vivian-Smith and Koltunow, 1999). In general the application of GA$_3$ to naked carpels results in parthenocarpic siliques with tissue morphology and size most resembling those observed in fertilization-induced siliques. Another prerequisite for silique formation is functional *FRUITFULL* (*FUL*) activity, because elongated, dehiscent siliques do not form in *ful* mutants following fertilization or exogenous growth regulator application (Gu *et al.*, 1998; Vivian-Smith *et al.*, 2001).

One way to identify the molecular factors controlling the initiation of fruit development and parthenocarpy is to try to uncouple fruit development from fertilization in *Arabidopsis* by mutation. Two parthenocarpic mutants have been recovered using different mutagenesis strategies and their comparative features are discussed below.

7.6.1.1   FRUIT WITHOUT FERTILIZATION (FWF)

In the parthenocarpic *Arabidopsis* mutant *fruit without fertilization* (*fwf*; Vivian-Smith *et al.*, 2001), male and female gamete formation is normal. This mutant is facultative because seedless siliques form only when pollination and fertilization are prevented. Parthenocarpy is recessive in *fwf*, suggesting that *fwf* represents a sporophytic mutation that contrasts the gametophytic class of *fis* mutants. The recessive nature of *fwf* is consistent with the hypothesis that *FWF* is involved in processes that repress development of the carpel in the absence of fertilization. Parthenocarpic silique development in *fwf* is dependent upon carpel identity because the presence of the *ful-7* mutation does not permit silique elongation (Vivian-Smith *et al.*, 2001).

Surprisingly, the presence of floral whorls surrounding the carpel significantly reduces parthenocarpy in *fwf*, indicating that interfloral organ signals and functional FWF activity normally act in concert to prevent fruit initiation in the absence of pollination and fertilization (Vivian-Smith *et al.*, 2001). Pollination and fertilization might be required to disable these restrictions. Postfertilization induced senescence of the perianth in *Arabidopsis* might serve to remove some of these inhibitory signals in addition to enabling the greater allocation of nutritive resources to growing fruit structures. During fruit development, *FWF* appears to be involved in the modulation of growth in both mesocarp cells and also vascular tissues as they are morphologically altered in the *fwf* mutant (Vivian-Smith *et al.*, 2001).

7.6.1.2 The role of the ovule in regulating fruit development

Several lines of evidence suggest that the ovules play a critical role during parthenocarpic fruit formation in the *fwf* mutant. The period in which *fwf* ovules are receptive to fertilization coincides with the period of parthenocarpic silique growth, suggesting that receptive and functional ovules may be required to transduce a signal that allows silique formation to occur (Vivian-Smith *et al.*, 2001).

Increased parthenocarpic silique lengths were observed when *fwf* was combined with the ovule mutant, *aberrant testa shape* (*ats*; Léon-Kloosterziel *et al.*, 1994). In this mutant, altered cell division results in a functional but composite inner and outer integument with reduced total integument cell layers. The *ats* mutation also negates the inhibitory effects of the surrounding floral whorls on parthenocarpy because the removal of these organs was no longer required to observe siliques in a conditional male sterile *fwf* background (Vivian-Smith *et al.*, 2001). This suggested that structural determinants of the ovule might be producing molecules that modulate parthenocarpy in *fwf*. However, the identity of *ATS* and its expression in the flower are unknown and as such the possibility that *ATS* may also be expressed in organs other than the ovule cannot yet be discounted.

The possibility that events in specific ovule tissues might be influencing the initiation of parthenocarpy in *Arabidopsis* was examined by combining *fwf* with ovule mutants in which the genetic lesions were known. The selected ovule mutants shown in figure 7.4 lack one or more structural features found in normal *Arabidopsis* ovules. Combination of the ovule mutants individually with *fwf* showed that the presence of the female gametophyte, endothelium, inner integument and funiculus are required for parthenocarpy. These observations support a hypothesis, originally proposed by Nitsch (1952), that the ovule plays an important role in directing fruit development. Specific regions of the ovule may individually or collectively synthesize or transmit molecules that modulate

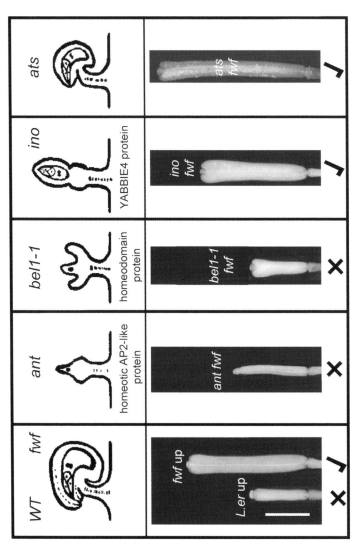

Figure 7.4 The effect of ovule structure on parthenocarpic fruit development in the *Arabidopsis fwf* mutant. Double mutants were generated between *fwf* and various *Arabidopsis* mutants with altered ovule structure (Schneitz *et al.*, 1997). The cartoon shows the structure of the ovule in each mutant and the panels below show the degree of silique development observed in the double mutant combination. (√) indicates parthenocarpic elongation, and (×) indicates absence of parthenocarpic development. *ant*, *aintegumenta*; *ats*, *aberrant testa shape*; *bel1-1*, *bell shaped ovules*; *fwf*, *fruit without fertilization*; *ino*, *inner no outer*; *up*, unpollinated silique; WT, wild type. Scale bar = 3 mm.

parthenocarpic silique development in the *fwf* background. In the absence of fertilization, *FWF* might act within these tissues to restrict the transduction of signals promoting fruit development. Once fertilization occurs *FWF* may also be required to modulate aspects of silique growth in mesocarp and vascular tissues. This hypothesis will be tested once the identity of *FWF* and its expression during flower and fruit development is known.

7.6.1.3 *Parthenocarpy in* 28-5

An activation-tagged parthenocarpic *Arabidopsis* mutant, *28-5* has been characterized that exhibits alterations in floral phenotype and vegetative structure. This mutant is male and female sterile and requires vegetative propagation *in vitro* (Ito and Meyerowitz, 2000). It produces siliques that are significantly increased in girth, despite the apparent absence of fertilization. The enhancer insertion activated *CYP78A9*, a cytochrome P450 gene whose function is unknown. Cytochrome P450 proteins are involved in the synthesis or degradation of plant secondary metabolites, and *CYP78A9* was postulated to be involved in the production of a novel plant growth substance because increased concentrations of phytohormones induce parthenocarpy (Ito and Meyerowitz, 2000). Unlike *fwf*, *28-5* displays elongated siliques despite the presence of anthers, petals and sepals. The basis for reproductive sterility in *28-5* has not yet been characterized and it is therefore uncertain whether aspects of ovule signalling are altered, eliminating the need to remove floral whorl organs, or whether floral whorl clues that inhibit silique development are related to aspects of pollen development.

7.6.2 *Genes conferring parthenocarpy in crops*

Parthenocarpy is exploited in breeding programmes aimed at producing seedless fruit and vegetables. Banana, grape, citrus, pineapple, tomato, capsicum, watermelon, pear and fig all provide examples of diverse parthenocarpic produce in the marketplace. These plants are either vegetatively propagated or hybrid seed is bulked from crosses between fertile parental lines bearing parthenocarpy.

In cultivated banana a single locus controls parthenocarpy however, a number of modifiers influence the extent of parthenocarpic development (Dodds and Simmonds, 1948). A number of loci influence parthenocarpy in citrus (Vardi *et al.*, 2000) but it is not certain how many are essential for fertilization-independent initiation of fruit growth and which are modifiers of the trait. Two distinct loci, *pat-1* and *pat-2* are known to confer parthenocarpy in tomato (George *et al.*, 1984), but in the tomato cultivar RP75/79 parthenocarpy is conferred by three recessive genes with additive effects (Nuez *et al.*, 1986; Pecaut and Philouze, 1978; Vardy *et al.*, 1989a). Efforts to map the *pat-2* locus in tomato have shown that it is located on chromosome 3, between the visible

markers *sf* and *bls*, and that the *pat-2* phenotype is also modified by another
locus *mp* (Vardy *et al.*, 1989b).

Parthenocarpy was recently investigated at the molecular level in apple (Yao
et al., 2001). Parthenocarpic apple varieties are rare, but characterized examples
exhibit floral defects reminiscent of those seen in mutants of the *Arabidopsis* B
class MADS-box gene *PISTILLATA* (*PI*), in that they display deformities in petal
and stamen development. Floral deformities are also evident in parthenocarpic
tomato (Mazzucato *et al.*, 1998) and other parthenocarpic cultivars (Nitsch,
1952). Mutations blocking the function of the apple *PI* orthologue, *MdPI*, were
observed in each of three seedless parthenocarpic apple varieties compared with
normal seeded varieties. Yao *et al.* (2001) concluded that this gene controlled
parthenocarpy in apple.

Mutations in *Arabidopsis PI* do not permit parthenocarpic fruit formation
(Chaudhury *et al.*, 1997; Vivian-Smith *et al.*, 2001). The difference in requirement for functional *PI* activity for parthenocarpy in *Arabidopsis* and apple
might relate to an altered balance of endogenous factors that contribute to the
two different fruit structures. Unlike *Arabidopsis*, the apple fruit forms from
the fused base of the sepal, petal and stamen rather than from just the carpel
tissues (figure 7.2). Alternatively, if floral whorl inhibitory cues are relevant
for preventing fruit growth in apple, then lesions in the apple *PI* orthologue
might overcome inhibitory cues, allowing parthenocarpy to be manifested. In
this sense, the loss in function of apple *PI* would enable the observation of
parthenocarpy caused by another gene.

7.6.3   Modifiers of parthenocarpic fruit development
are also members of the GRAS gene family

Diversity in fruit form occurs because the temporal patterns of cell division,
expansion and differentiation occur in a variety of spatial regions. A logi-
cal expectation is that the proteins involved with the initiation of fruit development form functional partnerships with genes and proteins that influence
cell division, growth and differentiation. There is now some indication of the
types of genes that might be involved in the initiation of fruit patterning
events.

Mutations in the gene, *LATERAL SUPRESSOR* (*LS*; Schumacher *et al.*, 1999)
suppress secondary meristem initiation in tomato, and *ls* mutants form flowers
that do not initiate petal formation (Szymkowiak and Sussex, 1993). Philouze
(1983) showed that *ls* also blocks parthenocarpic fruit development in tomato
lines containing *pat-2*, implying that *LS* is epistatic to *pat-2* and is a significant
modulator of parthenocarpy. *LS* is a member of the large *GRAS* family of genes
(figure 7.5). Two other members of the *GRAS* family, *SCARECROW* (*SCR*) and
SHORT-ROOT (*SHR*), play roles in the specification of asymmetric cell division,
the establishment of cellular pattern and the control of transmissible growth

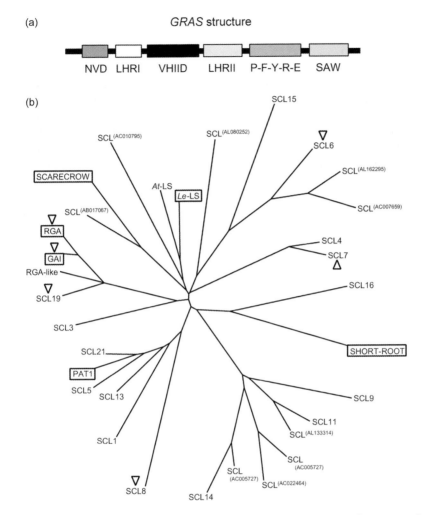

Figure 7.5 The *GRAS* gene family. (a) GRAS polypeptide consensus structure. GRAS genes comprise a variable N-terminal domain with homopolymeric repeat regions (NVD), a leucine heptad region (LHRI), a conserved VHIID motif, a second leucine heptad region (LHRII) and a P-F-Y-R-E and a SAW motif (Pysh *et al.*, 1999). (b) A neighbor-joining tree constructed for *GRAS* (*SCARECROW-LIKE*, *SCL*) members in *Arabidopsis* and *LATERAL SUPPRESSOR* from tomato (*Le-LS*; Schumacher *et al.*, 1999). Distances between neighboring polypeptides were determined using the Jukes–Cantor algorithm (Rice, 1994) and displayed using the TreeView program (Page, 1996). Members of the family cloned and functionally characterized on the basis of a mutant phenotype are indicated in boxes. *SCL1, 3, 5, 7* and *14* are expressed in silique tissues (Pysh *et al.*, 1999). The unnumbered members were found in the *Arabidopsis* genome sequence. Accession numbers of their coding sequence are indicated in superscript. Triangles represent *GRAS* members potentially mutated with *Ds* insertion elements (Parinov *et al.*, 1999).

signals in root and shoot organs (Fukaki *et al.*, 1998; Helariutta *et al.*, 2000; Wysocka-Diller *et al.*, 2000). In the root *SHR* functions upstream of *SCR* and controls the expression of *SCR* by non-cell-autonomous factors (Di Laurenzio *et al.*, 1996; Helariutta *et al.*, 2000). Histological analysis of tomato fruit development in *ls* and *ls pat-2* mutants is required to determine whether *LS* affects cell division in specific fruit tissues.

Another member of the *GRAS* gene family appears to play a role in asymmetric divisions in mesocarp cells during *Arabidopsis* silique development. Alterations in the function of *GAI*, a *GRAS* family member involved in GA perception, block anticlinal cell division in developing mesocarp cells in both parthenocarpic (Vivian-Smith *et al.*, 2001) and pollination-induced siliques (Vivian-Smith and Koltunow, 1999). In *gai-1 fwf* double mutants, parthenocarpic siliques form but silique growth occurs by cell expansion and shorter siliques are observed compared to the *fwf* background. The *gai-1* mutant allele used in these experiments is a gain-of-function allele that acts to constitutively repress GA perception. However, it is not possible to conclude that normal *GAI* activity is directly required for establishing appropriate cell planes in the mesocarp layer because another *GRAS* family member, *RGA* could have an overlapping function (Sun, 2000; figure 7.5). Nevertheless, the observation that anticlinal mesocarp cell division is blocked in the *gai-1* background, together with the requirement for functional *LS* in tomato, suggest that at least two *GRAS* family members are important for parthenocarpic fruit growth.

GAI and *RGA* in *Arabidopsis* negatively regulate the abundance of GA 20-oxidase, which catalyses multiple steps in GA biosynthesis (Peng *et al.*, 1997; Silverstone *et al.*, 1998) and has specific expression domains within flowers and siliques (Phillips *et al.*, 1995; Sponsel *et al.*, 1997). Therefore, the involvement of either of the *GRAS* family members *GAI* or *RGA* in silique development provides a potential feedback loop between GA signal transduction and the biosynthesis of GA, which is known to be essential for growth and differentiation in *Arabidopsis* siliques (Barendse *et al.*, 1986; Vivian-Smith and Koltunow, 1999; Vivian-Smith *et al.*, 2001).

It is tempting to speculate that other *GRAS* family members known to be expressed in *Arabidopsis* siliques (see figure 7.5) might have cell division and cell fate functions resembling their counterparts *SCR* and *SHR*. The *ats* ovule mutation has been found to restore anticlinal mesocarp cell division in parthenocarpic *gai-1 ats fwf* triple mutants, suggesting that *ATS* may also play a functional role in coordinating and communicating cell division processes in ovules and carpels (Vivian-Smith *et al.*, 2001). Further mutational analysis of the *GRAS* family members might provide clues to their function and role in cell determination events in early fruit development. This may provide a better understanding of the linkage between hormonal cues and patterning events.

7.7   Conclusions

Seed and fruit development require distinct sets of pattern-forming events. Basic pattern elements are established in the ovules but fruit form is highly variable and is often determined by localized growth changes in specific floral cell types following anthesis. The balance between fruit and seed growth and floral senescence appears to be maintained by long-range signalling involving phytohormones such as auxin and ethylene. However, other hormones such as gibberellins are also involved in fruit and seed set and growth.

Genetic and molecular evidence suggests that both seed and fruit development is actively restricted by mechanisms that prevent the initiation of continued pattern-forming events in the absence of fertilization. In *Arabidopsis*, *FWF* appears to be involved in processes that maintain a check on the initiation of fruit development and part of this process involves inhibitory cues from surrounding floral whorl tissues. *FWF* activity is required in the later events of fruit growth to modulate mesocarp and vascular cell growth. The *FIS*-class genes appear to collaborate in the embryo sac to restrict the initiation of endosperm development and they are subsequently required for endosperm pattern formation.

Fertilization-dependent and fertilization-independent fruit development are both reliant upon signal transduction events involving the ovule. How the events in the ovule are transduced to surrounding carpel tissues to enable the initiation of fruit growth remains unknown, but plant hormones may play a role in this signalling process. Fruit growth in tomato and also *Arabidopsis* is dependent on the activity of members of the *GRAS* family of genes. Genes of this type appear to be required before or after the initiation of fruit growth, and in the case of *GAI* or its substitute *RGA*, can influence development of specific tissue layers in *Arabidopsis*. In summary, the transition between flower maturity and the initiation of seed and fruit development is dependent upon the removal of inhibitory factors, long-distance communication between organs to coordinate seed and fruit growth, and the induction of genes required for local patterning events to effect appropriate form.

References

Adams, S., Vinkenoog, R., Spielman, M., Dickinson, H.G. and Scott, R.J. (2000) Parent-of-origin effects on seed development in *Arabidopsis thaliana* require DNA methylation. *Development*, **127**, 2493-2502.

Alleman, M. and Doctor, J. (2000) Genomic imprinting in plants: observations and evolutionary implications. *Plant Mol. Biol.*, **43**, 147-161.

Alvarez, J. and Smyth, D.R. (1999) *CRABS CLAW* and *SPATULA*, two *Arabidopsis* genes that control carpel development in parallel with *AGAMOUS*. *Development*, **126**, 2377-2386.

Arthur, L., Ozias-akins, P. and Hanna, W.W. (1993) Female sterile mutant in pearl-millet—evidence for initiation of apospory. *J. Hered.*, **84**, 112-115.

Barendse, G.W.M., Kepczynski, J., Karssen, C.M. and Koornneef, M. (1986) The role of endogenous gibberellins during fruit and seed development: studies on gibberellin-deficient genotypes of *Arabidopsis thaliana. Physiol. Plant.*, **67**, 315-319.

Berleth, T. and Mattsson, J. (2000) Vascular development: tracing signals along veins. *Curr. Opin. Plant Biol.*, **3**, 406-411.

Berleth, T. and Sachs, T. (2001) Plant morphogenesis: long distance coordination and local patterning. *Curr. Opin. Plant Biol.*, **4**, 57-62.

Bicknell, R.A. (1997) Isolation of a diploid, apomictic plant of *Hieracium aurantiacum. Sex. Plant Reprod.*, **10**, 168-172.

Bicknell, R.A., Borst, N.K. and Koltunow, A.M. (2000) Monogenic inheritance of apomixis in two *Hieracium* species with distinct developmental mechanisms. *Heredity*, **84**, 228-237.

Bicknell, R., Podivinsky, E., Catanach, A. and Erasmuson, S. (2001) Mutant analysis in *Hieracium. Proceedings of the 2nd International Apomxis Conference*, Como, Italy (Supported by the European Commission, Rockefeller Foundation, FAO, IRD), p. 30.

Bowman, J.L., Drews, G.N. and Meyerowitz, E.M. (1991) Expression of the *Arabidopsis* floral homeotic gene *AGAMOUS* is restricted to specific cell types late in flower development. *Plant Cell*, **3**, 749-758.

Bowman, J.L., Baum, S.F., Eshed, Y., Putterill, J. and Alvarez, J. (1999) Molecular genetics of gynoecium development in *Arabidopsis. Curr. Top. Dev. Biol.*, **45**, 155-205.

Carman, J.G. (1997) Asynchronous expression of duplicate genes in angiosperms may cause apomixis, bispory, tetraspory, and polyembryony. *Biol. J. Linn. Soc.*, **61**, 51-94.

Carman, J.G., Crane, C.F. and Riera-Lizarazu, O. (1991) Comparative histology of cell walls during meiotic and apomeiotic megasporogenesis in two hexaploid Australasian *Elymus* species. *Crop Sci.*, **31**, 1527-1532.

Chaudhury, A.M., Ming, L., Miller, C., Craig, S., Dennis, E.S. and Peacock, W.J. (1997) Fertilization-independent seed development in *Arabidopsis thaliana. Proc. Natl. Acad. Sci.*, **94**, 4223-4228.

Coombe, B.G. (1975) The development of fleshy fruits. *Annu. Rev. Plant Physiol.*, **27**, 507-528.

de Jong, H., van Dijk, P. and van Baarlen, P. (2001) Genetic dissection of diplosporous automomous apomixis in *Taraxacum. Proceedings of the 2nd International Apomixis Conference*, Como, Italy (Supported by the European Commission, Rockefeller Foundation, FAO, IRD), p. 24.

de Martinis, D. and Mariani, C. (1999) Silencing gene expression of the ethylene forming enzyme results in a reversible inhibition of ovule development in transgenic tobacco plants. *Plant Cell*, **11**, 1061-1072.

Denny, O.J. (1992) Xenia includes metaxenia. *HortScience*, **27**, 722-728.

Dilcher, D. (2000) Toward a new synthesis: major evolutionary trends in the angiosperm fossil record. *Proc. Natl. Acad. Sci. USA*, **97**, 7030-7036.

Di Laurenzio, L., Wysocka-Diller, J., Malamy, J.E., *et al.* (1996) The *SCARECROW* gene regulates an asymmetric cell division that is essential for generating the radial organization of the *Arabidopsis* root. *Cell*, **86**, 423-433.

Dodds, K.S. and Simmonds, N.W. (1948) Sterility and parthenocarpy in diploid hybrids of *Musa. Heredity*, **2**, 101-117.

do Valle, C.B. and Savidan, Y. (1996) Genetics, cytogenetics, and reproductive biology of *Brachiaria*, in *Brachiaria: Biology, Agronomy and Improvement* (eds. J.W. Miles, B.L. Maass and C.B. do Valle), CIAT-EMBRAPA, Cali, Colombia, pp. 147-163.

Drews, G.N., Lee, D. and Christensen, C.A. (1998) The genetic analysis of female gametophyte development and function. *Plant Cell*, **10**, 5-17.

Dujardin, M. and Hanna, W. (1984) Microsporogenesis, reproductive behaviour and fertility in 5 *Pennisetum* species. *Theor. Appl. Genet.*, **67**, 197-201.

Ferrándiz, C., Pelaz, S. and Yanofsky, M.F. (1999) Control of carpel and fruit development in *Arabidopsis. Annu. Rev. Biochem.*, **68**, 321-354.

Ficcadenti, N., Sestili, S., Pandolfini, T., Cirillo, C., Rotino, G.L. and Spena, A. (1999) Genetic engineering of parthenocarpic fruit development in tomato. *Mol. Breeding*, **5**, 463-470.

Fos, M., Nuez, F. and García-Martínez, J.L. (2000) The gene *pat-2*, which induces natural parthenocarpy, alters the gibberellin content in unpollinated tomato ovaries. *Plant Physiol.*, **122**, 471-479.

Friedman, W.E. (1994). The evolution of embryogeny in seed plants and the developmental origin and early history of endosperm. *Am. J. Bot.*, **81**, 1468-1486.

Friedman, W.E. (2001) Comparative embryology of basal angiosperms. *Curr. Opin. Plant Biol.*, **4**, 14-20.

Fukaki, H., Wysocka-Diller, J., Kato, T., Fujisawa, H., Benfey, P.N. and Tasaka, M. (1998) Genetic evidence that the endodermis is essential for shoot gravitropism in *Arabidopsis thaliana*. *Plant J.*, **14**, 425-430.

García-Martínez, J.L., Sponsel, V.M. and Gaskin, P. (1987) Gibberellins in developing fruits of *Pisum sativum* cv. Alaska: studies on their role in pod growth and seed development. *Planta*, **170**, 130-137.

García-Martínez, J.L., Santes, C., Croker, S.J. and Hedden, P. (1991) Identification, quantification and distribution of gibberellins in fruits of *Pisum sativum* L. cv. Alaska during pod development. *Planta*, **184**, 53-60.

García-Martínez, J.L., López-Diaz, I., Sánchez-Beltrán, M.J., Philllips, A.L., Ward, D.A., Gaskin, P. and Hedden, P. (1997) Isolation and transcript analysis of gibberellin 20-oxidase genes in pea and bean in relation to fruit development. *Plant Mol. Biol.*, **33**, 1073-1084.

Gasser, C.S., Broadhvest, J. and Hauser, B.A. (1998) Genetic analysis of ovule development. *Annu. Rev. Plant Physiol. Plant Mol. Biol.*, **49**, 1-24.

Geldner, N., Hamann, T. and Jurgens, G. (2000) Is there a role for auxin in early embryogenesis? *Plant Growth Regul.*, **32**, 187-191.

George, W.L., Scott, J.W. and Splittstoesser, W.E. (1984) Parthenocarpy in tomato. *Hortic. Rev.*, **6**, 65-84.

Gillaspy, G., Ben-David, H. and Gruissem, W. (1993) Fruits: a developmental perspective. *Plant Cell*, **5**, 1439-1451.

Grimanelli, D., Leblanc, O., Espinosa, E., Perotti, E., De Leon, D.G. and Savidan, Y. (1998) Mapping diplosporous apomixis in tetraploid *Tripsacum*: one gene or several genes? *Heredity*, **80**, 33-39.

Grimanelli, D., Tohme, J. and Gonzalez de Leon., D. (2001) Applications of molecular genetics in apomixis research, in *The Flowering of Apomixis: From Mechanisms to Genetic Engineering* (eds. Y. Savidan, J.G. Carman and T. Dresselhaus). CIMMYT and IRD, Mexico, pp. 83-93.

Grossniklaus, U., Koltunow, A.M. and van Lookeren Campagne, M. (1998a) A bright future for apomixis. *Trends Plant Sci.*, **3**, 415-416.

Grossniklaus, U., Vielle-Calzada, J.P., Hoeppner, M.A. and Gagliano, W.B. (1998b) Maternal control of embryogenesis by *MEDEA*, a Polycomb group gene in *Arabidopsis*. *Science*, **280**, 446-450.

Grossniklaus, U., Spillane, C., Page, D.R. and Köhler, C. (2001) Genomic imprinting and seed development: endosperm formation with and without sex. *Curr. Opin. Plant Biol.*, **4**, 21-27.

Gu, Q., Ferrándiz, C., Yanofsky, M. and Martiensson, R. (1998) The *FRUITFULL* MADS-box gene mediates cell differentiation during *Arabidopsis* fruit development. *Development*, **125**, 1509-1517.

Guerin, J., Rossel, J.B., Robert, S., Tsuchiya, T. and Koltunow, A. (2000) A *DEFICIENS* homologue is down-regulated during apomictic initiation in ovules of *Hieracium*. *Planta*, **210**, 914-920.

Gustafson, F.G. (1936) Inducement of fruit development by growth promoting chemicals. *Proc. Natl. Acad. Sci. USA*, **22**, 629-636.

Gustafson, F.G. (1939) The natural cause of parthenocarpy. *Am. J. Bot.*, **26**, 135-138.

Gustafson, F.G. (1942) Parthenocarpy: artificial and natural. *Bot. Rev.*, **8**, 599-654.

Hagberg, A. and Hagberg, G. (1980) High frequency of spontaneous haploids in the progeny of an induced mutation in barley. *Hereditas*, **93**, 341-343.

Hanna, W.W. and Bashaw, E.C. (1987) Apomixis: Its identification and use in plant breeding. *Crop Sci.*, **27**, 1136-1139.

Hanna, W.W. and Powell, J. B. (1973) *stubby head*, an induced facultative apomict in pearl millet. *Crop Sci.*, **13**, 726-728.

Hanna, W.W. and Powell, J.B. (1974) Radiation-induced female-sterile mutant in pearl millet. *J. Hered.*, **65**, 247-249.

Helariutta, Y., Fukaki, H., Wysocka-Diller, J., *et al.* (2000) The *SHORT-ROOT* gene controls radial patterning of the *Arabidopsis* root through radial signaling. *Cell*, **101**, 555-567.

Herr, J.M. (1995) The origin of the ovule. *Am. J. Bot.*, **82**, 547-564.

Hulme, A.C. (ed.) (1970) *The Biochemistry of Fruits and Their Products*, vol. 1, Academic Press, London.

Hulme, A.C. (ed.) (1971) *The Biochemistry of Fruits and Their Products*, vol. 2, Academic Press, London.

Ito, T. and Meyerowitz, E.M. (2000) Overexpression of a gene encoding a cytochrome P450, *CYP78A9*, induces large and seedless fruit in *Arabidopsis*. *Plant Cell*, **12**, 1541-1550.

Iwamasa, M., Ueno, I. and Nishiura, M. (1967) Inheritance of nucellar embryony in *Citrus*. *Bul. Hort. Res. Sta. Japan Ser. B*, **7**, 1-8.

Jassem, B. (1990) Apomixis in the genus *Beta*. *Apomixis Newsletter*, **2**, 7-23.

Jensen, P.J. and Bandurski, R.S. (1994) Metabolism and synthesis of indole-3-acetic acid (IAA) in *Zea mays*. Levels of IAA during kernel development and the use of *in vitro* endosperm systems for studying IAA biosynthesis. *Plant Physiol.*, **106**, 343-351.

Johnson, P.R. and Ecker, J.R. (1998) The ethylene gas signal transduction pathway: a molecular perspective. *Annu. Rev. Genet.*, **32**, 227-254.

Jongedijk, E. (1985) The pettern of megasporogenesis and megagametogenesis in diploid *Solanum* species hybrids: its relevance to the origin of 2n-eggs and the induction of apomixis. *Euphytica*, **34**, 599-611.

Kinoshita, T., Yadegari, R., Harada, J.J., Goldberg, R.B. and Fischer, R.L. (1999) Imprinting of the *MEDEA* polycomb gene in the *Arabidopsis* endosperm. *Plant Cell*, **11**, 1945-1952.

Kiyosue, T., Ohad, N., Yadegari, R., *et al.* (1999) Control of fertilization-independent endosperm development by the *MEDEA* polycomb gene *Arabidopsis*. *Proc. Natl. Acad. Sci. USA*, **96**, 4186-4191.

Knox, R.B., Gadget, M. and Dumas, C. (1987) Mentor pollen techniques. *Int. Rev. Cytol.*, **107**, 315-332.

Koltunow, A.M. (1993) Apomixis: embryo sacs and embryos formed without meiosis or fertilization in ovules. *Plant Cell*, **5**, 1425-1437.

Koltunow, A.M., Bicknell, R.A. and Chaudhury, A.M. (1995a) Apomixis: strategies for the generation of genetically identical seeds without fertilization. *Plant Physiol.*, **108**, 1345-1352.

Koltunow, A.M., Soltys, K., Nito, N. and Mcclure, S. (1995b) Anther, ovule seed, and nucellar embryo development in *Citrus sinensis* cv Valencia. *Can. J. Bot.*, **73**, 1567-1582.

Koltunow, A.M., Johnson, S.D. and Bicknell, R.A. (1998) Sexual and apomictic development in *Hieracium*. *Sex. Plant Reprod.*, **11**, 213-230.

Koltunow, A.M., Johnson, S.D. and Bicknell, R.A. (2000) Apomixis is not developmentally conserved in related, genetically characterized *Hieracium* plants of varying ploidy. *Sex. Plant Reprod.*, **12**, 253-266.

Koltunow, A.M., Johnson, S.D., Lynch, M., Yoshihara, T. and Costantino, P. (2001) Expression of *rolB* in apomictic *Hieracium piloselloides* VIll. Causes ectopic meristems *in planta* and changes in ovule formation, where apomixis initiates at higher frequency. *Planta*, in press.

Leblanc, O., Armstead, I., Pessino, S., *et al.* (1997) Non-radioactive mRNA fingerprinting to visualise gene expression in mature ovaries of *Brachiaria* hybrids derived from *B. brizantha*, an apomictic tropical forage. *Plant Sci.*, **126**, 49-58.

Léon-Kloosterziel, K.M., Keijzer, C.J. and Koornneef, M. (1994) A seed shape mutant of *Arabidopsis* that is affected in integument development. *Plant Cell*, **6**, 385-392.

Luo, M., Bilodeau, P., Koltunow, A., Dennis, E.S., Peacock, W.J. and Chaudhury, A.M. (1999) Genes controlling fertilization-independent seed development in *Arabidopsis thaliana*. *Proc. Natl. Acad. Sci. USA*, **96**, 296-301.

Luo, M., Bilodeau, P., Dennis, E.S., Peacock, W.J. and Chaudhury, A. (2000) Expression and parent-of-origin effects for *FIS2*, *MEA*, and *FIE* in the endosperm and embryo of developing *Arabidopsis* seeds. *Proc. Natl. Acad. Sci. USA*, **97**, 10637-10642.

Lyndon, R.F. (1990) *Plant Development*. Unwin Hyman, London.

Mann, J.R., Szabo, P.E., Reed, M.R. and Singer-Sam, J. (2000) Methylated DNA sequences in genomic imprinting. *Crit. Rev. Euk. Gene. Expression*, **10**(3-4), 241-257.

Mazzucato, A., Taddei, A.R. and Soressi, G.P. (1998) The parthenocarpic fruit (*pat*) mutant of tomato (*Lycopersicon esculentum* Mill.) sets seedless fruits and has aberrant anther and ovule development. *Development*, **125**, 107-114.

Mazzucato, A., Testa, G., Biancari, T. and Soressi, G.P. (1999) Effect of gibberellic acid treatments, environmental conditions, and genetic background on the expression of the parthenocarpic fruit mutation in tomato. *Protoplasma*, **208**, 18-25.

Mehouachi, J., Iglesias, D.J., Tadeo, F.R., Agusti, M., Primo-millo, E. and Talon, M. (2000) The role of leaves in citrus fruitlet abscission: effects on endogenous gibberellin levels and carbohydrate content. *J. Hort. Sci. Biotech.*, **75**, 79-85.

Meinke, D.W. and Sussex, I.M. (1979) Embryo lethal mutants of *Arabidopsis thaliana*. A model system for genetic analysis of plant embryo development. *Dev. Biol.*, **12**, 50-61.

Morgan, R.N., Alvernaz, J., Arthur, L., Hanna, W.W. and Ozias-Akins, P. (1997) Genetic characterization and floral development of *female sterile* and *stubby head*, two aposporous mutants of pearl millet. *Sex. Plant Reprod.*, **10**, 127-135.

Naumova, T.N. (1993) *Apomixis in angiosperms. Nucellar and Integumentary Embryony*, CRC Press, Boca Raton, FL, pp. 17-21.

Nemhauser, J.L., Feldman, L.J. and Zambryski, P.C. (2000) Auxin and *ETTIN* in gynoecium morphogenesis. *Development*, **127**, 3877-3888.

Nitsch, J.P. (1950) Growth and morphogenesis of the strawberry as related to auxin. *Am. J. Bot.*, **37**, 211-215.

Nitsch, J.P. (1952) Plant hormones in the development of fruits. *Q. Rev. Biol.*, **27**, 33-57.

Nitsch, J.P. (1970) Hormonal factors in growth and development, in *The Biochemistry of Fruits and Their Products* (ed. A.C. Hulme), vol. 1, Academic Press, London, pp. 427-472.

Nogler, G.A. (1984a) Gametophytic apomixis, in *Embryology of Angiosperms* (ed. B.M. Johri), Springer, Berlin.

Nogler, G.A. (1984b) Genetics of apomixis in *Ranunculus auricomus*. V. Conclusion. *Bot. Helv.*, **94**, 411-422.

Nogler, G.A. (1995) Genetics of apomixis in *Ranunculus auricomus*. VI. Epilogue. *Bot. Helv.*, **105**, 111-115.

Nuez, F., Costa, J. and Cuartero, J. (1986) Genetics of the parthenocarpy for tomato varieties 'Sub-Arctic Plenty', '75/59' and 'Severianin'. *Z. Pflanzenzuecht*, **96**, 200-206.

Nybom, H. (1988) Apomixis versus sexuality in blackberries (*Rubus* subgen. *Rubus*, Rosaceae). *Plant System. Evol.*, **160**, 207-218.

Ohad, N., Yadegari, R., Margossian, L., *et al.* (1999) Mutations in *FIE*, a WD polycomb group gene, allow endosperm development without fertilization. *Plant Cell*, **11**, 407-415.

O'Neill, S.D. (1997) Pollination and regulation of flower development. *Ann. Rev. Plant Physiol. Plant Mol. Biol.*, **48**, 547-574.

O'Neill, S.D. and Nadeau, J.A. (1997) Post-pollination flower development. *Horti. Rev.*, **19**, 1-58.

Ozias-Akins, P., Roche, D. and Hanna, W.W. (1998) Tight clustering and hemizygosity of apomixis-linked molecular markers in *Pennisetum squamulatum* genetic control of apospory by a divergent locus that may have no allelic form in sexual genotypes. *Proc. Natl. Acad. Sci. USA*, **95**, 5127-5132.

Page, R.D.M. (1996) TreeView: an application to display phylogenetic trees on personal computers. *Comput. Appl. Biosci.*, **12**, 357-358.

Parlevliet, J.E. and Cameron, J.W. (1959) Evidence on the inheritance of nucellar embryony in *citrus*. *Proc. Am. Soc. Hort. Sci.*, **74**, 252-260.

Parinov, S., Sevugan, M., De, Y., Yang, W.C., Kumaran, M. and Sundaresan, V. (1999) Analysis of flanking sequences from dissociation insertion lines. A database for reverse genetics in *Arabidopsis. Plant Cell*, **11**, 2263-2270.

Pecaut, P. and Philouze, J. (1978) A *sha pat* line obtained by natural mutation. *Rep. Tom. Genet. Coop.*, **28**, 12.

Peel, M.D., Carman, J.G. and Leblanc, O. (1997) Megasporocyte callose in apomictic buffelgrass, Kentucky bluegrass, *Pennisetum squamulatum* Fresen, *Tripsacum* L, and weeping lovegrass. *Crop Sci.*, **37**, 724-732.

Peng, J., Carol, P., Richards, D.E., *et al.* (1997) The *Arabidopsis GAI* gene defines a signaling pathway that negatively regulates gibberellin responses. *Genes Dev.*, **11**, 3194-3205.

Pennel, R.I. and Roberts, K. (1990) Sexual development in the pea is presaged by the altered expression of arabinogalactan protein. *Nature*, **344**, 547-549.

Pessino, S.C., Ortiz, J.P. A., Leblanc, O., doValle, C.B., Evans, C. and Hayward, M.D. (1997) Identification of a maize linkage group related to apomixis in Brachiaria. *Theor. Appl. Genet.*, **94**, 439-444.

Phillips, A.L., Ward, D.A., Uknes, S., *et al.* (1995) Isolation and expression of three gibberellin 20-oxidase cDNA clones from *Arabidopsis*. *Plant Physiol.*, **108**, 1049-1057.

Philouze, J. (1983) Epistatic relations between *ls* and *pat-2*. *Rep. Tom. Genet. Coop.*, **33**, 9-12.

Pysh, L.D., Wysocka-Diller, J.W., Camilleri, C., Bouchez, D. and Benfey, P.N. (1999) The *GRAS* gene family in *Arabidopsis*: sequence characterization and basic expression analysis of the *SCARECROW-LIKE* genes. *Plant J.*, **18**, 111-119.

Quarin, C.L. (1999) Effect of pollen source and pollen ploidy on endosperm formation and seed set in pseudogamous apomictic *Paspalum notatum*. *Sex. Plant Reprod.*, **11**, 331-335.

Quarin, C.L., Norrmann, G.A. and Espinoza, F. (1998) Evidence for autoploidy in apomictic *Paspalum rufum*. *Hereditas*, **129**, 119-124.

Rice, P. (1994) *Genetics Computer Group EGCG Program Manual*, The Sanger Centre, Cambridge.

Rotino, G.L., Perri, E., Zottini, M., Sommer, H. and Spena, A. (1997) Genetic engineering of parthenocarpic plants. *Nature Biotechnology*, **15**, 1398-1401.

Savidan, Y. (1982) Nature et heredite de l'apomixie chez *Panicum maximum* Jacq. *Trav. Doc. ORSTOM*, **153**, 1-159.

Savidan, Y. (2000) Apomixis: genetics and breeding, in *Plant Breeding Reviews* (ed. J. Janick), vol. 18, Wiley, New York, pp. 13-86.

Scott, R.J., Spielman, M., Bailey, J. and Dickinson, H.G. (1998). Parent-of-origin effects on seed development in *Arabidopsis thaliana*. *Development*, **125**, 3329-3341.

Schneitz, K., Hülskamp, M., Kopczak, S. and Pruitt, R.E. (1997). Dissection of sexual organ ontogenesis: a genetic analysis of ovule development in *Arabidopsis thaliana*. *Development*, **124**, 1367-1376.

Schumacher, K., Schmitt, T., Rossberg, M., Schmitz, G. and Theres, K. (1999) The *LATERAL SUPPRESSOR* (*LS*) gene of tomato encodes a new member of the VHIID protein family. *Proc. Natl. Acad. Sci. USA*, **96**, 290-295.

Schwabe, W.W. and Mills, J.J. (1981) Hormones and parthenocarpic fruit set: a literature survey. *Hort. Abstr.*, **51**, 661-699.

Sedgley, M. and Griffin, A.R. (1989) *Sexual Reproduction of Tree Crops*, Academic, London.

Silverstone, A.L., Ciampaglio, C.N. and Sun, T.P. (1998) The *Arabidopsis RGA* gene encodes a transcriptional regulator repressing the gibberellin signal transduction pathway. *Plant Cell*, **10**, 155-169.

Skalinska, M. (1970) Further cytological studies in natural populations of *Hieracium aurantiacum* L. *Acta. Biol. Cracov. Ser. Bot.*, **13**, 111-118.

Skalinska, M. (1971) Experimental and embryological studies in facultative apomixis of *Hieracium aurantiacum* L. *Acta. Biol. Cracov. Ser. Bot.*, **16**, 139-155.

Skalinska, M. (1973) Further studies in facultative apomixis of *Hieracium aurantiacum* L. *Acta. Biol. Cracov. Ser. Bot.*, **16**, 121-137.

Sørenson, M.B., Chaudhury, A.M., Robert, H., Bancharel, E. and Berger, F. (2001) Polycomb group genes control pattern formation in plant seed. *Curr. Biol.*, **11**, 1-20.

Spillane, C., MacDougall, C., Stock, C., *et al.* (2000) Interaction of the *Arabidopsis* Polycomb group proteins *FIE* and *MEA* mediates their common phenotypes. *Curr. Biol.*, **10**, 1535-1538.

Sponsel, V.M., Schmidt, F.W., Porter, S.G., Nakayama, M., Kohlstruk, S. and Estelle, M. (1997) Characterization of new gibberellin-responsive semi-dwarf mutants of *Arabidopsis*. *Plant Physiol.*, **115**, 1009-1020.

Strahl, B.D. and Allis, C.D. (2000) The language of covalent histone modifications. *Nature*, **403**, 41-45.

Sun, G., Dilcher, D.L., Zeng, S. and Zhou, Z. (1998) In search of the first flower: a Jurassic angiosperm, *Archaefructus*, from northeast China. *Science*, **282**, 1692-1695.

Sun, T. (2000) Gibberellin signal transduction. *Curr. Opin. Plant. Biol.*, **3**, 374-380.

Swain, S.M., Ross, J.R. and Kamiya, Y. (1995a) Gibberellins and pea seed development. *Planta*, **195**, 426-433.

Swain, S.M., Ross, J.J., Reid, J.B. and Kamiya, Y. (1995b) Gibberellin and pea seed development: expression of the *lhi*, *ls* and *le5839* mutations. *Planta*, **195**, 426-433.

Szymkowiak, E.J. and Sussex, I.M. (1993) Effect of *lateral suppressor* on petal initiation in tomato. *Plant J.*, **4**, 1-7.

Theissen, G., Becker, A., Di Rosa, A., *et al.* (2000) A short history of MADS-box genes in plants. *Plant. Mol. Biol.*, **42**, 115-149.

Traveset, A. (1993) Deceptive fruits reduce seed predation by insects in *Pistacia terebinthus* L. (Anacardiaceae). *Evolution and Ecology*, **7**, 357-361.

Tucker, M.R., Paech, N.A., Willemse, M.T.M. and Koltunow, A.M.G. (2001) Dynamics of callose deposition and β-1,3-glucanase expression during reproductive events in sexual and apomictic *Hieracium. Planta*, **212**, 487-498.

van Baarlen, P., van Dijk, P., Hoekstra, R.F. and de Jong, J.H. (2000) Meiotic recombination in sexual diploid and apomictic triploid dandelions (*Taraxacum officinale* L.). *Genome*, **43**, 827-835.

van Overbeek, J., Conklin, M.E. and Blakeslee, A.F. (1941) Factors in coconut milk are essential for growth and development of very young *Datura* embryos. *Science*, **94**, 350-351.

Vardi, A., Neumann, H., Frydman-Shani, Y., Yaniv, Y. and Spiegel-Roy, P. (2000) Tentative model on the inheritance of juvenility, self-incompatibility and parthenocarpy. *Acta Hort.*, **535**, 199-205.

Vardy, E., Lapushner, D., Genizi, A. and Hewitt, J. (1989a) Genetics of parthenocarpy in tomato under a low temperature regime: I. Line RP75/79. *Euphytica*, **41**, 1-8.

Vardy, E., Lapushner, D., Genizi, A. and Hewitt, J. (1989b) Genetics of parthenocarpy in tomato under a low temperature regime: II. Cultivar 'Severianin'. *Euphytica*, **41**, 9-15.

Varoquaux, F., Blanvillain R., Delseny, M. and Gallois, P. (2000) Less is better: new approaches for seedless fruit production. *TIBTECH*, **18**, 233-242.

Vielle-Calzada, J.P., Nuccio, M.L., Budiman, M.A., *et al.* (1996) Comparative gene expression in sexual and apomictic ovaries of *Pennisetum ciliare* (L) Link. *Plant Mol. Biol.*, **32**, 1085-1092.

Vielle-Calzada, J.P., Thomas, J., Spillane, C., Coluccio, A., Hoeppner, M.A. and Grossniklaus, U. (1999) Maintenance of genomic imprinting at the *Arabidopsis MEDEA* locus requires zygotic DDM1 activity. *Genes Dev.*, **13**, 2971-2982.

Vinkenoog, R., Spielman, M., Adams, S., Fischer, R.L., Dickinson, H.G. and Scott, R.J. (2000) Hypomethylation promotes autonomous endosperm development and rescues post-fertilization lethality in *fie* mutants. *Plant Cell*, **12**, 2271-2282.

Vivian-Smith, A. and Koltunow, A.M. (1999) Genetic analysis of growth-regulator-induced parthenocarpy in *Arabidopsis. Plant Physiol.*, **121**, 437-452.

Vivian-Smith, A., Luo, M., Chaudhury, A. and Koltunow, A.M. (2001) Fruit development is actively restricted in the absence of fertilization in *Arabidopsis. Dev.*, **128**, 2321-2331.

Wang, W.-C. and Sundaresan, V. (2000) Genetics of gametophyte biogenesis in *Arabidopsis. Curr. Opin. Plant Biol.*, **3**, 53-57.

Western, T.L. and Haughn, G.W. (1999) *BELL1* and *AGAMOUS* genes promote ovule identity in *Arabidopsis thaliana. Plant J.*, **18**, 329-336.

Willemse, M.T.M. and van Went, J.L. (1984) The female gametophyte, in *Embryology of Angiosperms* (ed. B.M. Johri), Springer-Verlag, Berlin, pp. 159-191.

Williams, L. and Grafi, G. (2000) The retinoblastoma protein—a bridge to heterochromatin. *Trends Plant Sci.*, **5**, 239-240.

Wysocka-Diller, J.W., Helariutta, Y., Fukaki, H., Malamy, J.E. and Benfey, P.N. (2000) Molecular analysis of *SCARECROW* function reveals a radial patterning mechanism common to root and shoot. *Development*, **127**, 595-603.

Yadegari, R., Kinoshita, T., Lotan, O., *et al.* (2000) Mutations in the *FIE* and *MEA* genes that encode interacting polycomb proteins cause parent-of-origin effects on seed development by distinct mechanisms. *Plant Cell*, **12**, 2367-2381.

Yanofsky, M.F., Ma, H., Bowman, J.L., Drews, G.N., Feldmann, K.A. and Meyerowitz, E.M. (1990) The protein encoded by the *Arabidopsis* homeotic gene *AGAMOUS* resembles transcription factors. *Nature*, **346**, 35-39.

Yao, J.L., Dong, Y.H. and Morris, B.A.M. (2001) Parthenocarpic apple fruit production conferred by transposon insertion mutations in a MADS-box transcription factor. *Proc. Natl. Acad. Sci. USA*, **98**, 1306-1311.

Yasuda, S. (1930) Parthenocarpy induced by the stimulus of pollination in some plants of the *Solanaceae*. *Agric. Hort.*, **5**, 287-294.

Yasuda, S. (1935) Parthenocarpy caused by the stimulation of pollination in some plants of the *Cucurbitaceae*. *Agric. Horti.*, **10**, 1385-1390.

Zangerl, A.R., Berenbaum, M.R. and Nitao, J.K. (1991) Parthenocarpic fruits in wild parsnip: decoy defence against a specialist herbivore. *Evol. Ecol.*, **5**, 136-145.

Zarembinski, T.I. and Theologis, A. (1994) Ethylene biosynthesis and action: a case of conservation. *Plant Mol. Biol.*, **26**, 1579-1597.

Zhang, X.S. and O'Neill, S.D. (1993) Ovary and gametophyte development are coordinately regulated by auxin and ethylene following pollination. *Plant Cell*, **5**, 403-418.

Zimmermann, W. (1952) The main results of the 'telome theory'. *Paleobotanist*, **1**, 456-470.

8 Self-incompatibility in flowering plants: the ribonuclease-based systems

Jinhong Li and Ed Newbigin

8.1 An overview of the various self-incompatibility systems

Self-incompatibility (SI) is a genetic barrier to self-fertilisation that has arisen many times during the evolution of flowering plants. Self-fertilisation can occur following the transfer of pollen within a flower (self-pollination) or between flowers on an individual plant (geitonogamous pollination). By allowing pistils to distinguish the plant's own pollen (self-pollen) from pollen from another plant (non-self pollen), and only allowing non-self pollen to fertilise an ovule, SI prevents fertilisation after both self- and geitonogamous pollinations. This makes SI plants obligate outcrossers.

Another feature of these systems is that they regulate the outcome of pollinations between plants of the same species as well. Pollen recognition in most SI systems is controlled by a single genetic locus with many alleles: pistils will reject pollen grains that have the same allele of this locus as themselves. An SI species is thus divided into groups of plants that cannot fertilise each other because they have the same allele. Within a population, plants that either are siblings or are otherwise closely related will usually belong to the same incompatibility group. This helps SI species avoid the negative consequences of inbreeding that occur when closely related plants cross. Avoidance of inbreeding, either by preventing self-fertilisation or reducing the chances of fertilisation between closely related plants, is thought to have been a crucial factor in the evolution of SI systems and in their persistence in angiosperms.

SI systems also influence the geographical distribution of species. Because they are obligate outcrossers that require the presence of nearby compatible plants for reproduction, species with an SI system are less suited to long-distance dispersal than species capable of self-fertilisation, where a single individual can found a sexually reproducing colony. For this reason, SI species are found chiefly near the centre of distribution for the genera to which they belong, whereas related self-fertilising species are more common near the periphery of the distribution. This also helps explain why so few SI species are indigenous to remote oceanic islands like the Galapagos (McMullen, 1987).

SI systems have been found in 60 to 90 of the 320-odd angiosperm families (Charlesworth, 1985). All SI species from the same family have the same system. Systems from different families vary in features such as the way SI is controlled genetically, the manner in which incompatible pollen is rejected, and whether the members of different incompatibility groups can be distinguished by differences in floral form. In most families, the SI system is described as homomorphic because plants from different incompatibility groups are morphologically identical. In heteromorphic SI systems, differences in the way the stamens and styles are arranged within the flower identify the different incompatibility groups.

Heteromorphic systems have so far been found in 25 to 27 plant families and can be further classified according to whether there are two or three incompatibility groups (distyly and tristyl, respectively; Darwin, 1877; Ganders, 1979). In distylous systems, incompatibility is controlled by a single locus with two alleles, S and s (Gibbs, 1986). This system is typified by *Primula* in which the S allele is associated with short styles and long anther filaments ('thrum' form) and is dominant over the s allele, which is associated with long styles and short anther filaments ('pin' form). Figure 8.1a illustrates the basic principles of this system. The tristylous system, which is found in plants such as *Lythrum salicaria* (purple loosestrife), is less common than the distylous system. Two loci, each with two alleles, control expression of incompatibility types in most cases (Barrett, 1992).

Homomorphic SI systems are more widespread than heteromorphic systems. Many are monogenic systems controlled by a single locus, usually called S. A few are polygenic systems under the control of two or more loci. Figures 8.1b and c show the genetics controlling pollen rejection in the two types of monogenic homomorphic systems. Figure 8.1b shows rejection in a sporophytic system, where the haploid pollen grain expresses both the S alleles present in the diploid pollen-producing plant. Figure 8.1c shows rejection in a gametophytic system, where the pollen grain expresses one or other of the S alleles present in the diploid plant. In both cases, a haploid pollen grain alighting on a diploid pistil that expresses the same S allele is rejected. Incompatible pollen grains either do not germinate (stigmatic inhibition), or germinate but produce tubes that grow slowly and fail to reach the ovary before the flower senesces (stylar inhibition; Newbigin *et al.*, 1993; Nasrallah and Nasrallah, 1993). Pollen in gametophytic systems is usually rejected in the style and on the stigma in a sporophytic system. But this is not always so, and what distinguishes the two systems is the outcome of crosses between plants that have one S allele in common (middle panel in figures 8.1b and c). In a sporophytic system, such crosses are incompatible. To slightly complicate matters, some sporophytic systems have dominant and recessive S alleles, and these relationships also determine whether a cross is compatible or incompatible (Thompson and Taylor, 1966; Ockendon, 1974).

Reviews of the various SI systems have been published at regular intervals over the years (see, for example, Thompson and Kirch, 1992; Nasrallah and Nasrallah, 1993; Newbigin *et al.*, 1993; Barrett and Cruzan, 1994; Kao and McCubbin, 1996; Nasrallah, 2000; Jordan *et al.*, 2000; but particularly de Nettancourt, 2000). Here we review the strikingly similar gametophytic

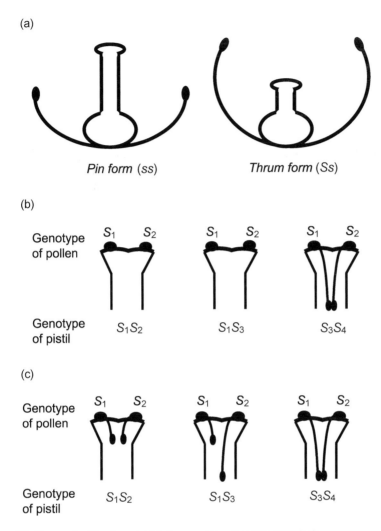

Figure 8.1 Three major SI systems. (a) A heteromorphic system. The example shown represents distyly in species such as *Primula*. Compatible pollinations occur between anthers and styles at the same level. Control is by an *S* locus with two alleles (*S* and *s*). The thrum form arises when the *S* allele is present. This theoretically can only exist in the heterozygous condition. The pin form arises in plants homozygous for the *s* allele. (b) A homomorphic sporophytic system. The example shown represents SI as seen in some species of *Brassica*. The pollen parent genotype is $S_1 S_2$. When an allele in the pollen parent matches that of the pistil (e.g., $S_1 S_2$ or $S_1 S_3$), pollen germination is arrested at the stigma surface. Where there is no match ($S_3 S_4$), the pollen may germinate and grow through the style towards the ovules. The central panel applies only if the S_1 allele is dominant to or codominant with the S_2 allele in the pollen and if S_1 is dominant to or codominant with S_3 in the style. If S_3 is dominant S_1 in the style, or if S_2 is dominant to S_1 in the pollen, pollen from the $S_1 S_2$ parent will be compatible. (c) A homomorphic gametophytic system. The example shown represents SI as seen in some species of *Nicotiana*. The pollen parent genotype is $S_1 S_2$. When an allele in an individual haploid pollen grain matches either allele in the diploid pistil, the pollen tube grows slowly within the style. For example, growth of both S_1 pollen and S_2 pollen is inhibited in an $S_1 S_2$ style, whereas S_2 pollen will grow successfully through the $S_1 S_3$ style. When there is no match of alleles (e.g., pollen grains from an $S_1 S_2$ plant on an $S_3 S_4$ pistil), all the pollen tubes can grow through the style toward the ovules.

systems found in three families of plants—the Solanaceae (nightshade family), the Rosaceae (cherries and apples) and the Scrophulariaceae (snapdragons and foxgloves)—paying particular attention to what is currently known about how these systems work. Gametophytic SI controlled by a single *S* locus has also been studied in the Papaveraceae. Although the genetics are the same, the Papaveraceae system and the systems found in the other three families work in different ways and are not evolutionarily related. Readers wanting to learn more about the cell and molecular biology of the Papaveraceae system are directed to the reviews recently published by Snowman *et al.* (2000) and Jordan *et al.* (2000).

8.2   Gametophytic self-incompatibility in three families

8.2.1   *Observations on compatible and incompatible pollinations*

Gametophytic SI (GSI) in the Solanaceae has long been viewed as a simple model in which to study cell recognition and consequently has been much studied at the cellular level. Fewer descriptions of pollen rejection in the Scrophulariaceae and Rosaceae have been published (for examples see Modlibowska, 1945; Tupy, 1959; Yoder, 1998; Austin *et al.*, 1998), but those that are available indicate this happens in a manner superficially similar to that in the Solanaceae. As other similarities described below support this suggestion, it appears likely that pollen rejection in the three families occurs by identical means.

When studied by light or electron microscopy, the compatible and incompatible pollen tubes of most solanaceous species show similar morphologies. Both grow normally through the stigma, but incompatible pollen tubes grow about 20 times more slowly than compatible pollen tubes after they enter the style. Although most incompatible tubes do not get beyond the upper third of the style (Cresti *et al.*, 1980), some do continue growing, albeit slowly, but still do not reach the ovules (Herrero and Dickinson, 1980; Lush and Clarke, 1997). *Anthocersis gracilis*, a species native to Australia, is an interesting exception to this general picture. After both self- and cross-pollinations, the placenta fills with pollen tubes that enter the micropyles of many ovules. Only the ovules of flowers that have been cross-pollinated will subsequently enlarge and develop into seeds (Stace, 1995)

Beside growing more slowly, the walls of incompatible pollen tubes are thicker than those of compatible tubes and have irregular deposits of the polysaccharide callose that accumulate behind the growing tip (de Nettancourt *et al.*, 1974; Lush and Clarke, 1997). Wall thickening is probably a secondary effect of incompatibility that arises because the tubes are growing slowly but are still depositing cell wall material at a rate similar to that of a compatible pollen tube. The primary effect of incompatibility therefore seems to be to reduce the rate at which pollen tubes grow. The incompatibility mechanism is reversible, as grafting experiments with the large styles of *Nicotiana alata* have shown that pollen tubes growing through an incompatible style can 'recover' (i.e. grow at the same rate as compatible pollen tubes) when they enter a section of compatible style (Lush and Clarke, 1997).

The most striking feature of incompatible pollen tubes when viewed with the electron microscope is the stacking of cisternae in the rough endoplasmic reticulum (de Nettancourt *et al.*, 1973; Walles and Han, 1998a). The intricate patterns of longitudinal and concentric stacks that form only a few hours after pollination are the earliest ultrastructural change seen in incompatible *Lycopersicon peruvianum* and *Brugmansia suaveolens* pollen tubes. As the formation of stacks is coincident with the onset of the slower growth rate, de Nettancourt *et al.* (1974) considered them relevant to an incompatibility mechanism that they suggested was probably based on a 'general cessation of protein synthesis'.

8.2.2   Genes and proteins associated with self-incompatibility in the Solanaceae, Rosaceae and Scrophulariaceae

The *S* phenotype of the pistil in SI plants from the Solanaceae, Rosaceae and Scrophulariaceae appears to be determined by an extracellular glycoprotein with ribonuclease activity known as the *S*-RNase (see Newbigin, 1996, Richman *et al.*, 1997, and references therein). The first *S*-RNase characterised in detail came from the solanaceous plant, *N. alata* (Anderson *et al.*, 1986). Similar proteins were subsequently found in self-incompatible members of the Rosaceae (*Pyrus serotina*; Sassa *et al.*, 1993) and Scrophulariaceae (*Antirrhinum*; Xue *et al.*, 1996). *S*-RNases are the only products of *S* locus so far identified in all three families.

In the Solanaceae, *S*-RNases are responsible for recognising incompatible pollen tubes and inhibiting their growth. This has been shown through transgenic experiments in *Petunia* and *Nicotiana* that indicate that a high level of *S*-RNase expression in the style is both necessary and sufficient for the SI response (Lee *et al.*, 1994; Murfett *et al.*, 1994). Studies with transgenic *Petunia* plants and self-compatible accessions of the self-incompatible species *L. peruvianum* have also shown that the ribonuclease activity of *S*-RNases is required for pollen rejection (Huang *et al.*, 1994; Royo *et al.*, 1994). Because *S*-RNases are expressed at high levels by mature styles and at low levels during pollen development, potentially they are the only product of the *S* locus (Dodds *et al.*, 1993; Clark and Sims, 1994). However, this appears unlikely, as raising and lowering the expression of *S*-RNases in pollen does not alter the SI response of transgenic plants (Dodds *et al.*, 1999). From this and other findings, it can be concluded that the solanaceous *S* locus has a bipartite structure with separate genes encoding the pollen and stylar products. Lewis (1949a) had earlier inferred the presence of separate pollen and stylar genes from observations of mutations that affected

SI in another plant family (the Onagraceae). The gene(s) controlling the pollen phenotype, usually called *pollen-S*, has still to be identified in any gametophytic SI system, and the way its product and the *S*-RNase interact to initiate the SI response is unknown.

Evidence for the involvement of *S*-RNases in the rosaceous SI system comes primarily from an analysis of a self-compatible mutant of Japanese pear, Osa-Nijisseiki (Sassa *et al.*, 1993, 1997). This plant is derived from Nijisseiki, a self-incompatible variety with the genotype $S_2 S_4$. Self-compatibility in Osa-Nijisseiki is thought to result from deletion of the S_4-RNase gene. Because this mutation first took place in a bud growing on a Nijisseiki tree, Osa-Nijisseiki is a periclinal chimera with the mutant S_4 allele in the L1 and L2 cell layers, and a wild-type S_4 allele in the L3 layer (Sassa *et al.*, 1997). Cells from the L1 layer form the transmitting tract, and Sassa *et al.* (1997) argue that the reason the styles of Osa-Nijisseiki cannot reject pollen bearing the S_4 allele is that these cells lack the S_4-RNase gene. However, Hiratsuka *et al.* (1999) have recently questioned whether the S_4-RNase gene has been deleted, as they could find a low amount of S_4-RNase in Osa-Nijisseiki styles. The S_4-RNase gene must be missing from cells of the L2 layer as well, as cells from this layer give rise to the gametes (pollen and egg cells) and the mutant phenotype is heritable (Sassa *et al.*, 1997; Ishimizu *et al.*, 1999a). Sassa *et al.* (1997) concluded that the rosaceous *S*-RNase gene, like the solanaceous gene, controls the style's SI response but not the pollen's SI response, because reciprocal crosses between Nijisseiki and Osa-Nijisseiki indicated the mutation only affects the SI phenotype of the style (this is described as a style-part mutation).

The only paper describing *S*-RNases from the Scrophulariaceae published so far is by Xue *et al.* (1996). Using primers based on the solanaceous sequences, three cDNAs were identified in *Antirrhinum* styles encoding proteins related to the *S*-RNases. Classical genetic mapping showed that the cDNAs represent different alleles of a gene linked to the *S* locus. *In situ* hybridisation showed that the gene is expressed in *Antirrhinum* styles in the same way that *S*-RNase genes are expressed in styles of solanaceous plants such as *N. alata*. The Scrophulariaceae and Solanaceae are closely related families and belong to a subclass of flowering plants called the Asteridae. Genealogical analysis of *S*-RNase sequences indicates that the SI systems in these two families have a single evolutionary origin (Xue *et al.*, 1996; Richman *et al.*, 1997). The Rosaceae belong to a different subclass (the Rosidae) and are hence only distantly related to the Scrophulariaceae and Solanaceae. It is unclear whether the three RNase-based SI systems are homologous in an evolutionary sense, or whether RNases have been recruited for this function many times (Richman *et al.*, 1997). Indeed, the time since divergence is so great that it may be impossible to determine this unequivocally. However, if the three systems are homologous, then the RNase-based system is likely to be widespread, as roughly 80% of all dicot families are descendants of the most recent common ancestor of the Rosidae and Asteridae.

Surface features are likely sites for interactions between stylar S-RNases and their pollen counterparts. Amino acid identity between S-RNases is low and very few residues are common to sequences from the Solanaceae, Rosaceae and Scrophulariaceae (Richman *et al.*, 1997; Ishimizu *et al.*, 1998). Alignments reveal five short blocks of conserved sequence that are interspersed by longer blocks of more variable sequence (Ioerger *et al.*, 1991; Tsai *et al.*, 1992; Ishimizu *et al.*, 1998). Among the conserved amino acids are Cys residues that are involved in disulfide bonds (Parry *et al.*, 1998; Ishimizu *et al.*, 1998) and His residues that are essential for RNase activity (Ishimizu *et al.*, 1995; Parry *et al.*, 1997). The two or three most variable regions of sequence are called hypervariable domains (HV domains) and characteristically contain hydrophilic residues that are presumably on the protein's surface. S-RNases are all glycosylated and carbohydrate side-chains must also contribute significantly to the protein's overall shape (Oxley and Bacic, 1995; Ishimizu *et al.*, 1999b). Thus, despite their sequence diversity, Anderson *et al.* (1989) concluded that S-RNases have similar structures composed of a framework of conserved regions held together by disulfide bonds between the conserved Cys residues, with N-glycosyl chains and hypervariable domains exposed on the surface. Crystallographic analyses that are currently in progress to determine the three-dimensional structure of S-RNases from the Solanaceae and Rosaceae should resolve remaining questions as to the shape of these molecules (Ida *et al.*, 2001; Matsuura *et al.*, 2001).

Given their high degree of sequence diversity and probable surface location, HV domains have often been suggested as determining allelic identity of the S-RNase through interactions with pollen-S. This idea has been tested by a number of groups using a 'domain-swap' approach. In these experiments, short stretches of sequence from one S-RNase gene are replaced with the corresponding sequence from another gene. The chimeric gene, containing sequences from two different S-RNase alleles, is then expressed in styles of transgenic plants, which are then tested to see whether they can reject pollen bearing either of the two S alleles used as sources of DNA. Domain-swap experiments are interesting, as they have not only helped us understand how allele identity is specified in the sequence of an S-RNase but also provided important insights into how S alleles evolve.

Contrary to expectations, most domain-swap experiments have resulted in S-RNases with no apparent allele identity. For example, Zurek *et al.* (1997) performed an extensive set of domain-swap experiments using different lengths of sequence from two *N. alata* S-RNases. Although the S-RNases used (from the S_{A2} and S_{C10} alleles) had quite divergent sequences, all the hybrid proteins were active ribonucleases and so were likely to have been folded correctly. Transgenic *Nicotiana* hybrids expressing either the S_{A2}-RNase or S_{C10}-RNase, reject pollen grains carrying the S_{A2} and S_{C10} alleles, respectively. However, transgenic plants expressing hybrid proteins composed of sequences from the S_{A2}-RNase and S_{C10}-RNase rejected neither S_{A2} nor S_{C10} pollen. This includes plants that expressed a protein differing from S_{A2}-RNase by only eight amino acids at the N-terminal end! Kao and McCubbin (1996) cite similar results from their domain-swap experiments with a pair of *P. inflata* S-RNases. Although S-RNases determine the style's allele identity, the way identity is specified within the sequence is clearly very complex.

Other domain-swap experiments involving pairs of closely related S-RNases show that identity can be exchanged between alleles. Matton *et al.* (1997) used two S-RNases from *Solanum chacoense* (from the S_{11} and S_{13} alleles) that differ in sequence by only ten amino acids, four of which are in two of the HV domains (called HVa and HVb). By replacing the HVa and HVb domains from the S_{11}-RNase with the corresponding region from the S_{13}-RNase, Matton *et al.* (1997) made a hybrid protein, S_{11}-HVab. Transgenic *S. chacoense* plants express-ing S_{11}-RNase in their styles rejected S_{11} pollen and accepted S_{13} pollen. Trans-genic plants that expressed S_{11}-HVab at a high level rejected S_{13} pollen and accepted S_{11} pollen. That is, the four amino-acid differences in the HVa and HVb domains determine the allelic specificity of S_{13}-RNase.

In summarising the domain-swap experiments described thus far, Verica *et al.* (1998) pointed out that the results are strongly influenced by the particular combination of S-RNase genes used. The *Solanum* results show that a few scattered differences in the HV domains can determine allele identity in pairs of highly similar S-RNases, while the *Petunia* and *Nicotiana* results show that residues outside the HV domains also have a role. Considered together, the results suggest that interactions with pollen-S probably occur over much of the S-RNase surface. This includes, but is not necessarily limited to, the HV domains. Not all surface features determine the allelic specificity of the S-RNase, however. For instance, modifying a single amino acid in a *Petunia* S-RNase so that a carbohydrate side-chain can no longer be attached at that position does not affect the protein's ability to cause incompatible pollen to be rejected (Karunanandaa *et al.*, 1994).

The domain-swap experiments have recently been taken a step further to provide an interesting evolutionary insight. Using the transgenic *S. chacoense* system already described, Matton *et al.* (1999) investigated which of the four amino acid differences in the S_{11}-HVab construct specified allele identity. Three of the differences are in the HVa domain and the other difference is in the HVb domain. Plants expressing an S_{11}-RNase with the S_{13}-RNase HVa domain sequence rejected neither S_{11} nor S_{13} pollen. Clearly, the single difference in the HVb domain is involved in allele identity. Another hybrid protein had substitutions at two of the three HVa sites and at the HVb site. Surprisingly, this caused styles to reject *both* S_{11} and S_{13} pollen, which led Matton *et al.* (1999) to describe the chimeric protein as a 'dual-specificity' S-RNase. This may be very significant in evolutionary terms because dual-specificity stylar factors could help solve one of the biggest puzzles presented by any S locus with a bipartite structure: how new S alleles arise.

On the face of it, this seems a fairly straightforward problem. New S alleles can arise from pre-existing S alleles by mutation, or from two different S alleles by recombination. However, attempts to create new alleles from existing alleles have so far been unsuccessful (Lewis, 1949b), presumably because, for the new allele to function, any change that alters identity must also preserve recognition between all S locus products. This means that coordinate changes must occur in both the pollen and stylar genes, something that is unlikely to happen simultaneously but could happen sequentially. The conceptual challenge has therefore been to describe a pathway to a new S allele that involves a change in one gene and then the other, while maintaining allelic recognition and thus preserving SI in all intermediates.

Matton *et al.*'s (1999) two-step pathway begins with a mutation that lets the S-RNase recognise both the existing S allele and a new pollen specificity not yet present in the population (that is, the S-RNase has dual specificity). In the next step, a change to *pollen-S* creates a new pollen specificity. In the final step, further mutation to the S-RNase restricts rejection to the new pollen specificity alone. Figure 8.2 illustrates these steps. Readers interested in an extended discussion on dual-specificity alleles and their implications should see Charlesworth (2000a,b) and Uyenoyama and Newbigin (2000). For an analysis of pathways to new S alleles that permit self-compatibility at an intermediate step, see Uyenoyama *et al.* (2001).

The identity of *pollen-S* and its product are clearly necessary to understanding of the molecular basis of SI in the Solanaceae, Scrophulariaceae and Rosaceae, and to testing ideas about how these systems might have evolved. Cloning this gene has, however, been more difficult than might be expected, given the ease with which S-RNases can be seen on protein gels of stylar extracts (Bredemeijer and Blass, 1981; Sassa *et al.*, 1993). Since similarly abundant and polymorphic proteins have not yet been identified in pollen extracts, researchers have been forced to take other experimental approaches to find *pollen-S*. These searches are perhaps most advanced in the Solanaceae, where both protein and genetic approaches have been tried. The genetic approach usually uses one of the many variants of the polymerase chain reaction to identify pollen-expressed genes that are differentially expressed in plants with different S alleles (Li *et al.*, 2000; McCubbin *et al.*, 2000a). In the protein approach, a technique such as affinity chromatography or the yeast two-hybrid screen is used to identify proteins that can to bind to the S-RNase (Dowd *et al.*, 2000). Of the two, the genetic approach has so far been the more successful.

A third approach, still untried in any of the GSI systems, is to sequence large insert clones containing DNA from an S allele. This can be done, for example, by isolating and sequencing S-RNase-containing clones from a yeast or bacterial artificial chromosome library, such as those from the SI species potato (Leister *et al.*, 1997) and *Petunia inflata* (McCubbin *et al.*, 2000b). Genomic DNA sequences can then be scanned with software developed by the various

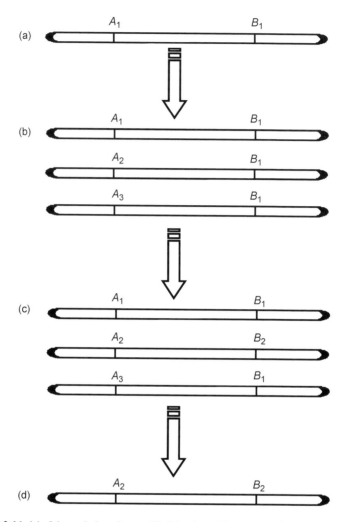

Figure 8.2 Model of the evolution of a new SI allele. A_1 and B_1 are respectively the style and pollen specificity genes of the S_1 allele. The initial state of the allele is shown in (a). In (b), neutral mutations in the style specificity gene increase its genetic diversity. The new forms of A (A_2 and A_3) are functionally the same as A_1. In (c), a rare mutation in B_1 leads to the formation of B_2, which can recognise A_2 but not A_1. This creates a new allele (S_2) that, owing to its selective advantage over S_1, can spread through the population and eventually replace the older allele (d).

sequencing projects to find potential genes. Suzuki *et al.* (1999) describe this type of analysis for a 75 kb segment of DNA from the *Brassica campestris* S locus, a species with a sporophytic type of SI system. One of the genes they found was subsequently shown to be the *pollen-S* gene from this system (Schopfer *et al.*, 1999). The main problem with this approach, at least for the

Solanaceae, is that the *S* locus in many species is near a centromere (Bernacchi and Tanksley, 1997; Entani *et al.*, 1999; Golz *et al.*, 2001). Gene density in this region of a chromosome is low, meaning that genes at or near the *S* locus may be difficult to find because they are separated by large amounts of mostly repetitive DNA. Indeed, one solanaceous *S* locus is estimated to contain more than a megabase of DNA (McCubbin and Kao, 1999). This is so much DNA that even a set of overlapping *S*-RNase clones cannot be guaranteed to cover it completely. Sequencing may therefore be better suited to species where a smaller genome means less DNA at the *S* locus. Examples can be found in the Rosaceae, where the *Prunus* genome is only about twice the size of the *Arabidopsis* genome. This compares to the *N. alata* genome, which is more than twenty times the size of the *Arabidopsis* genome.

Regardless of how the search for *pollen-S* is done, all approaches should provide a crop of candidate molecules whose involvement in SI needs to be assessed somehow. One way is to sort through the candidates and only consider those genes that have the features expected of *pollen-S*. That is, only to analyse those genes that are polymorphic (each *S* allele should have a different allele of *pollen-S*), expressed late in pollen development, and genetically close to the *S*-RNase gene. In this regard, it is instructive to reflect on the experiments to date to see how often these features have been observed.

Li *et al.* (2000) pooled RNA samples based on *S* genotype and used differential display to identify three pollen-expressed cDNA fragments that were encoded by genes closely linked to the *N. alata S* locus. Using a similar approach, McCubbin *et al.* (2000a) identified 10 pollen cDNA fragments from the *P. inflata S* locus. The genes detected by all these fragments are polymorphic. Perhaps the best-characterised so far is a gene called *48A* that has all the hallmarks of *pollen-S*, such as a high degree of polymorphism, cosegregation with the *S* locus, and abundant expression in mature pollen and anthers that contain postmeiotic pollen (Li *et al.*, 2000). *48A* also appears not to be expressed in vegetative tissues. As satisfying as this is, we must now consider whether genes other than *pollen-S* also have these features.

Large numbers of genes are expressed in pollen. Mascarenhas (1993) estimated the pool of RNA present in the pollen grain at anthesis is the product of around 20 000 different genes. By chance, some of these 20 000 genes may be near the *S* locus. Owing to its involvement in SI, the *S*-RNase gene is under a form of diversifying selection called 'balancing selection'. This form of selection will also affect *pollen-S*. Another possible source of variation at the *S* locus is an effect called 'hitchhiking' (Kreitman and Akashi, 1995). This term describes changes in the frequency of a selectively neutral mutation in a population that are brought about through linkage to a nearby mutation that is selected. In the case of the *S* locus, a lack of recombination (well known at the *S* locus, see Li *et al.*, 2000 for an example) and the extreme age of alleles (lineages often predate speciation events, see Ioerger *et al.*, 1990) also act to increase levels of DNA variation. Consequently, all the genes at this locus, not just those involved in SI, can be polymorphic. Each S allele (or S allele lineage) may therefore have its own unique set of variant forms of these genes. Schierup *et al.* (2000) investigated hitchhiking at the S locus further using computer simulations and described ways of studying this effect in wild populations. However, determining whether variation in a gene is due to balancing selection or hitchhiking still poses problems for investigators searching for *pollen-S*. Thus, although a useful guide, the criteria used currently are not particularly restrictive and may leave a large field of candidates that need to be further tested in a transgenic plant. The question then becomes 'What phenotype will a *pollen-S* transgene cause?' How will an S_1 pollen grain with a *pollen-*S_2 transgene behave? Will it behave like an S_2 pollen grain? Will it express both the S_1 and S_2 alleles? Or will it appear not to express any S alleles at all? A possible answer appears *inter alia* in the next section and comes from an understanding of how S-RNases are thought to enter pollen tubes and retard their growth.

8.2.3   *Reviewing the current models of self-incompatibility*

Figure 8.3 shows two models of pollen rejection in the Solanaceae that may also reflect events in the Rosaceae and Scrophulariaceae as well. Although the models differ in the way self- and non-self-pollen tubes are distinguished (see below), both assume that S-RNases act as intracellular toxins within incompatible pollen tubes. Because it is currently thought that ribosomal rRNA genes are not transcribed in the pollen of flowering plants after germination (Mascarenhas, 1993), rRNA has been proposed as the S-RNase substrate. Retarded growth of incompatible pollen tubes due to a loss of rRNA, and hence a loss of functional ribosomes (McClure *et al.*, 1990), conforms to de Nettancourt *et al.*'s (1974) earlier suggestion that a 'general cessation of protein synthesis' is the cause of pollen tube rejection.

While central to the models, it has not been easy to show that S-RNases are cytotoxins or that RNA hydrolysis plays a pivotal role in pollen tube rejection. *N. alata* S-RNases are toxic to *Nicotiana* suspension-culture cells (Lush and Clarke, 1995), but surprisingly do not markedly inhibit *N. alata* pollen tubes growing in culture (Jahnen *et al.*, 1989). *In vitro*, S-RNases can degrade any RNA and show no marked preference for nucleotide sequence. However, S-RNases degrade RNA at rates that vary from high to barely discernible (McClure *et al.*, 1989; Parry *et al.*, 1998). The rRNA of incompatible, but not compatible, pollen tubes is degraded during growth in *N. alata* styles (McClure *et al.*, 1990). However, it remains to be seen whether RNA degradation in an incompatible pollen tube is a cause or consequence of rejection. Nor is there any marked breakdown of ribosomes during an incompatible response, as Walles and Han (1998b)

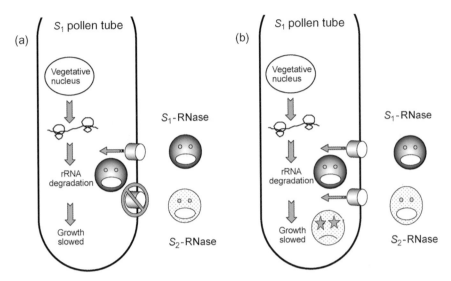

Figure 8.3 Two models of the RNase-based, gametophytic SI system. The style in both panels is $S_1 S_2$ and the pollen tube is S_1. (a) shows the translocator model and (b) shows the inhibitor model. See text for details.

recently showed that the number of ribosomes per unit of rough endoplasmic reticulum membrane in the incompatible pollen tubes of *Brugmansia suaveolens* and *N. alata* does not change over time as expected.

This leads us to ask whether pollen tubes can synthesise rRNA. Although pollen tubes are generally thought to rely on rRNA synthesised during pollen development (Mascarenhas, 1993), Tupy *et al.* (1977) found that *N. tabacum* pollen tubes can synthesise rRNA. Moreover, the grafting experiments reported by Lush and Clarke (1997) show that incompatible pollen tubes can also 'recover' when allowed to grow into a compatible style, indicating that rejection is reversible in at least some pollen tubes. So perhaps *Nicotiana* pollen tubes can synthesise rRNA after all. This might account for both the recovery and lack of ribosome loss seen in incompatible tubes. But what if rRNA hydrolysis were not central to the mechanism of pollen tube rejection? What if incompatibility relied instead on the hydrolysis of some other type of RNA molecule?

One recent suggestion is that rejection could be mediated by a change in membrane permeability or stability, a possibility that could be consistent with the stacking of membranes seen in incompatible pollen tubes. This suggestion comes from the observations made by MacIntosh *et al.* (2001) on *RNY1*, the only gene in the yeast genome coding for an extracellular RNase of the *S*-RNase type. Yeast cells that cannot make Rny1 because of mutations are larger than wild-type cells and have bigger vacuoles; they also cannot grow in high salt or at elevated temperatures. These are features indicative of cells that are osmotically sensitive. Mutations in *RNY1* are complemented by Rny1 itself and by any one of several related RNase genes from plants. Surprisingly, a gene for an unrelated mammalian RNase (RNase A) also complements the mutation, but a version of this gene that encodes an enzymatically inactive form of RNase A does not. This demonstrates that it is the ability to hydrolyse RNA and not some feature of the protein's structure that is key to Rny1 function in a wild-type cell. The substrate that Rny1 degrades to regulate membrane permeability is not known (membrane-associated RNA has been put forward as one possibility; MacIntosh *et al.*, 2001), but the speculation that *S*-RNases and the related yeast enzyme act on similar substrates is particularly interesting and worth further investigation.

Although the pollen substrate for *S*-RNase is still not clear, more progress has been made on the mechanism used to distinguish incompatible and compatible pollen tubes. The two models referred to above are the receptor or translocator model and the inhibitor model (figure 8.3). The receptor model proposes that the *pollen-S* product specifically recognises an *S*-RNase from a matching *S* allele and allows it to enter the pollen tube. Other *S*-RNases are prevented from entering the pollen tube and therefore cannot act as intracellular cytotoxins. The inhibitor model proposes that *S*-RNases enter the pollen tube nonspecifically. Most are inactivated in the cytoplasm by the *pollen-S* product, which is an *S*-RNase inhibitor. However, pollen-*S* cannot inhibit *S*-RNases derived from the same *S* allele, which therefore remain active in the pollen tube cytoplasm (Dodds *et al.*, 1996).

The two models make different predictions about the accumulation of *S*-RNases by pollen tubes and the mutability of *pollen-S*. According to the receptor model, *S*-RNases cannot enter compatible pollen tubes and pollen tubes that do not make pollen-*S* cannot be rejected by an incompatible style. Hence the *pollen-S* gene is not required for pollen tube survival and can be deleted. The inhibitor model predicts that *S*-RNases accumulate in both compatible and incompatible pollen tubes, and that because they cannot detoxify *S*-RNases, pollen tubes that cannot make pollen-*S* are rejected by any style. Hence, according to the inhibitor model, deleting the *pollen-S* gene is lethal.

Recent immunocytochemistry experiments showing that *S*-RNases enter both compatible and incompatible pollen tubes provides strong support for the inhibitor model (Luu *et al.*, 2000). Using an antibody raised to the HV region of an *S*-RNase, Luu *et al.* (2000) found heavy labelling inside both types of tubes in *S. chacoense*. Labelling was mostly in the cytoplasm near the tips of pollen tubes, and was absent from regions farther back in the tube that are devoid of cytoplasm. Intriguingly, approximately ten times more label was seen in pollen tubes than in the extracellular spaces of the style where *S*-RNases accumulate. This may indicate that *S*-RNases do not just diffuse into the pollen tube but are actively taken up. It also poses a problem, because pollen-*S* does not appear to be very abundant, or at least probably has not been seen using any of the techniques tried so far. Yet in its current form the inhibitor model proposes that at least one pollen-*S* molecule is needed to bind to and inactivate each *S*-RNase molecule. Unless binding is transient and inactivation is irreversible, this would require at least as many pollen-*S* molecules as *S*-RNase molecules in the pollen tube.

The question also needs to be raised of how the roughly 30 kDa *S*-RNase enters the pollen tube. Although the data show this to happen, the route used is not known. Endocytosis allows molecules as large as dextrans of M_r 9400 (but not larger) to get into tobacco pollen tubes growing in culture (O'Driscoll *et al.*, 1993), but if this route were used the *S*-RNase would be enclosed in a membrane-bound vesicle and unable to access potential substrates in the cytoplasm.

Mutational studies also support the inhibitor model. Mutations likely to cause the deletion of a gene can be made by exposing cells to a source of ionising radiation. Golz *et al.* (1999) exposed immature anthers of *N. alata* to ionising radiation, then collected pollen at anthesis and used it to pollinate an incompatible style. The vast majority of the pollen was rejected, but a few of the irradiated grains produced tubes that grew through the incompatible styles and fertilised ovules. Plants from the resulting seeds were analysed for changes to their SI phenotype and *S* genotype. Most were self-compatible owing to mutations affecting the pollen function of the *S* locus (these plants are called pollen-part mutants or PPMs and are described in detail by Golz *et al.*, 2000). Inheritance of the mutant phenotypes was followed through a series of crosses to unmutated plants. DNA blotting with *S*-RNase cDNA probes and the *S*-linked cDNA fragments identified by Li *et al.* (2000) was used to determine the *S* genotypes of parents and progeny (Golz *et al.*, 1999, 2001).

It was clear from these experiments that the PPM phenotype in all plants was caused by the presence of an extra *S* allele in the genomes of some pollen grains. Most of the PPMs had an entire *S* allele, while a few had only part of an extra allele; this part included *pollen-S* but did not include the *S*-RNase gene (Golz *et al.*, 1999, 2001). Some of the extra *S* alleles were carried on small additional chromosomes called 'centric fragments', while others could not be observed cytologically because they were part of a chromosome. Even though more than 5 million irradiated pollen grains were examined, none of the PPMs had a deletion of *pollen-S*. This makes deletions of *pollen-S* either exceedingly rare or lethal to the pollen grain. Moreover, as only one cause for the PPM phenotype was found, either the pathway leading to pollen rejection requires no other pollen gene products except pollen-*S*, or the other products cannot be mutated because they too are essential to the pollen grain.

The way the duplicated *S* alleles and the PPM phenotype were inherited is also consistent with the inhibitor model (Golz *et al.*, 2001). Pollen tubes could only grow through an otherwise incompatible style if the *S* allele at the

S locus and the extra *S* allele were different. This is known as a competitive interaction, and a similar phenomenon is also seen in self-compatible tetraploid plants derived from self-incompatible diploid plants (de Nettancourt, 2000; Chawla *et al.*, 1997). Pollen tubes carrying two different *S* alleles will produce two different types of pollen-*S* inhibitor. In combination, these can inactivate *S*-RNase encoded by any *S* allele and hence allow the pollen tube to grow through any style. To answer the question posed earlier: So long as *pollen-S* transgenes behave like an extra *S* allele, plants with such transgenes should behave like PPMs.

The inhibitor model does not currently delineate some aspects of SI. These aspects are how pollen-*S* and an *S*-RNase from the same *S* allele interact, and the role played by a class of genes known as modifiers. The inhibitor model proposes that pollen-*S* can inhibit the *S*-RNases encoded by other *S* alleles. However, as the domain-swap experiments have already shown, interactions also take place between products of the same *S* allele. Thus it appears that pollen-*S* can interact with an *S*-RNase in either an inhibitory way or an allele-specific way. This may mean that *S*-RNases have on their surface sites for two types of interaction with pollen-*S*. The avidity of pollen-*S* for these sites may be what determines where binding occurs. Pollen-*S* will preferentially bind to the allele-specific site if its avidity for this site is higher than for the inhibitory site. But what happens following an allele-specific interaction? *S*-RNase activation has been put forward as one possibility (Luu *et al.*, 2000), but the presence of extra copies of *S* alleles in the PPMs appears to this rule out.

If there are two sites on the *S*-RNase, which residues make up in the inhibitory site and which the allele-specific site? Although domain-swap experiments have been extremely helpful in identifying some of the residues in the allele-specific site, this approach cannot be used to examine the inhibitory site, because this site must be on all *S*-RNases. Domain-swap experiments can only be used to study those features that are allele-specific.

The inhibitor model does not mention the role of modifiers in SI. These are factors required for SI but not the specificity of pollen rejection. At least some are stylar factors, as implied by observations such as the inability of pollen tubes growing in culture to respond in an allele-specific manner to the addition of *S*-RNases to the medium (Jahnen *et al.*, 1989). Genetic evidence for modifiers also exists (Ai *et al.*, 1991).

Recently a gene for a small, asparagine- and aspartic acid-rich protein called HT was cloned from the styles of *N. alata* (McClure *et al.*, 1999). The HT protein by itself is not sufficient for allele-specific rejection, but *Nicotiana* plants that accumulate less-than-normal amounts of HT owing to an antisense construct cannot reject incompatible pollen tubes. The HT protein is acidic and so may interact with a basic protein like an *S*-RNase. It appears, however, that HT/*S*-RNase complexes do not form and HT possibly interacts with pollen tubes instead (McClure *et al.*, 1999).

8.3   Conclusions—the utility of models

If the arguments in the foregoing section have been convincingly presented, then the reader should be in little doubt that the translocator model is the poorer of the two SI models. But what of the alternative model, which also has limitation that have been noted here? Since the inhibitor model also is not an adequate description of the SI mechanism, would it not be better if we just left off all discussion of models until we know enough to develop one that is satisfactory?

Models are not authoritative doctrines; they are our attempt to bring disparate observations together into a working draft of what is going on inside a pollinated style. Models let us think in different ways and help us to design experiments that test the model's basic assumptions. Experiments that are not guided by a clear model are often quite difficult to interpret and important observations may be overlooked simply because we do not understand what they mean. Alternatively, we can also run the risk of being overimaginative if we attempt to fit each observation, no matter how unlikely, to a particular model or scheme.

Some features of the SI response predicted by the models have not yet been explored, although the tools needed for these experiments either are already available or could be made. For example, in the translocator model, allele recognition occurs outside the pollen tube, whereas it occurs inside the pollen tube in the inhibitor model. The style in both cases can be considered merely a source of S-RNases. So what happens when the pollen tube itself expresses an S-RNase? According to the translocator model any S-RNase can cause an incompatibility response once it is in the pollen tube, but according to the inhibitor model this can only occur when the S-RNase allele matches the pollen grain's allele. One way to find out what actually happens would be to produce a number of stably transformed plants and see whether a pollen-expressed S-RNase transgene can be transmitted through the male line. The translocator model says that that it cannot, while the inhibitor model says that it can so long as the S genotype of the pollen grain and S-RNase are not the same. A quicker way to do this would be to transiently express S-RNase constructs in pollen and study the growth of tubes in culture. Such experiments are at least conceivable, as Wang *et al.* (1998) have recently described a method to express genes in deexined *Nicotiana* pollen grains. Likewise, the current models predict that one possible way to block the incompatibility response is to make a protein in the pollen tube's cytoplasm that can specifically bind to the S-RNase. No natural proteins with this interesting property are known at present, although advances in immunology suggest ways in which they could be made synthetically (Whitelam and Cockburn, 1996).

Note that both experiments can be done without knowing the sequence of *pollen-S*. So, although it is an important piece of the puzzle and a goal of several research groups, not all experiments on SI need this sequence. Indeed, while waiting for this long-awaited discovery to be made, we can continue to advance our understanding of the mechanism of pollen rejection by testing ideas of how pollen-*S* works.

Acknowledgements

We thank colleagues at the School of Botany and elsewhere who gladly provided their suggestions and comments on this paper. J.H.L. was supported by an Australian Development Cooperation Scholarship. SI research in the Plant Cell Biology Research Centre is supported by grants from the Australian Research Council.

References

Ai, Y., Kron, E. and Kao, T.-h. (1991) *S*-alleles are retained and expressed in a self-compatible cultivar of *Petunia hybrida*. *Mol. Gen. Genet.*, **230**, 353-358.

Anderson, M.A., Cornish, E.C., Mau, S.L., *et al.* (1986) Cloning of cDNA for a stylar glycoprotein associated with expression of self-incompatibility in *Nicotiana alata* Link et Otto. *Nature*, **321**, 38-44.

Anderson, M.A., McFadden, G.I., Bernatzky, R., *et al.* (1989) Sequence variability of three alleles of the self-incompatibility gene of *Nicotiana alata*. *Plant Cell*, **1**, 483-491.

Austin, P.T., Hewett, E.W., Noiton, D. and Plummer, J.A. (1998) Self-incompatibility and temperature affect pollen tube growth in 'Sundrop' apricot (*Prunus armeniaca* L.) *J. Hort. Sci. Biotech.*, **73**, 375-386.

Barrett, S.C.H. (1992) Heterostylous genetic polymorphisms: model systems for evolutionary analysis, in *Monographs on Theoretical and Applied Genetics 15. Evolution and Function of Heterostyly* (ed. S.C.H. Barrett), Springer-Verlag, Berlin, pp. 1-29.

Barrett, S.C.H. and Cruzan, M.B. (1994) Incompatibility in heterostylous plants, in *Genetic Control of Self-Incompatibility and Reproductive Development in Flowering Plants* (eds. E.G. Williams, A.E. Clarke and R.B. Knox), Kluwer Academic, Dordrecht, pp. 189-219.

Bernacchi, D. and Tanksley, S.D. (1997) An interspecific backcross of *Lycopersicon esculentum* × *L. hirsutum*: linkage analysis and a QTL study of sexual compatibility factors and floral traits. *Genetics*, **147**, 861-877.

Bredemeijer, G.M.M. and Blass, J. (1981) *S*-specific proteins in styles of self-incompatible *Nicotiana alata*. *Theor. Appl. Genet.*, **59**, 185-190.

Charlesworth, D. (1985) Distribution of dioecy and self-incompatibility in angiosperms, in *Evolution— Essays in Honour of John Maynard Smith* (eds. J.J. Greenwood and M. Slatkin), Cambridge University Press, Cambridge, pp. 237-268.

Charlesworth, D. (2000a) How can two-gene models of self-incompatibility generate new specificities? *Plant Cell*, **12**, 309-310.

Charlesworth, D. (2000b) Unlocking the secrets of self-incompatibility. *Curr. Biol.*, **10**, R184-R186.

Chawla, B., Bernatzky, R., Liang, W. and Marcotrigiano, M. (1997) Breakdown of self-incompatibility in tetraploid *Lycopersicon peruvianum*: Inheritance and expression of *S*-related proteins. *Theor. Appl. Genet.*, **95**, 992-996.

Clark, K.R. and Sims, T. (1994) The S-ribonuclease gene of *Petunia hybrida* is expressed in nonstylar tissue, including immature anthers. *Plant Physiol.*, **106**, 25-36.

Cresti, M., Ciampolini, F. and Sarfatti, G. (1980) Ultrastructural investigations on *Lycopersicum peruvianum* pollen activation and pollen tube organization after self- and cross-pollination. *Planta*, **150**, 211-217.

Darwin, C. (1877) *The Different Forms of Flowers on Plants of the Same Species*, John Murray, London.

de Nettancourt, D. (2000) *Incompatibility and Incongruity in Wild and Cultivated Plants*, Springer Verlag, Berlin.

de Nettancourt, D., Devreux, M., Bozzini, A., Cresti, M., Pacini, E. and Sarfatti, G. (1974) Genetical and ultrastructural aspects of self- and cross-incompatibility in interspecific hybrids between self-compatible *Lycopersicum esculentum* and self-incompatible *L. peruvianum. Theor. Appl. Genet.*, **44**, 278-288.

de Nettancourt, D., Devreux, M., Bozzini, A., Cresti, M., Pacini, E. and Sarfatti, G. (1973) Ultrastructural aspects of the self-incompatibility mechanism in *Lycopersicum peruvianum* Mill. *J. Cell Sci.*, **12**, 403-419.

Dodds, P.N., Bönig, I., Du, H., *et al.* (1993) The S-RNase gene of *Nicotiana alata* is expressed in developing pollen. *Plant Cell*, **5**, 1771-1782.

Dodds, P.N., Clarke, A.E. and Newbigin, E. (1996) A molecular perspective on pollination in flowering plants. *Cell*, **85**, 141-144.

Dodds, P.N., Ferguson, C., Clarke, A.E. and Newbigin, E. (1999) Pollen-expressed S-RNases are not involved in self-incompatibility in *Lycopersicon peruvianum. Sex. Plant Reprod.*, **12**, 76-87.

Dowd, P.E., McCubbin, A.G., Wang, X., *et al.* (2000) Use of *Petunia inflata* as a model for the study of solanaceous type self-incompatibility. *Ann. Bot.*, **85** (suppl. A), 87-93.

Entani, T., Iwano, M., Shiba, H., Takayama, S., Fukui, K. and Isogai, A. (1999) Centromeric localization of an *S*-RNase gene in *Petunia hybrida* Vilm. *Theor. Appl. Genet.*, **99**, 391-397.

Ganders, F.R. (1979) The biology of heterostyly. *NZ J. Bot.*, **17**, 607-635.

Gibbs, P.E. (1986) Do homomorphic and heteromorphic self-incompatibility systems have the same sporophytic mechanism? *Plant Syst. Evol.*, **154**, 285-323.

Golz, J.F., Su, V., Clarke, A.E. and Newbigin, E. (1999) A molecular description of mutations affecting the pollen component of the *Nicotiana alata S* locus. *Genetics*, **152**, 1123-1135.

Golz, J.F., Clarke, A.E. and Newbigin, E. (2000) Mutational approaches to the study of self-incompatibility: Revisiting the pollen-part mutants. *Ann. Bot.*, **85** (suppl. A), 95-103.

Golz, J.F., Oh, H.Y., Su, V., Kusaba, M. and Newbigin E. (in press) Genetic evidence for an *S*-ribonuclease inhibitor at the *S* locus of *Nicotiana alata, Proc. Natl. Acad. Sci. USA*, in press.

Herrero, M. and Dickinson, H.G. (1980) Pollen tube growth following compatible and incompatible intraspecific pollinations in *Petunia hybrida. Planta*, **148**, 217-221.

Hiratsuka, S., Nakashima, M., Kamasaki, K., Kubo, T. and Kawai, Y. (1999) Comparison of an S-protein expression between self-compatible and -incompatible Japanese pear cultivars. *Sex. Plant Reprod.*, **12**, 88-93.

Huang, S., Lee, H.-S., Karunanandaa, B. and Kao, T.-h. (1994) Ribonuclease activity of *Petunia inflata* S proteins is essential for rejection of self-pollen. *Plant Cell*, **6**, 1021-1028.

Ida, K., Shinkawa, T., Norioka, S., *et al.* (2001) Crystallization and preliminary X-ray crystallographic analysis of S-allelic glycoprotein S_{F11}-RNase from *Nicotiana alata. Acta Crystallogr. Sect. D*, **57**, 143-144.

Ioerger, T.R., Clarke, A.G. and Kao, T.-h. (1990) Polymorphism at the self-incompatibility locus in Solanaceae predates speciation. *Proc. Natl. Acad. Sci. USA*, **87**, 9732-9735.

Ioerger, T.R., Gohlke, J.R., Xu, B. and Kao, T.h. (1991) Primary structural features of the self-incompatibility protein in Solanaceae. *Sex. Plant Reprod.*, **4**, 81-87.

Ishimizu, T., Miyagi, M., Norioka, S., Liu, Y.-H., Clarke, A.E. and Sakiyama, F. (1995) Identification of histidine 31 and cysteine 95 in the active site of self-incompatibility associated S_6 RNase in *Nicotiana alata. J. Biochem.*, **118**, 1007-1013.

Ishimizu, T., Endo, T., Yamaguchi-Kabata, Y., Nakamura, K.T., Sakiyama, F. and Norioka, S. (1998) Identification of regions in which positive selection may operate in *S*-RNase of

Rosaceae: implication for *S*-allele-specific recognition sites in *S*-RNase. *FEBS Lett.*, **440**, 337-342.

Ishimizu, T., Inoue, K., Shimonaka, M., Saito, T., Terai, O. and Norioka, S. (1999a) PCR-based method for identifying the *S*-genotypes of Japanese pear cultivars. *Theor. Appl. Genet.*, **98**, 961-967.

Ishimizu, T., Mitsukami, Y., Shinkawa, T., *et al.* (1999b) Presence of asparagine-linked *N*-acetylglucosamine and chitobiose in *Pyrus pyrifolia* *S*-RNases associated with gametophytic self-incompatibility. *Eur. J. Biochem.*, **263**, 624-634.

Jahnen, W., Lush, W.M. and Clarke, A.E. (1989) Inhibition of *in vitro* pollen tube growth by isolated *S*-glycoproteins of *Nicotiana alata*. *Plant Cell*, **1**, 501-510.

Jordan, N.D., Ride, J.P., Rudd, J.J., Davies, E.M., Franklin-Tong, V.E. and Franklin, F.C.H. (2000) Inhibition of self-incompatible pollen in *Papaver rhoeas* involves a complex series of cellular events. *Ann. Bot. (Suppl. A)*, **85**, 197-202.

Kao, T.-h. and McCubbin, A.G. (1996) How flowering plants discriminate between self and non-self pollen to prevent inbreeding. *Proc. Natl. Acad. Sci. USA*, **93**, 12059-12065.

Karunanandaa, B., Huang, S. and Kao, T.-h. (1994) Carbohydrate moiety of the *Petunia inflata* S$_3$ protein is not required for self-incompatibility interactions between pollen and pistil. *Plant Cell*, **6**, 1933-1940.

Kreitman, M. and Akashi, H. (1995) Molecular evidence for natural selection. *Annu. Rev. Ecol. Syst.*, **26**, 403-422.

Lee, H.S., Huang, S. and Kao, T.-h. (1994) S proteins control rejection of incompatible pollen in *Petunia inflata*. *Nature*, **367**, 560-563.

Leister, D., Berger, A., Thelen, H., Lehmann, W., Salamini, F. and Gebhardt, C. (1997) Construction of a potato YAC library and identification of clones linked to the disease resistance loci R1 and GRO1. *Theor. Appl. Genet.*, **95**, 954-960.

Lewis, D. (1949a) Structure of the incompatibility gene. II. Induced mutation rate. *Heredity*, **3**, 339-355.

Lewis, D. (1949b) Structure of the incompatibility gene. III. Types of spontaneous and induced mutations. *Heredity*, **5**, 399-414.

Li, J.-H., Nass, N., Kusaba, M., *et al.* (2000) A genetic map of the *Nicotiana alata* *S* locus that includes three pollen-expressed genes. *Theor. Appl. Genet.*, **100**, 956-964.

Lush, W.M. and Clarke, A.E. (1995) Growth inhibition of suspension cultured plant cells by ribonucleases. *J. Plant Res.*, **108**, 305-312.

Lush, W.M. and Clarke, A.E. (1997) Observations of pollen tube growth in *Nicotiana alata* and their implications for the mechanism of self-incompatibility. *Sex. Plant Reprod.*, **10**, 27-35.

Luu, D.-T., Qin, X., Morse, D. and Cappadocia, M. (2000) *S*-RNase uptake by compatible pollen tubes in gametophytic self-incompatibility. *Nature*, **407**, 649-651.

MacIntosh, G.C., Bariola, P.A., Newbigin, E. and Green, P.J. (2001) Characterization of Rny1, the *Saccharomyces cerevisiae* member of the T$_2$ RNase family of RNases: unexpected functions for ancient enzymes? *Proc. Natl. Acad. Sci. USA*, **98**, 1018-1023.

Mascarenhas, J.P. (1993) Molecular mechanisms of pollen tube growth and differentiation. *Plant Cell*, **5**, 1303-1314.

Matsuura, T., Unno, M., Sakai, H., Tsukihara, T. and Norioka, S. (2001) Purification and crystallization of Japanese pear S-RNase associated with gametophytic self-incompatibility. *Acta Crystallogr. Sect. D*, **57**, 172-173.

Matton, D.P., Maes, O., Laublin, G., *et al.* (1997) Hypervariable domains of self-incompatibility RNases mediate allele-specific pollen recognition. *Plant Cell*, **9**, 1757-1766.

Matton, D.P., Luu, D.T., Qin, X., *et al.* (1999) Production of an *S*-RNase with dual specificity suggests a novel hypothesis for the generation of new *S* alleles. *Plant Cell*, **11**, 2087-2097.

McClure, B.A., Haring, V., Ebert, P.R., *et al.* (1989) Style self-incompatibility gene products are ribonucleases. *Nature*, **342**, 955-957.

McClure, B.A., Gray, J.E., Anderson, M.A. and Clarke, A.E. (1990) Self-incompatibility in *Nicotiana alata* involves degradation of pollen rRNA. *Nature*, **347**, 757-760.

McClure, B.A., Mou, B., Canevascini, S. and Bernatzky, R. (1999) A small asparagine-rich protein required for *S*-allele-specific rejection in *Nicotiana*. *Proc. Natl. Acad. Sci. USA*, **96**, 13548-13553.

McCubbin, A.G. and Kao, T.-h. (1999) The emerging complexity of self-incompatibility (*S*-) loci. *Sex. Plant Reprod.*, **12**, 1-5.

McCubbin, A.G., Wang, X. and Kao, T.-h. (2000a) Identification of self-incompatibility (*S*-) locus linked pollen cDNA markers in *Petunia inflata*. *Genome*, **43**, 619-627.

McCubbin, A.G., Zuniga, C. and Kao, T.-h. (2000b) Construction of a binary bacterial artificial chromosome library of *Petunia inflata* and the isolation of large genomic fragments linked to the self-incompatibility (*S*-) locus. *Genome*, **43**, 820-826.

McMullen, C.K. (1987) Breeding systems of selected Galapagos Islands angiosperms. *Am. J. Bot.*, **74**, 1694-1705.

Modlibowska, I. (1945) Pollen tube growth and embryo sac development in apples and pears. *J. Pomol. Hortic. Sci.*, **2**(1), 57-89.

Murfett, J., Atherton, T.L., Mou, B., Gasser, C.S. and McClure, B.A. (1994) S-RNase expression in transgenic *Nicotiana* causes S-allele-specific pollen rejection. *Nature*, **367**, 563-566.

Nasrallah, J.B. (2000) Cell–cell signaling in the self-incompatibility response. *Curr. Opin. Plant Biol.*, **3**, 368-373.

Nasrallah, J.B. and Nasrallah, M.E. (1993) Pollen-stigma signalling in the sporophytic self-incompatibility response. *Plant Cell*, **5**, 1325-1335.

Newbigin, E. (1996) The evolution of self-incompatibility: a molecular voyeur's perspective. *Sex. Plant Reprod.*, **9**, 357-361.

Newbigin, E., Anderson, M.A. and Clarke, A.E. (1993) Gametophytic self-incompatibility systems. *Plant Cell*, **5**, 1315-1324.

Ockendon, D.J. (1974) Distribution of self-incompatibility alleles and breeding structure of open-pollinated cultivars of Brussels sprouts. *Heredity*, **33**, 159-171.

O'Driscoll, D., Read, S.M. and Steer, M.W. (1993) Determination of cell-wall porosity by microscopy: walls of cultured cells and pollen tubes. *Acta Bot. Neerl.*, **42**, 237-244.

Oxley, D. and Bacic, A. (1995) Microheterogeneity of *N*-glycosylation on a stylar self-incompatibility glycoprotein of *Nicotiana alata*. *Glycobiology*, **5**, 517-523.

Parry, S., Newbigin, E., Currie, G., Bacic, A. and Oxley, D. (1997) Identification of active-site histidine residues of a self-incompatibility ribonuclease from a wild tomato. *Plant Physiol.*, **115**, 1421-1429.

Parry, S.K., Newbigin, E., Craik, D., Nakamura, K.T., Bacic, A. and Oxley, D. (1998) Structural analysis and molecular model of a self-incompatibility ribonuclease from *Lycopersicon peruvianum*. *Plant Physiol.*, **116**, 463-469.

Richman, A.D., Broothaerts, W. and Kohn, J.R. (1997) Self-incompatibility RNases from three plant families—homology or convergence? *Am. J. Bot.*, **84**, 912-917.

Royo, J., Kunz, C., Kowyama, Y., Anderson, M., Clarke, A.E. and Newbigin, E. (1994) Loss of a histidine residue at the active site of the *S*-locus ribonuclease is associated with self-compatibility in *Lycopersicon peruvianum*. *Proc. Natl. Acad. Sci. USA*, **91**, 6511-6514.

Sassa, H., Hirano, H. and Ikehashi, H. (1993) Identification and characterization of stylar glycoproteins associated with self-incompatibility genes of Japanese pear, *Pyrus serotina* Rehd. *Mol. Gen. Genet.*, **241**, 17-25.

Sassa, H., Hirano, H., Nishio, T. and Koba, T. (1997) Style-specific self-incompatibility mutation caused by deletion of the *S*-RNase gene in Japanese pear (*Pyrus serotina*). *Plant J.*, **12**, 223-227.

Schierup, M.H., Charlesworth, D. and Vekemans, V. (2000) The effect of hitch-hiking on genes linked to a balanced polymorphism in a subdivided population. *Genet. Res.*, **76**, 63-73.

Schopfer, C.R., Nasrallah, M.E. and Nasrallah, J.B. (1999) The male determinant of self-incompatibility in *Brassica*. *Science*, **286**, 1697-1700.

Snowman, B.N., Geitmann, A., Clarke, S.R., *et al.* (2000) Signalling and the cytoskeleton of pollen tubes in Papaver rhoeas. *Ann. Bot.*, **85** (suppl. A), 49-57.

Stace, H.M. (1995) Protogyny, self-incompatibility and pollination in *Anthocersis gracilis* (Solanaceae). *Aust. J. Bot.*, **43**, 451-459.

Suzuki, G., Kai, N., Hirose, T., *et al.* (1999) Genomic organization of the *S* locus: identification and characterization of genes in *SLG/SRK* region of S^9 haplotype of *Brassica* campestris (syn. *rapa*). *Genetics*, **153**, 391-400.

Thompson, K.F. and Taylor, J.P. (1966) Non-linear dominance relationships between *S* alleles. *Heredity*, **21**, 345-362.

Thompson, R.D. and Kirch, H.-H. (1992) The *S*-locus of flowering plants: when self-rejection is self interest. *Trends Genet.*, **8**, 383-387.

Tsai, D.S., Lee, H.S., Post, L.C., Kreiling, K.M. and Kao, T.h. (1992) Sequence of an S-protein of *Lycopersicon Peruvianum* and comparison with other solanaceous S-proteins. *Sex. Plant Reprod.*, **5**, 256-263.

Tupy, J. (1959) Callose formation in pollen tubes and incompatibility. *Biol. Plant. (Praha)*, **1**, 192-198.

Tupy, J., Hrabetova, E. and Balatkova, V. (1977) Evidence for ribosomal RNA synthesis in pollen tubes in culture. *Biol. Plant.*, **19**, 226-230.

Uyenoyama, M. and Newbigin, E. (2000) Evolutionary dynamics of dual-specificity self-incompatibility alleles. *Plant Cell*, **12**, 310-312.

Uyenoyama, M.K., Zhang, Y. and Newbigin, E. (2001) On the origin of self-incompatibility haplotypes: Transition through self-compatible intermediates. *Genetics*, **157**, 1805-1817.

Verica, J.A., McCubbin, A.G. and Kao, T.-h. (1998) Are the hypervariable regions of S-RNases sufficient for allele-specific recognition of pollen? *Plant Cell*, **10**, 314-317.

Walles, B. and Han, S.-P. (1998a) Development of the rough endoplasmic reticulum in incompatible pollen tubes of *Brugmansia suaveolens* (Solanaceae). *Int. J. Plant Sci.*, **159**, 738-743.

Walles, B. and Han, S.-P. (1998b) Ribosomes in incompatible pollen tubes in the Solanaceae. *Physiol. Plantarum.*, **103**, 461-465.

Wang, J., Shi, H.-Z., Zhou, C., Yang, H.-Y., Zhang, X.-L. and Zhang, R.-D. (1998) Beta-glucuronidase gene and green fluorescent protein gene expression in de-exined pollen of *Nicotiana tabacum* by microprojectile bombardment. *Sex. Plant Reprod.*, **11**, 159-162.

Whitelam, G.C. and Cockburn, W. (1996) Antibody expression in transgenic plants. *Trends Plant Sci.*, **1**, 268-272.

Xue, Y., Carpenter, R., Dickinson, H.G. and Coen, R.S. (1996) Origin of allelic diversity in *Antirrhinum* S-locus RNases. *Plant Cell*, **8**, 805-814.

Yoder, J.I. (1998) Self and cross compatibility in three species of the hemiparasite *Triphysaria* (Scrophulariaceae). *Environ. Exp. Bot.*, **39**, 77-83.

Zurek, D.M., Mou, B., Beecher, B. and McClure, B. (1997) Exchanging domains between S-RNases from *Nicotiana alata* disrupts pollen recognition. *Plant J.*, **11**, 797-808.

9 Pollination signals and flower senescence

William R. Woodson

9.1 Introduction

The angiosperm flower is composed of a number of tissue types, each of which performs a unique function in reproduction. Flower petals, typically the most prominent floral organ, function to attract pollinators. Successful pollination sets in motion a series of biochemical and developmental events that often culminate in the senescence and/or abscission of petals (Stead, 1992). Given the role of the petals as visual cues, pollination-induced senescence has likely evolved as a mechanism to deter further visits by pollinators. Furthermore, remobilization of cellular constituents from the senescing organ would likely facilitate development of the gynoecium, providing an advantage to the survival of the species. The regulation of developmental events in the corolla following pollination requires a mechanism for communication among the floral organs, signaling that a successful pollination has occurred. In many species, an increase in the production of the phytohormone ethylene is one of the earliest detectable biochemical events in the pollinated pistil. This often occurs within the first few hours after pollination, and in many cases proceeds the growth of pollen tubes (Nichols, 1977; Nichols et al., 1983; Hoekstra and Weges, 1986; Pech et al., 1987; Jones and Woodson, 1997, 1999a,b; Bui and O'Neill, 1998). This review will focus on the current state of knowledge regarding the biochemical processes and signals that regulate pollination-induced flower senescence. Pollination-regulated developmental processes have been reviewed recently (O'Neill, 1997; O'Neill and Nadeau, 1997).

9.2 Pollination-induced senescence

The premature senescence of flower petals is one of many developmental responses associated with pollination (Stead, 1992; O'Neill 1997; O'Neill and Nadeau, 1997). Other developmental processes that appear to be coordinated by pollination include changes in pigmentation and development of the female gametophyte. Collectively, these developmental events are thought to facilitate reproduction by preparing the female gametophyte for fertilization and removing organs that have fulfilled their function in the attraction of pollinators. The induction of corolla senescence by pollination has been studied extensively in the flowers of carnation (Nichols, 1977; Nichols et al., 1983; Whitehead *et al.*, 1983; Larsen *et al.*, 1995; Jones and Woodson, 1997, 1999a,b), petunia (Gilissen, 1976; Gilissen and Hoekstra, 1984; Hoekstra and Weges, 1986; Pech *et al.*, 1987) and orchids (Burg and Dijkman, 1967; O'Neill *et al.*, 1993; Porat *et al.*, 1995; Bui and O'Neill, 1998). In these species, pollination is associated with a rapid increase in ethylene production, first by the pistil and subsequently by the petals (Nichols *et al.*, 1983; Zhang and O'Neill, 1993; Larsen *et al.*, 1995; Jones and Woodson, 1997, 1999a,b). In these cases, it is clear that ethylene plays a critical role in the senescence of the petals, as will be described below.

9.3   Ethylene biosynthesis

Ethylene is synthesized in plant tissues from the amino acid precursor methionine (Adams and Yang, 1979; Kende, 1989). The first intermediate in the pathway is *S*-adenosylmethionine (SAM), which is converted to the immediate precursor of ethylene, 1-aminocyclopropane-1-carboxylic acid (ACC) by the enzyme ACC synthase (Boller *et al.*, 1979). Finally, ACC synthase is converted into ethylene by the action of ACC oxidase (Kende, 1993). The conversion of methionine to SAM is catalyzed by the ubiquitous enzyme SAM synthetase and is not thought to be rate limiting (Larsen and Woodson, 1991). In many tissues, the application of ACC results in the rapid production of ethylene, suggesting that ACC synthase is a rate-limiting step in the *in vivo* synthesis of ethylene (Yang and Hoffman, 1984). Genes encoding ACC synthase have now been isolated from many species (for review see Zarembinski and Theologis, 1994) and have been shown to be a large, divergent multigene family. ACC synthase genes are subject to differential regulation, often in a tissue-specific manner.

The ACC oxidase enzyme was identified through the functional expression of a ripening-related cDNA clone from tomato fruit. Initially, it was shown that expression of an antisense RNA from the *pTOM13* cDNA inhibited ethylene production in transgenic tomatoes (Hamilton *et al.*, 1990). This led to the speculation that *pTOM13* encoded an enzyme in the ethylene biosynthetic pathway. This was subsequently shown to be the case, as expression of *pTOM13* homologues in *Xenopus oocytes* (Spanu *et al.*, 1991) and yeast (Hamilton *et al.*, 1991) resulted in the capacity to convert ACC to ethylene. Sequence homology between ACC oxidase and flavanone 3-hydroxylase, initially reported by Hamilton *et al.* (1990), led to the development of an *in vitro* assay for ACC oxidase (Ververidis and John, 1991). This revealed that ACC oxidase requires CO_2, ascorbate and Fe(II).

ACC oxidase is constitutively expressed in many tissues and is generally not regarded as rate limiting in the biosynthesis of ethylene (Kende, 1989, 1993). However, ACC oxidase transcripts have been shown to increase in response to a number of stimuli such as wounding (Hamilton *et al.*, 1990; Kim and Yang, 1994) and during senescence (Woodson *et al.*, 1992; Tang *et al.*, 1994) and ripening (Hamilton *et al.*, 1990; Balague *et al.*, 1993). ACC oxidase is encoded by a multigene family that exhibits differential patterns of regulation. In petunia, four ACC oxidase genes have been identified (Tang *et al.*, 1993, 1994). Three of these genes are transcriptionally active and expressed in a tissue-specific manner (Tang *et al.*, 1994). The ACC oxidase gene family in tomato includes at least three members (Barry *et al.*, 1996).

9.3.1   Ethylene and flower senescence

Ethylene production has long been known to be associated with the senescence of flower petals (for review see Borochov and Woodson, 1989). Presenescent petals of carnation, for example, exhibit very low levels of ethylene production and limited activities of both ACC synthase and ACC oxidase (Woodson *et al.*, 1992). The onset of senescence is accompanied by a significant increase in the production of ethylene (for review see Borochov and Woodson, 1989). The onset of ethylene production was shown to be associated with a concomitant increase in the expression of ACC synthase and ACC oxidase mRNAs and enzyme activity, suggesting that this was regulated at the levels of both transcription and translation (Woodson *et al.*, 1992).

Analagously to ripening in climacteric fruit, ethylene production in senescing carnation petals appears to be subject to autocatalytic regulation. Exposure of presenescent petals to ethylene induces ethylene production and petal senescence (Nichols, 1968, 1971; Woodson and Lawton, 1988). Ethylene plays a critical role in the regulation of ethylene production and senescence (Borochov and Woodson, 1989). A number of genes shown to be upregulated during flower senescence are under the regulation of ethylene (Lawton *et al.*, 1990). These include a thiol protease (Jones *et al.*, 1995), a glutathione-*S*-transferase (Meyer *et al.*, 1991; Itzhaki *et al.*, 1994; Maxson and Woodson, 1996), and a number of genes of unknown function (Lawton *et al.*, 1989; Wang *et al.*, 1993). Treatment of carnation flowers with chemical inhibitors of ethylene synthesis or action prevent the increase in ethylene production, the expression of senescence-related genes, and the premature onset of senescence (Bufler *et al.*, 1980; Reid *et al.*, 1980; Wang and Woodson, 1989; Lawton *et al.*, 1990). These treatments are standard in the floral industry, where delayed senescence is sought to prolong the marketable life of cut carnation flowers. Further evidence for central role of ethylene in carnation petal senescence comes from experiments in transgenic plants. Savin *et al.* (1995) expressed an antisense ACC oxidase transcript in transgenic carnations, which led to a reduction in ethylene production and a delay in the onset of petal senescence. This approach promises to improve the longevitiy of flowers without chemical treatments currently employed in the industry.

9.4   Pollination-induced ethylene biosynthesis

In many species, an increase in the production of ethylene is one of the earliest detectable biochemical events in the pollinated pistil. This often occurs within the first few minutes or hours following pollination (Nichols, 1977; Hoekstra and Weges, 1986; Pech *et al.*, 1987; O'Neill *et al.*, 1993; Larsen *et al.*, 1995; Jones and Woodson, 1999b). In petunia (Hoekstra and Weges, 1986; Singh *et al.*, 1992; Tang and Woodson, 1996) and carnation (Nichols *et al.*, 1983; Larsen *et al.*, 1995; Jones and Woodson, 1997), increased ethylene biosynthesis by the style is detected within the first 30 minutes following pollination. This increased ethylene precedes the germination and growth of the pollen tube, suggesting that penetration of the stigmatic surface is not required for the induction of ethylene biosynthesis. Application of synthetic beads to the stigma of orchids failed to elicit an increase in ethylene production, indicating that physical contact alone is insufficient to elicit a response (Zhang and O'Neill, 1993). Pollination-induced ethylene in the stigma is often followed by increased ethylene in other attached organs. In carnations, ethylene is first detected in the pollinated style within 1 hour, and subsequently increases in the ovary by 8 hours, and finally in the petals by 24 hours (Larsen *et al.*, 1995; Jones and Woodson, 1999a,b). Similar patterns have been observed in petunia (Pech *et al.*, 1987) and orchid (O'Neill *et al.*, 1993). This induction of ethylene in floral organs occurs before fertilization of the ovules, and in the case of orchid triggers the further development of the ovules, making them competent to be fertilized (Zhang and O'Neill, 1993).

9.4.1   *Pollination-induced ACC synthase*

The synthesis of ACC from methionine is often the rate-limiting step in the biosynthesis of ethylene. Because of this, ACC synthase activity and transcript abundance have been studied extensively in pollinated flowers. In carnation, the wave of ethylene production that initiates in the style, progressing to the ovary and finally the petals, is associated with a concomitant increase in ACC content, suggesting that the synthesis of ACC is critical to the regulation of ethylene production (Nichols *et al.*, 1983; Jones and Woodson, 1999b). This increase in ACC content was shown to be associated with increased ACC synthase activity in styles and petals (Jones and Woodson, 1999b). In striking contrast, the ovary, which was shown to produce significant amounts of ethylene following pollination, and to contain increased amounts of ACC, did not exhibit significant ACC synthase activity (Jones and Woodson, 1999b). ACC synthase is encoded by a multigene family in all plants studied to date. In carnation, three unique ACC synthase genes have been identified and characterized (Park *et al.*, 1992; ten Have and Woltering, 1997; Jones and Woodson, 1999a). *DC-ACS1* was shown to be expressed primarily in senescing petals and styles during the final stages of the wave of increased ethylene that follows pollination (Woodson *et al.*, 1992; Jones and Woodson, 1997, 1999a). The expression of *DC-ACS1* was shown to be under the regulation of ethylene, as it was blocked by inhibitors of ethylene action such as 2,5-norbornadiene (Jones and Woodson, 1997). In striking contrast, *DC-ACS2* and *DC-ACS3* were shown to be expressed in pollinated styles beginning 1 hour after pollination (Jones and Woodson, 1997, 1999a). The expression of both *DC-ACS2* and *DC-ACS3* is independent of ethylene and appears to be related to primary signals associated with the interaction of pollen with the pistil. Consistent with this observation, treatment with ethylene action inhibitors fails to prevent the early increase in ethylene by pollinated styles, but prevents the final wave of ethylene produced by both styles and petals (Jones and Woodson, 1997). This final wave of ethylene is primarily associated with the expression of *DC-ACS1*, which is inhibited by 2,5-norbornadiene. *DC-ACS1* appears to play a role in amplifying the ethylene signal by increasing the capacity to synthesize ACC in response to early pollination-induced ethylene.

The regulation of ethylene biosynthesis in pollinated orchid flowers has also been studied extensively. In this flower, three ACC synthase cDNAs have been characterized (O'Neill *et al.*, 1993; Bui and O'Neill, 1998). As in carnation, these genes appear to respond to both primary and secondary signals following pollination in a coordinated manner. One ACC synthase gene (*Phal-ACS1*) is under the regulation of ethylene and appears to lead to an amplification of the ethylene signal like DC-ACS1 in carnation (Bui and O'Neill, 1998). Two additional ACC synthase genes (*Phal-ACS2* and *Phal-ACS3*) were shown to be expressed primarily in the pistil in response to primary pollination signals. In contrast to carnation, where ACC accumulates in the ovary without the expression of ACC synthase genes or ACC synthase activity, orchid flowers fail to exhibit increased ACC synthase activity or ACC synthase gene expression in Petals (Bui and O'Neill, 1998). This is in spite of the fact that petals account for much of the ethylene produced from a pollinated orchid flower.

Tomato has become an important genetic and molecular model for studying ethylene biosynthesis and perception in plants. Recently, Llop-Tous (2000) reported that pollination of tomato flowers leads to increased ethylene production by the pistil and petals as in other flowers. This ethylene was shown to play an important role in senescence, as the never ripe (Nr) mutant failed to exhibit premature senescence following pollination. The Nr mutant displays ethylene insensitivity (Lanahan *et al.*, 1994) and is defective in a member of the ethylene receptor gene family (Wilkinson *et al.*, 1995). Expression of the LEACS1A gene was correlated with the production of ethylene by the pistil in both wild-type and Nr flowers following pollination. Like carnation and orchid, early ethylene production by the pistil appears to be independent of ethylene in tomato flowers. A second ACC synthase gene (*LEACS6*) was shown to be expression in petals of pollinated flowers, and this expression was blocked in Nr flowers.

The role of ACC synthesis in pollination-induced ethylene has also been studied in the diploid geranium (Clark *et al.*, 1997). In this species, pollination is associated with a very rapid increase in ethylene production, leading to petal abscission within 4 hours of pollination. A single ACC synthase cDNA was identified in this species, but the pattern of expression did not correlate with ethylene production. In geranium, posttranscriptional regulation of ACC synthase activity may play a role in pollination-induced ethylene.

9.4.2   *Pollination-induced ACC oxidase*

The final step in the ethylene biosynthetic pathway is catalyzed by ACC oxidase. In contrast to ACC synthase, this enzyme is often constitutive in plant tissues. In carnation, it was shown that petals did not exhibit significant activity of ACC oxidase until the onset of petal senescence (Woodson *et al.*, 1992). This increased ACC oxidase activity was associated with the expression of ACC oxidase mRNA encoded by the *DC-ACO1* gene. The expression of ACC oxidase mRNA in carnation petals is under strict regulation by ethylene. Inhibitors of ethylene action prevent the expression of the *DC-ASO1* gene in petals and ethylene stimulates expression prior to the onset of senescence. In contrast to petals, styles of mature carnation flowers are capable of converting ACC to ethylene and ACC oxidase mRNA is abundant in this tissue (Woodson *et al.*, 1992; Jones and Woodson, 1997). While pollination stimulates the expression of ACC oxidase, and inhibitors of ethylene action prevents this increase, carnation styles contain constitutive levels of ACC oxidase mRNA that lead to high levels of ACC oxidase activity independent of pollination. In this species, ACC oxidase expression plays an important role in amplifying the ethylene signal in styles and petals.

The ACC oxidase gene family has been studied extensively in petunia. Four ACC oxidase genes were identified and shown to be arranged in two unlinked clusters of tandemly arranged genes (Tang *et al.*, 1993). Three members of this gene family were shown to be expressed (Tang *et al.*, 1994). As in carnation, petunia petals contain very low levels of ACC oxidase mRNA. The onset of senescence is associated with a significant increase in ACC oxidase mRNA levels and ACC oxidase enzyme activity (Tang *et al.*, 1994). The *PH-ACO1* gene was shown to be responsible for this activity in petals, whereas *PH-ACO3* and *PH-ACO4* were found to be expressed primarily in pistil tissue. Pistils from immature flowers contained no ACC oxidase mRNA, but the levels increased with flower opening. Localization of ACC oxidase mRNA in pistils revealed that much of the expression was in the stigmatic region of the style. Consistent with this, mature styles were shown to be capable of converting ACC to ethylene when applied to the stigma (Pech *et al.*, 1987). Pollination of mature petunia flowers leads to increased ethylene within 1 hour. In contrast, pollination-induced ethylene in floral buds was delayed by several hours and was associated with increased expression of ACC oxidase (Tang and Woodson, 1996). This delay in ethylene production was not associated with a delay in pollen germination or tube growth, suggesting that the low levels of ACC oxidase in pistils from flower buds was rate limiting. The induction of ACC oxidase in pollinated stigmas from flower buds was not prevented by treatment with ethylene action inhibitors, indicating this was a primary pollination response. Similarly, mature tomato flowers exhibit constitutive levels of ACC oxidase mRNA in pistils of both wild-type and the ethylene insensitive Nr mutant. In mature carnation, petunia and tomato flowers, pollination-induced ethylene in the pistil appears to be under the primary control of ACC synthase.

In contrast to petunia and carnation, the stigmas of orchid flowers exhibit limited ACC oxidase activity prior to pollination (Nadeau *et al.*, 1993). This activity increases dramatically following pollination and is associated with increased expression of ACC oxidase mRNA. In this species, ACC oxidase appears to play a significant role in regulating pollination-induced ethylene.

9.5   Primary pollination signals

In the flowers of carnation (Nichols *et al.*, 1983; Jones and Woodson, 1999a,b) and petunia (Pech *et al.*, 1987; Singh *et al.*, 1992; Tang and Woodson, 1996), increased ethylene is apparent within the first few minutes of pollination. In orchids (Bui and O'Neill, 1998), increased stylar ethylene is detectable within the first 2 hours after applying pollen to the stigma. In all of these situations, the initial burst of ethylene precedes the germination and growth of the pollen tubes, suggesting that the initial pollen–pistil interaction signals the increase in ethylene production. The nature of these interactions has been the subject of numerous investigations.

The initial burst of ethylene is believed to be a response to chemicals derived from the pollen grain itself rather than physical interaction. While wounding stimulates the synthesis of ethylene in petunia stigmas (Hoekstra and Weges, 1986), similar responses are not seen in carnation (Woodson *et al.*, 1992; Larsen *et al.*, 1995), geranium (Clark *et al.*, 1997) or orchid (O'Neill *et al.*, 1993). Mock pollination with latex beads or brushing the stigma to mimic pollination failed to elicit an ethylene response in orchids (Zhang *et al.*, 1993) or carnations (Larsen *et al.*, 1995), respectively. In contrast to physical perturbation of the stigma, application of foreign pollen or pollen that has been killed by heat often results in a transient increase in ethylene production (Whitehead *et al.*, 1983; Hoekstra and Weges, 1986). Taken together, the preponderance of results suggests that the pollen–pistil interactions that lead to an early increase in ethylene are chemical in nature and may involve factors borne in the pollen grain itself.

9.5.1 Auxin

The phytohormone auxin was implicated as a pollen-borne chemical capable of eliciting a pollination-like response in orchid flowers (for review of early literature see O'Neill, 1997; O'Neill and Nadeau, 1997). Auxin has been shown to stimulate ethylene production and mimic pollination when applied to the stigma of orchids (Zhang and O'Neill, 1993; O'Neill, 1997; O'Neill and Nadeau, 1997). O'Neill *et al.* (1993) and Bui and O'Neill (1998) reported that auxin stimulated the production of ethylene when applied to orchid stigmas and induced the expression of ACC synthase genes. ACC synthase genes responded differentially to auxin (Bui and O'Neill, 1998). The *Phal-ACS2* gene apparently responded to auxin directly, as its expression was not inhibited by treatment with norbornadiene, an ethylene action inhibitor. In contrast, the *Phal-ACS1* gene product increased in abundance in auxin-treated stigmas, apparently in response to the ethylene produced rather than directly to the application of auxin. In contrast to the response in orchids, application of auxin to carnation and petunia stigmas did not result in increased ethylene production or any other postpollination phenomena (Reid *et al.*, 1984; Pech *et al.*, 1987).

In an effort to identify the pollen-borne substance(s) that is responsible for pollination-induced ethylene in orchids, Porat *et al.* (1998) performed a water extraction of the pollinia from *Phalaenopsis* flowers. This study revealed a low molecular weight, water-soluble component that was capable of eliciting the postpollination response when applied to the stigma of orchid flowers. Furthermore, this substance was unaffected by proteinase K or boiling, suggesting that it was not proteinaceous in nature. Preliminary purification experiments revealed the pollen-borne substances were separated into a number of distinct fractions that, when applied to the stigma, elicited a pollination response. One of these fractions was shown to contain the ethylene precursor ACC. None of the active fractions from HPLC separation co-eluted with free indole acetic acid (IAA). This preliminary report holds much promise for identifying the major constituents present in pollen that serve as primary pollination signals.

9.5.2 ACC

Pollen from a number of plant species has been shown to contain significant levels of the ethylene precursor ACC (Whitehead *et al.*, 1983; Hill *et al.*, 1987). Petunia pollen contains as much as 1500 nmol ACC/g (Hoekstra and Weges, 1986; Singh *et al.*, 1992). This observation led to the proposal that pollen-borne ACC could account for the ethylene produced by styles immediately following pollination. Singh *et al.* (1992) reported that the endogenous ACC content of pollen in petunia correlated with the amount of ethylene produced by petunia styles immediately after pollination. They concluded that this early ethylene was due to the conversion of pollen-borne ACC to ethylene. In support of this conclusion, unpollinated petunia stigmas exhibit high levels of ACC oxidase activity and are capable of converting applied ACC to ethylene (Pech *et al.*, 1987; Tang *et al.*, 1994; Tang and Woodson, 1996).

The synthesis of ACC in developing pollen has been investigated. Lindstrom *et al.* (1999) reported that the accumulation of ACC in pollen was a rather late developmental event, occurring just as the flower began to open. Furthermore, they showed that this ACC was the product of an ACC synthase gene, *PH-ACS2*, that was expressed specifically in the male gametophyte late in development. The *PH-ACS2* gene and its promoter were analyzed further, and sequences flanking the 5'-region of the gene were shown to direct pollen-specific expression of the GUS reporter gene. A role for ACC or ethylene in pollen development has not been identified to date.

While pollen from petunia contains significant levels of ACC and this ACC could contribute to the ethylene produced following pollination, a number of experiments have brought this into question. The application of aminoethoxy-vinylglycine (AVG), an inhibitor of ACC synthase, to petunia and carnation styles has been shown to effectively block pollination-induced ethylene production by the style, indicating that this early ethylene is dependent on ACC synthase activity (Hoekstra and Weges, 1986; Woltering *et al.*, 1993). The rapid induction of ACC synthase activity in the styles of petunia following pollination also provides evidence that ACC synthase and the synthesis of ACC in the style rather than pollen-borne ACC is the precursor of this pollination-induced ethylene (Pech *et al.*, 1987). In carnations, pollen-borne ACC is very low (25 nmol/g) and would appear insufficient to elicit the level of ethylene produced by pollinated styles immediately following the application of pollen. When this is taken together with the early induction of ACC synthase genes in the pistil of carnation (Jones and Woodson, 1997) and tomato (Llop-Tous *et al.*, 2000), it would appear that *de novo* synthesis of ACC is needed for the burst of ethylene immediately following pollination.

9.6 Secondary pollination signals and interorgan communication

The induction of ethylene and petal senescence by pollination suggests that signals originating in the gynoecium are translocated to the petal. The presence of a mobile 'wilting factor' was proposed by Gilissen and Hoekstra (1984), who reported that removal of the pollinated style from the flower within 6 hours of pollination prevented the subsequent senescence of the petals. Similarly, removal of the styles from carnation flowers within 10 hours of pollination prevented senescence of the corolla (Jones and Woodson, 1999b). In both petunia and carnation, ethylene production by the gynoecium increased prior to the critical time of removal. By placing removed styles back in the flower, it was shown that diffusion of ethylene through the atmosphere was unlikely to account for the subsequent induction of ethylene in petals (Gilissen and Hoekstra, 1984; Jones and Woodson, 1999b). The nature of the translocated signal responsible for coordinating postpollination development, including petal senescence, has been the subject of numerous investigations.

9.6.1   Role of auxin

Auxin has been recognized as a plant hormone for many years and is known to move basipetally in plants. Burg and Dijkman (1967) proposed that pollen-derived auxin was the primary signal responsible for pollination-induced ethylene and senescence in orchid flowers. Strauss and Arditti (1982) subsequently reported that ^{14}C-labeled IAA was largely immobile when applied to the stigma. This report did indicate that auxin moved very slowly in the flower, but did ultimately accumulate in the ovary. This is consistent with the more recent work of Zhang and O'Neill (1993), where auxin was shown to play a role in pollination-induced ovule differentiation in orchids. In contrast to orchids, application of auxin to carnation and petunia stigmas did not induce significant ethylene production or result in premature petal senescence (Reid et al., 1984; Pech et al., 1987).

9.6.2   Role of ACC and ethylene

The sequential nature of increased ethylene by styles, ovary and finally petals following pollination suggests that ethylene or its precursor ACC may play a role in interorgan communication. In pollinated carnation and orchid flowers, the level of ACC increases in each of the floral organs concomitantly with increased ethylene production (Nichols et al., 1983, 1993; Larsen et al., 1995; Jones and Woodson, 1997). Also, in a recent study, Jones and Woodson (1999b) reported that ACC content and ethylene production began in the tip of the style (stigma) and proceeded basipetally toward the base of the style following pollination.

   As a soluble hormone precursor, it is reasonable to suggest that ACC would be more amenable to targeted translocation within the flower than the gaseous molecule ethylene. Translocation of ACC has been demonstrated to occur in flooded plants, in which it is synthesized in roots and translocated to shoots where ACC is subsequently oxidized to ethylene (Bradford and Yang, 1980). Evidence for movement of ACC in flowers after pollination can be found in both carnations and orchids. In carnations, increased ACC content and ethylene production are detected in the ovary within 12 hours of pollination (Jones and Woodson, 1999b). However, ovaries exhibit no concomitant increases in ACC synthase mRNA (Jones and Woodson, 1997) or enzyme activity (Jones and Woodson, 1999b). Similarly, in orchids (O'Neill et al., 1993; Bui and

O'Neill, 1998), ACC and ethylene are detected in petals without evidence for either ACC synthase mRNA or enzyme activity. When petals were removed from the pollinated orchid flower, ethylene evolution decreased, indicating that ethylene biosynthesis in petals was dependent on the translocation of ACC from the gynoecium (O'Neill et al., 1993). Taken together, these experiments clearly point to ACC being synthesized in one organ and translocated to another following pollination.

More direct evidence for the translocation of ACC within flowers came from experiments using isotope-labeled ACC. Reid et al. (1984) applied ^{14}C-labeled ACC to the stigma of carnation flowers and detected the production of (^{14}C)ethylene by the petals. In contrast, Woltering et al. (1995) reported that ^{14}C-ACC and the ACC analogue α-aminoisobutyric acid were immobile when applied to the central column or stigma of Cymbidium orchid flowers. While it is clear that ACC accumulates in organs that exhibit no apparent capacity to synthesize this compound, the role of ACC translocation in interorgan signaling following pollination remains to be determined.

Ethylene has also been implicated as the translocated signal in pollination-induced senescence. This could occur through the movement of dissolved ethylene or through diffusion of ethylene within intercellular spaces. Internal concentrations of ethylene have been shown to be quite high (Woltering, 1990). The diffusion of gaseous ethylene from styles to petals could offer another explanation for the evolution of (^{14}C)ethylene from carnation petals when styles were treated with ^{14}C-ACC (Reid et al., 1984). Ethylene and its analogue propylene have been shown to diffuse from the gynoecium to the petals in orchids (Woltering, 1990; Woltering et al., 1995) and carnations (Jones and Woodson, 1999b).

While there is considerable evidence for the translocation of both ACC and ethylene in pollinated flowers, these results do not clearly define either as the translocated pollination signal. It is clear ethylene plays a critical role in pollination-induced petal senescence, as inhibitors of ethylene action prevent this response (O'Neill et al., 1993; Tang and Woodson, 1996; Jones and Woodson, 1997; Bui and O'Neill, 1998). The expression of ACC oxidase genes in the ovary and petals of carnation is completely dependent on ethylene (Jones and Woodson, 1999a), suggesting that ethylene produced in response to primary pollination signals in the gynoecium is critical to propagate the signal throughout the flower. Similar results were reported in orchid flowers (Nadeau et al., 1993; O'Neill et al., 1993). Consistent with the role of stylar ethylene in propagating the pollination signal, inhibition of ethylene action specifically in the style by treatment with diazocyclopentadiene prevented induction of ethylene in ovaries and petals, preventing pollination-induced senescence (Jones and Woodson, 1997). This points to a role for gynoecium-produced ethylene in the regulation of ethylene production in other floral organs following pollination.

9.6.3 Role of sexual compatibility and pollen tube growth

The early increase in ethylene in orchid, petunia and carnation flowers precedes the germination and growth of pollen tubes (Pech *et al.*, 1987; O'Neill *et al.*, 1993; Larsen *et al.*, 1995). Heat-killed and incompatible pollen elicited increased ethylene in the stigma/style of petunia and carnation (Gilissen, 1976; Hoekstra and Weges, 1986; Singh *et al.*, 1992; Larsen *et al.*, 1995). However, only compatible pollen induced sustained ethylene production by the gynoecium, which in turn led to increased ethylene in the petals and petal senescence. This observation led to the suggestion that mechanical wounding or other signals associated with the growth of pollen tubes induced sustained ethylene production and petal senescence (Hoekstra and Weges, 1996). Excision or wounding of petunia styles induced ethylene production of petunia styles, which led to the acceleration of corolla senescence (Gilissen, 1976, 1977). In contrast, wounding failed to elicit ethylene production or petal senescence in unpollinated carnation flowers (Larsen *et al.*, 1995).

The role of pollen tube growth in pollination-induced ethylene has been called into question by sexual compatibility studies in carnation (Larsen *et al.*, 1995). Similar to petunia, they reported that incompatible pollen induced early stylar ethylene in carnation, but did not result in sustained ethylene production or petal senescence. Surprisingly, this pollen exhibited ovular incompatibility, but pollen tubes germinated and grew the length of the style. Therefore, in this species, growth of pollen tubes alone is insufficient to induce sustained ethylene production by the style or other floral organs, including the petals.

9.6.4 Role of ethylene sensitivity

The capacity to respond to ethylene has been implicated in the regulation of a number of processes including pollination-induced petal senescence. Halevy *et al.* (1994) demonstrated that pollination-induced abscission of cyclamen petals was prevented by treatment with the ethylene action inhibitor silver thiosulfate. In contrast, ethylene did not induce abscission of petals from unpollinated flowers. This result suggested that pollination resulted in the flowers becoming 'sensitive' to ethylene. Pollination-induced ethylene sensitivity increased within a few hours of pollination in the flowers of *Phalaenopsis* orchids (Porat *et al.*, 1994, 1995). This increase in ethylene sensitivity was associated with a significant increase in short-chain saturated free fatty acids in the column and perianth of pollinated flowers (Halevy *et al.*, 1996). These compounds had previously been identified in the eluates from pollinated petunia styles and shown to exhibit senescence-inducing properties (Whitehead and Halevy, 1989). Other reports have failed to confirm the effects of these short-chain fatty acids on flower senescence in petunia, carnation and orchids

(Woltering *et al.*, 1993), calling into question the role of these compounds in postpollination signaling.

9.7   Model for interorgan communication in carnation flower

A model for interorgan signaling in pollinated carnation flowers has been developed based on the data published to date. In this model, pollen provides the primary signal that leads to the induction of ACC synthase transcripts encoded by *DC-ACS3* in the stigma. The resulting ACC synthase activity in the style leads to production of ACC, which in turn is converted to ethylene by the basal constitutive levels of ACC oxidase mRNA and enzyme activity. Increased production of ethylene in the style results in the induction of *DC-ACS2* mRNA and increases the expression of *DC-ACS3* and *DC-ACO1* mRNAs. A secondary signal is translocated through the style to the ovary and petals, specifically in response to compatible pollination. Ethylene and ACC are both implicated as playing this role. For example, ethylene production by the ovary following pollination is completely dependent on ACC from other floral organs, as there is no evidence for either ACC synthase transcripts or enzyme activity. Further, the induction of ACC oxidase mRNA is under the regulation of ethylene. Therefore, ethylene produced in the base of the style after pollination likely induces *DC-ACO1* mRNA, and ACC translocated from the style is then oxidized to ethylene. The initial signal perceived by petals following pollination of carnation flowers is likely to be ethylene itself, as the accumulation of ethylene in the intercelluar spaces of the ovary and receptacle is quite significant. Furthermore, the ethylene analogue has been shown to diffuse from the ovary to the petals. The perception of this ethylene by the petals leads to the induction of both *DC-ACS1* and *DC-ACO1* mRNAs and the enzymes they encode. The autocatalytic nature of this signaling pathway results in the sustained production of high levels of ethylene gas, leading ultimately to petal senescence.

A number of very important questions need to be addressed in future research. First, the nature of the initial interaction between the pollen and pistil that leads to ethylene biosynthesis remains to be determined. There is evidence for pollen-borne substances that are capable of eliciting this response; however, the identification of these compounds remains to be accomplished (Porat *et al.*, 1993, 1998). Also, the signals associated with compatible and incompatible pollination that affect postpollination ethylene and development remain to be determined. For example, early ethylene production in petunia and carnation is independent of compatibility signals, as incompatible pollen elicits a similar response (Singh *et al.*, 1992; Larsen *et al.*, 1995; Hoesktra and Weges, 1996). This ethylene appears to be insufficient to elicit a complete postpollination response, as sustained ethylene production and petal senescence depend on a compatible pollination. Therefore, there must be signals associated with compatible pollen that are perceived by the gynoecium, leading in turn to increased ethylene production.

Acknowledgements

I thank former graduate students and postdoctoral associates for their many discussions and research contributions that framed this topic over the past decade. These include Drs. Michelle Jones, Paul Larsen, Jon Lindstrom, and Xiaoyan Tang. I am grateful to the USDA-National Research Initiative, the National Science Foundation, the Gloeckner Foundation and the American Floral Endowment for providing support to my laboratory to address questions related to ethylene and pollination signaling in petunia and carnation flowers.

References

Adams, D.O. and Yang, S.F. (1979) Ethylene biosynthesis: identification of 1-aminocyclopropane-1-carboxylic acid as an intermediate in the conversion of methionine to ethylene. *Proc. Natl. Acad. Sci. USA*, **83**, 7755-7759.

Balague, C., Watson, C.F. and Turner, A.J., *et al.* (1993) Isolation of a ripening and wound-induced cDNA from *Cucumis melo* L. encoding a protein with homology to the ethylene-forming enzyme. *Eur. J. Biochem.*, **212**, 27-34.

Barry, C.S., Blume, B., Bouzayen, M., Cooper, W., Hamilton, A.J. and Grierson, D. (1996) Differential expression of the 1-aminocyclopropane-1-carboxylate oxidase gene family of tomato. *Plant J.*, **9**, 525-535.

Boller, T., Herner, R.C. and Kende, H. (1979) Assay for and enzymatic formation of an ethylene precursor, 1-aminocyclopropane-1-carboxylic acid. *Planta*, **145**, 293-303.

Borochov, A. and Woodson, W.R. (1989) Physiology and biochemistry of flower petal senescence. *Hort. Rev.*, **11**, 15-43.

Bradford, K.J. and Yang, S.F. (1980) Xylem transport of 1-aminocyclopropane-1-carboxylic acid, an ethylene precursor, in waterlogged tomato plants. *Plant Physiol.*, **65**, 322-326.

Bufler, G., Mor, Y., Reid, M.S. and Yang, S.F. (1980) Changes in 1-aminocyclopropane-1-carboxylic acid content of cut carnation flowers in relation to their senescence. *Planta*, **150**, 439-442.

Bui, A. and O'Neill, S.D. (1998) Three 1-aminocyclopropane-1-carboxylate synthase genes regulated by primary and secondary pollination signals in orchid flowers. *Plant Physiol.*, **116**, 419-428.

Burg, S.P. and Dijkman, M.J. (1967) Ethylene and auxin participation in pollen induced fading of *Vanda* orchid blossoms. *Plant Physiol.*, **42**, 1648-1650.

Clark, D.G., Richards, C., Hilioti, Z., Lind-Iversen, S. and Brown, K. (1997) Effect of pollination on accumulation of ACC synthase and ACC oxidase transcripts, ethylene production and flower petal abscission in geranium (*Pelargonium* × *hortorum* L.H. Bailey). *Plant Mol. Biol.*, **34**, 855-865.

Gilissen, L.J.W. (1976) The role of the style as a sense-organ in relation to wilting of the flower. *Planta*, **131**, 201-202.

Gilissen, L.J.W. (1977) Style-controlled wilting of the flower. *Planta*, **133**, 275-280.

Gilissen, L.J.W. and Hoekstra, F. (1984) Pollination-induced corolla wilting in *Petunia hybrida*: rapid transfer through the style of a wilting-inducing substance. *Plant Physiol.*, **75**, 496-498.

Halevy, A.H., Whitehead, C.S. and Kofranek, A.M. (1984) Does pollination induce corolla abscission of cyclamen flowers by promoting ethylene production? *Plant Physiol.*, **75**, 1090-1093.

Halevy, A.H., Porat, R., Spiegelstein, M., Borochov, A., Botha, L. and Whitehead, C.S. (1996) Short-chain saturated fatty acids in the regulation of pollination-induced ethylene sensitivity of *Phalaenopsis* flowers. *Physiol. Plant.*, **97**, 469-474.

Hamilton, A.J., Bouzayen, M. and Grierson, D. (1990) Antisense gene that inhibits synthesis of the hormone ethylene in transgenic plants. *Nature*, **346**, 284-287.

Hamilton, A.J., Bouzayen, M. and Grierson, D. (1991) Identification of a tomato gene for the ethylene-forming enzyme by expression in yeast. *Proc. Natl. Acad. Sci. USA*, **88**, 77434-77437.

Hill, S.E., Stead, A.D. and Nichols, R. (1987) Pollination-induced ethylene and production of 1-aminocyclopropane-1-carboxylic acid by pollen of *Nicotiana tabacum* cv White Burley. *J. Plant Growth Regul.*, **6**, 1-13.

Hoekstra, F.A. and Weges, R. (1986) Lack of control by early pistillate ethylene of the accelerated wilting of *Petunia hybrida* flowers. *Plant Physiol.*, **80**, 403-408.

Itzhaki, H., Maxson, J.M. and Woodson, W.R. (1994) An ethylene-reponsive enhancer element is involved in the senescence-related expression of the carnation glutathione-*S*-transferase (*GST1*) gene. *Proc. Natl. Acad. Sci. USA*, **99**, 8925-8929.

Jones, M.L. and Woodson, W.R. (1997) Pollination-induced ethylene in carnation: role of stylar ethylene in corolla senescence. *Plant Physiol.*, **115**, 205-212.

Jones, M.L. and Woodson, W.R. (1999a) Differential expression of three members of the 1-aminocyclopropane-1-carboxylate synthase gene family in carnation. *Plant Physiol.*, **119**, 755-764.

Jones, M.L. and Woodson, W.R. (1999b) Interorgan signaling following pollination in carnations. *J. Am. Soc. Hort. Sci.*, **124**, 598-604.

Jones, M.L., Larsen, P.B. and Woodson, W.R. (1995) Ethylene-regulated expression of a carnation cysteine proteinase during flower petal senescence. *Plant Mol. Biol.*, **28**, 505-512.

Kende, H. (1989) Enzymes of ethylene biosynthesis. *Plant Physiol.*, **91**, 1-4.

Kende, H. (1993) Ethylene biosynthesis. *Annu. Rev. Plant Physiol.*, **44**, 283-307.

Kim, W.T. and Yang, S.F. (1994) Structure and expression of cDNAs encoding 1-aminocyclopropane-1-carboxylate oxidase homologs from excised mung bean hypocotyls. *Planta*, **194**, 223-229.

Lanahan, M.B., Yen, H.-C., Giovannoni, J.J. and Klee, H.J. (1994) The *Never-ripe* mutation blocks ethylene perception in tomato. *Plant Cell*, **6**, 521-530.

Larsen, P.B. and Woodson, W.R. (1991) Cloning and nucleotide sequence of a *S*-adenosylmethionine synthetase cDNA from carnation. *Plant Physiol.*, **96**, 997-999.

Larsen, P.B., Woltering, E.J. and Woodson, W.R. (1993) Ethylene and interorgan signaling in flowers following pollination, in *Plant Signals in Interactions with Other Organisms* (eds. J. Schultz and I. Raskin), American Society of Plant Physiologists, Rockville, MD, pp. 171-181.

Larsen, P.B., Ashworth, E.N., Jones, M.L. and Woodson, W.R. (1995) Pollination-induced ethylene in carnation: role of pollen tube growth and sexual compatibility. *Plant Physiol.*, **108**, 1405-1412.

Lawton, K.A., Huang, B., Goldsbrough, P.B. and Woodson, W.R. (1989) Molecular cloning and characterization of senescence-related genes from carnation flower petals. *Plant Physiol.*, **90**, 690-696.

Lawton, K.A., Raghothama, K.G., Goldsbrough, P.B. and Woodson, W.R. (1990) Regulation of senescence-related gene expression in carnation flower petals by ethylene. *Plant Physiol.*, **93**, 1370-1375.

Llop-Tous, I., Barry, C.S. and Grierson, D. (2000) Regulation of ethylene biosynthesis in response to pollination in tomato flowers. *Plant Physiol.*, **123**, 971-978.

Lindstrom, J.T., Lei, C.-H., Jones, M.L. and Woodson, W.R. (1999) Accumulation of 1-aminocyclopropane-1-carboxylic acid (ACC) in petunia pollen is associated with expression of a pollen-specific ACC synthase late in development. *J. Am. Soc. Hort. Sci.*, **124**, 145-151.

Maxson, J.M. and Woodson, W.R. (1996) Cloning of a DNA-binding protein that interacts with the ethylene-responsive enhancer element of the carnation *GST1* gene. *Plant Mol. Biol.*, **31**, 751-759.

Meyer, R.C. Jr, Goldsbrough, P.B. and Woodson, W.R. (1991) An ethylene responsive flower senescence-related gene from carnation encodes a protein homologous to glutathione *S*-tranferase. *Plant Mol. Biol.*, **17**, 277-282.

Nadeau, J.A., Zhang, X.S. and O'Neill, S.D. (1993) Temporal and spatial regulation of 1-aminocyclopropane-1-carboxylate oxidase in the pollination-induced senescence of orchid flowers. *Plant Physiol.*, **103**, 31-39.

Nichols, R. (1968) The response of carnations (*Dianthus caryophyllus*) to ethylene. *J. Hort. Sci.*, **43**, 335-349.

Nichols, R. (1971) Induction of flower senescence and gynaecium development in the carnation by ethylene and 2-chloroethylphosphonic acid. *J. Hort. Sci.*, **46**, 323-332.

Nichols, R. (1977) Sites of ethylene production in the pollinated and unpollinated senescing carnation (*Dianthus caryophyllus*) inflorescence. *Planta*, **135**, 155-159.

Nichols, R., Bufler, G., Mor, Y., Fujino, D.W. and Reid, M.S. (1983) Changes in ethylene production and 1-aminocyclopropane-1-carboxylic acid content of pollinated carnation flowers. *J. Plant Growth Regul.*, **2**, 1-8.

O'Neill, S.D. (1997) Pollination regulation of flower development. *Annu. Rev. Plant Physiol. Plant Mol. Biol.*, **48**, 547-574.

O'Neill, S.D. and Nadeau, J.A. (1997) Postpollination flower development. *Hort. Rev.*, **19**, 1-58.

O'Neill, S.D., Nadeau, J.A., Zhang, X.S., Bui, A.Q. and Halevy, A.H. (1993) Interorgan regulation of ethylene biosynthetic genes by pollination. *Plant Cell*, **5**, 419-432.

Park, K.Y., Drory, A. and Woodson, W.R. (1992) Molecular cloning of a 1-aminocyclopropane-1-carboxylate synthase from senescing carnation flower petals. *Plant Mol. Biol.*, **18**, 377-386.

Pech, J.-C., Latche, A., Larrigaudiere, C. and Reid, M.S. (1987) Control of early ethylene synthesis in pollinated petunia flowers. *Plant Physiol. Biochem.*, **25**, 431-437.

Porat, R., Borochov, A. and Halevy, A.H. (1993) Enhancement of *Petunia* and *Dendrobium* flower senescence by jasmonic acid methyl ester is via the promotion of ethylene production. *Plant Growth Regul.*, **13**, 297-301.

Porat, R., Borochov, A., Halevy, A.H. and O'Neill, S.D. (1994) Pollination-induced senescence of *Phalaenopsis* petals: the wilting process, ethylene production and sensitivity to ethylene. *Plant Growth Regul.*, **15**, 129-136.

Porat, R., Halevy, A.H., Serek, M. and Borochov, A. (1995) An increase in ethylene sensitivity following pollination is the initial event triggering an increase in ethylene production and enhanced senescence of *Phalaenopsis* orchid flowers. *Physiol. Plant.*, **93**, 778-784.

Porat, R., Nadeau, J.A., Kirby, J.A., Sutter, E.G. and O'Neill, S.D. (1998) Characterization of the primary pollen signal in the postpollination syndrome of *Phalaenopsis* flowers. *Plant Growth Regul.*, **24**, 109-117.

Reid, M.S., Paul, J.L., Farhoomand, M.B., Kofranek, A.M. and Staby, G.L. (1980) Pulse treatments with silver thiosulfate complex extend the vase life of cut carnations. *J. Am. Soc. Hort. Sci.*, **105**, 25-27.

Reid, M.S., Fujino, D.W., Hoffman, N.E. and Whitehead, C.S. (1984) 1-aminocyclopropane-1-carboxylic acid (ACC). The transmitted signal in pollinated flowers? *J. Plant Growth Regul.*, **3**, 189-196.

Savin, K.W., Baudinette, S.C., Granham, M.W., *et al.* (1995) Antisense ACC oxidase RNA delays carnation petal senescence. *Hort. Sci.*, **30**, 970-972.

Singh, A., Evensen, K.B. and Kao, T. (1992) Ethylene synthesis and floral senescence following compatible and incompatible pollinations in *Petunia inflata*. *Plant Physiol.*, **99**, 38-45.

Spanu, P., Reinhardt, D. and Boller, T. (1991) Analysis and cloning of the ethylene-forming enzyme from tomato by functional expression of its mRNA in *Xenopus laevis* oocytes. *EMBO J.*, **10**, 2007-2013.

Stead, A.D. (1992) Pollination-induced flower senescence: a review. *Plant Growth Regul.*, **11**, 13-20.

Strauss, M. and Arditti, J. (1982) Postpollination phenomena in orchid flowers. X. Transport and fate of auxin. *Bot. Gaz.*, **143**, 286-293.

Tang, X. and Woodson, W.R. (1996) Temporal and spatial expression of 1-aminocyclopropane-1-carboxylate oxidase mRNA following pollination of immature and mature petunia flowers. *Plant Physiol.*, **112**, 503-511.

Tang, X., Wang, H., Brandt, A.S. and Woodson, W.R. (1993) Organization and structure of the 1-aminocyclopropane-1-carboxylate oxidase gene family from *Petunia hybrida*. *Plant Mol. Biol.*, **23**, 1151-1164.

Tang, X., Gomes, A.M.T.R., Bhatia, A. and Woodson, W.R. (1994) Pistil-specific and ethylene-regulated expression of 1-aminocyclopropane-1-carboxylate oxidase genes in petunia flowers. *Plant Cell*, **6**, 1227-1239.

ten Have, A. and Woltering, E. (1997) Ethylene biosynthetic pathway genes are differentially expressed during carnation (*Dianthus caryophyllus* L.) flower senescence. *Plant Mol. Biol.*, **34**, 89-97.

van Doorn, W.G. (1997) Effects of pollination on floral attraction and longevity. *J. Exp. Bot.*, **48**, 1615-1622.

Ververidis, P. and John, P. (1991) Complete recovery *in vitro* of ethylene-forming enzyme activity. *Phytochemistry*, **30**, 725-727.

Wang, H. and Woodson, W.R. (1989) Reversible inhibition of ethylene action and interruption of petal senescence in carnation flowers by norbornadiene. *Plant Physiol.*, **89**, 434-438.

Wang, H., Brandt, A.S. and Woodson, W.R. (1993) A flower senescence-related mRNA from carnation encodes a novel protein related to enzymes involved in phosphonate biosynthesis. *Plant Mol. Biol.*, **22**, 719-724.

Whitehead, C.S. and Halevy, A.H. (1989) Ethylene sensitivity: the role of short-chain saturated fatty acids in pollination-induced senescence of *Petunia hybrida* flowers. *Plant Growth Regul.*, **8**, 41-54.

Whitehead, C.S., Fujino, D.W. and Reid, M.S. (1983) Identification of the ethylene precursor 1-aminocyclopropane-1-carboxylic acid (ACC) in pollen. *Sci. Hort.*, **21**, 291-297.

Wilkinson, J.Q., Lanahan, M.B., Yen, H.-C., Giovannoni, J.J. and Klee, H.J. (1995) An ethylene-inducible component of signal transduction encoded by *Never-ripe*. *Science*, **270**, 1807-1809.

Woltering, E.J. (1990) Interorgan translocation of 1-aminocyclopropane-1-carboxylic acid and ethylene coordinates senescence in emasculated *Cymbidium* flowers. *Plant Physiol.*, **92**, 837-845.

Woltering, E.J., van Hout, M., Somhorst, D. and Harren, F. (1993) Roles of pollination and short-chain saturated fatty acids in flower senescence. *Plant Growth Regul.*, **12**, 1-10.

Woltering, E.J., Somhorst, D. and van der Veer, P. (1995) The role of ethylene in interorgan signaling during flower senescence. *Plant Physiol.*, **109**, 1219-1225.

Woltering, E.J., de Vrije, T., Harren, F. and Hoekstra, F.A. (1997) Pollination and stigma wounding: same response, different signal? *J. Exp. Bot.*, **48**, 1027-1033.

Woodson, W.R. and Lawton, K.A. (1988) Ethylene-induced gene expression in carnation petals. Relationship to autocatalytic ethylene production and senescence. *Plant Physiol.*, **87**, 498-503.

Woodson, W.R., Park, K.Y., Drory, A., Larsen, P.B. and Wang, H. (1992) Expression of ethylene biosynthetic pathway transcripts in senescing carnation flowers. *Plant Physiol.*, **99**, 526-532.

Xu, Y. and Hanson, M.R. (2000) Programmed cell death during pollination-induced petal senescence in petunia. *Plant Physiol.*, **122**, 1323-1333.

Yang, S.F. and Hoffman, N.E. (1984) Ethylene biosynthesis and its regulation in higher plants. *Annu. Rev. Plant Physiol.*, **35**, 155-189.

Zarembinski, T.I. and Theologis, A. (1994) Ethylene biosynthesis and action: a case of conservation. *Plant Mol. Biol.*, **26**, 1579-1597.

Zhang, X.S. and O'Neill, S.D. (1993) Ovary and gametophyte development are coordinately regulated by auxin and ethylene following pollination. *Plant Cell*, **5**, 403-418.

Index